Historical & Cultural Astronomy

The Historical & Cultural Astronomy series includes high-level monographs and edited volumes covering a broad range of subjects in the history of astronomy, including interdisciplinary contributions from historians, sociologists, horologists, archaeologists, and other humanities fields. The authors are distinguished specialists in their fields of expertise. Each title is carefully supervised and aims to provide an in-depth understanding by offering detailed research. Rather than focusing on the scientific findings alone, these volumes explain the context of astronomical and space science progress from the pre-modern world to the future. The interdisciplinary Historical & Cultural Astronomy series offers a home for books addressing astronomical progress from a humanities perspective, encompassing the influence of religion, politics, social movements, and more on the growth of astronomical knowledge over the centuries.

The Historical & Cultural Astronomy Series Editors are: Wayne Orchiston, Marc Rothenberg, and Cliff Cunningham.

Roger Kinns
Editor

The World of Visual Time Signals for Mariners

Time Balls, Time Guns, Time Lights and Other Signals

Springer

Editor
Roger Kinns
The Maudslay Society
Pembroke College
Cambridge, England

ISSN 2509-310X ISSN 2509-3118 (electronic)
Historical & Cultural Astronomy
ISBN 978-3-031-57333-0 ISBN 978-3-031-57334-7 (eBook)
https://doi.org/10.1007/978-3-031-57334-7

This Springer imprint is published by the registered company Springer Nature Switzerland AG
The registered company address is: Gewerbestrasse 11, 6330 Cham, Switzerland

Appreciating
Henry Maudslay (1771–1831), a great engineer whose memory is celebrated by the Maudslay Society and the Maudslay Scholarship Foundation at Pembroke College, Cambridge.

and

my wife Marion, children Jennifer and Malcolm, and grandchildren Isla, Harris and Isabella, who have lived with my obsession.

δός μοί ποῦ στῶ καὶ κινῶ τὴν γῆν.
Give me somewhere where I might stand and move the world
(Archimedes, according to Pappos of Alexandria)

Contents

About the Authors

Roger Kinns was born in Winchester, England, in 1944. He read Mechanical Sciences at Gonville and Caius College, Cambridge, and then took an MASc degree in control engineering at the University of Waterloo, Ontario, Canada, before returning to Cambridge to complete a PhD in unsteady aerodynamics. Roger was Maudslay Research Fellow of Pembroke College, Cambridge, from 1971 to 1975. He then joined YARD Ltd in Glasgow, Scotland, to lead the development and application of techniques for the acoustic design of ships and submarines and has worked as an independent consultant since 1999. Until 2019, he was a senior visiting research fellow in the School of Mechanical and Manufacturing Engineering at the University of New South Wales in Sydney, Australia. He is presently treasurer of the Maudslay Society and Maudslay Scholarship Foundation Trust, and chairman of the Younger Benmore Trust that has supported development of Benmore Botanic Garden since 1928. Roger has lived in Clynder, near Helensburgh, Scotland, since 1975. The Maudslay connection led to an enduring fascination with the history of engineering and particularly time signals worldwide. Roger has published a succession of research papers on time balls and other time-signalling devices and techniques in *The Journal of Astronomical History and Heritage* and *The International Journal for the History of Engineering and Technology*. He has a chapter on the time signals of Southeast Asia in *Exploring the History of Southeast Asian Astronomy: A Review of Current Projects and Future Prospects and Possibilities* (2021, Springer) and another on evolutions in the history of visual time signals for mariners in *Essays on Astronomical History and Heritage: A Tribute to Wayne Orchiston on his 80th Birthday* (2023, Springer). Coincidentally, he is co-owner of a racing yacht *Thalia*, built in 1924, which shares the name of the frigate that Robert Wauchope, inventor of the visual time signal, commanded in the 1830s.

Douglas Bateman was born in 1939 and spent most of his school years in Yorkshire. He moved south to join the then Royal Aircraft Establishment at Farnborough, eventually being sponsored for a physics degree at the City University, London. Work entailed research with lasers and infrared countermeasures. After a spell in London in the Ministry of Defence, he returned to Farnborough for project management and weapons support. An interest in horology began at an early age, and in 1972, Doug built a regulator clock with photoelectric control, temperature, and barometric compensation, which gave very good performance. This led to published papers and studies into what is important in timekeeping. Following a commission to build electronics for the Greenwich time ball, a major three-part series on its history was published in *Antiquarian Horology*. A parallel interest is in sundials, again having designed an unusual noon dial. He has lectured widely on horological and related topics.

Paul Fuller was born in Slough in 1960. He studied Geography at the University of Sheffield and graduated with an upper second class degree. He also obtained a postgraduate qualification in statistics from the University of Kent at Canterbury. He has been employed by British local government in producing key performance measures and statutory information. A member of the Society for the History of Astronomy, he is author of the paper "The Life and Times of John Rand Capron (1829–1888)", *The Antiquarian Astronomer*, 8, 21–45 (2014). His interest in horology was triggered by Douglas Bateman's lecture on the Greenwich Time Ball to this society in October 2014. He is also a member of the Society for the History of Meteorology and is co-author, with Jenny Randles, of the book *Crop Circles: A Mystery Solved* (1990, 1993, Robert Hale).

Part I
Development of Visual Time Signals

Chapter 1
Foundations

Roger Kinns

Abstract The foundations for this book lie in a search for time balls in Australia that might have been supplied by Maudslay, Sons & Field in London, who had made apparatus for Greenwich, Deal and Edinburgh. They built apparatus for Sydney, NSW in 1855, but no others in Australia could be attributed directly to the firm. Another was believed to have been supplied to the Cape of Good Hope in collaboration with Siemens. It was found, however, that the 1873 mechanical apparatus was instead shipped to Lyttelton, NZ.

Some extraordinary errors, such as an exaggerated mass of the Edinburgh time ball, were resolved, as well as the history of a short-lived time ball in Glasgow that was replaced by controlled clocks along the Clyde. The study broadened to include visual time signals of many different types worldwide, making extensive use of lists issued by the British Admiralty from 1880 onwards. Those outside the British Empire were by then in an increasing majority.

1 Introduction

The need for visual time signals in ports around the world emerged when marine chronometers could be made in sufficient numbers at a sufficiently low price to be supplied to ocean-going ships for determination of longitude. Chronometers were designed to indicate time at a chosen meridian on the Earth's surface but significant errors could accumulate during long voyages.

Time signals under control by an astronomical observatory provided a check on chronometer accuracy in service conditions. How far had the indicated time drifted since it was last checked and what was the daily rate of change? That rate could be determined by observing a time signal over a period of days. In principle, it was then a simple process to correct the indicated time after the ship had left port. Constancy of rate was the sought-after chronometer characteristic. Visual time

R. Kinns (✉)
The Maudslay Society, Pembroke College, Cambridge, England

R. Kinns (ed.), *The World of Visual Time Signals for Mariners*, Historical & Cultural Astronomy, https://doi.org/10.1007/978-3-031-57334-7_1

signals could take many different forms, but there was a wide preference for time balls. It will be shown later how they came into being, thanks to the persistence of Robert Wauchope (1788–1862), a Royal Navy officer.

This book stemmed from forensic engineering investigations into time balls in Australia and New Zealand. It evolved into a much wider study of the history of visual time signals and how they were distributed, bringing me into contact with specialists in the history of astronomy and time distribution around the world. The foundations for this two-part book are described first. Part I shows how early ideas for accurate visual time signals evolved into specific solutions and how visual signals proliferated for the general population. It includes detailed descriptions of early apparatus which set the standards for later arrangements. Part II contains descriptions of visual signals for mariners that existed in different geographical areas worldwide, derived particularly from lists of time signals for mariners that were published by the British Admiralty from 1880 onwards. The timescale extends until after the Second World War.

2 The Starting Point

My interest in the history of time signals started through my membership of the Maudslay Society, which celebrates the memory of Henry Maudslay (1771–1831), the pioneer of precision engineering, and the subsequent work of Maudslay, Sons & Field (Foster Petree, 1949). It led me in directions I had never anticipated.

A first stimulus came in 2006 from Colin Evans, a former President of the Society. I had become a regular visitor to Australia and New Zealand owing to family links and my professional interests in marine engineering and acoustics. He suggested that when I next visited Adelaide, I might try to find whether a time ball there had been supplied or influenced by Maudslay Sons & Field (abbreviated to Maudslays).

2.1 Contributions by Maudslay, Sons & Field

Maudslays had designed and erected the first public time ball at Greenwich, which became operational in October 1833. It set the standard for time ball arrangements and is still in operation after many modifications to the apparatus and almost two centuries of time signal development (Howse, 1997; Bateman, 2013). Maudslays were thought to have provided four later systems, all using a rack and pinion arrangement to hoist the ball. Detail was missing as the company was liquidated in 1900 and almost all the company records had been destroyed at that time. There was, however, one tantalising list of time balls in a surviving notebook. Apparatus had been ordered for Greenwich in 1833 and for Deal, Edinburgh and Sydney, NSW during the 1850s, confirmed by independent records in the UK and Australia.

Another appeared to have been ordered for the Cape of Good Hope in 1873, in some arrangement with Siemens. What happened to it and did Maudslays supply other time ball systems?

2.2 Initial Investigations in Australia and New Zealand

The first time ball apparatus I ever saw was at Lyttelton during a visit to friends in New Zealand. That was long before I started my historical research into time signals. It was demonstrated then by the late Bruce Carr (1930–2023), who had an infectious enthusiasm for the iconic time ball tower and restored apparatus at Lyttelton, and later became a generous correspondent. Its significance for my own research became clear in 2008.

The search for the 1873 time ball led to my first papers concerning time balls in Australia, working with historian Lesley Abell in Adelaide. We worked particularly on the Charles Todd correspondence, trying to establish whether Maudslays had any involvement in the time ball at Semaphore, near Port Adelaide. Todd had been responsible for operation of Maudslay time balls at Greenwich and Deal, before he left for South Australia in 1855. We found that the Semaphore arrangement had been designed by Todd, but was only constructed using local suppliers after he had led construction of the telegraph line between Adelaide and Darwin. The investigation then took an unexpected turn. Todd's correspondence with Henry Russell, the Government Astronomer in Sydney, showed how Russell had modified the original Sydney apparatus which Todd had seen under construction at Maudslays' works in 1855 (Kinns & Abell, 2009). It became apparent that there were extraordinary similarities between the apparatus at Sydney and Lyttleton. The differences were essentially Russell's modifications to the Sydney apparatus.

Lyttelton had the only time ball system ever supplied by Siemens (Kinns, 2009). It was now clear that it incorporated the missing mechanical apparatus made by Maudslays in 1873. It had been shipped to New Zealand, not the Cape of Good Hope and became operational in 1875 after a new tower had been constructed. Lyttelton had the third official time ball for mariners in New Zealand, those at Wellington and Port Chalmers having used apparatus made by other companies. The timing of that research was fortunate because the tower collapsed and the time ball apparatus was severely damaged during the 2011 earthquakes that devastated Christchurch. Later, the history of short-lived time balls at Auckland and the introduction of time lights at Wellington and Auckland completed an extraordinary picture of visual time signal development in a single country.

It was Nick Lomb, then Curator of Sydney Observatory, who suggested that I might consider publication of my work on time balls in the *Journal of Astronomical History and Heritage*. That led to a long and productive association with Wayne Orchiston, covering time signals of many different types worldwide.

2.3 Investigations in Scotland

Edinburgh has one of the finest early time balls, first operational in 1854 as an official signal, but many of the assumed dates concerning its origin were imprecise and there was an extraordinary myth that the ball itself had weighed over 750 kg, when its real mass was about 90 kg. Investigation of original documents and contemporary newspaper articles showed how errors had arisen (Kinns, 2011, 2014). Edinburgh also introduced an automated time gun service in 1861 that set new standards for signal accuracy. The topology of Edinburgh lends itself to the combination, with its time ball at high elevation and a time gun on Castle battlements above residential areas. George Robinson of the "One O'clock Gun Association" in Edinburgh was always helpful with references and information.

Glasgow has long been the largest city in Scotland, expanding rapidly during the Industrial Revolution with its access to rest of the world via the River Clyde. Glasgow had a privately-operated time ball from 1857 to 1863 when there was no telegraph link from Glasgow Observatory to the City centre. It had been demonstrated that a ball could easily be dropped in Glasgow by electric telegraph from Edinburgh Observatory and there were experiments with multiple time guns controlled from Edinburgh, but the relative status of observatories in the two cities was a sensitive political issue (Kinns, 2010; Clarke & Kinns, 2011). There were other important issues too. Increasing smoke pollution, as well as the construction of buildings, low water depth and the growing number of ships along extensive Glasgow wharves, led to poor visibility of a time ball near the city centre. The topology of the city did not lend itself to time guns either. Glasgow opted instead for controlled clocks at various locations along the Clyde. There was no Glasgow time ball after 1864 and none were erected further out along the Clyde. It was an anomaly for a major port, even though the accuracy of the clocks was highly regarded.

It was a good start for exploring time signals in the rest of the world.

3 The Broadening View

There has been extensive research into the history of the Royal Observatory at Greenwich and its role in time determination. *Greenwich Time and the Longitude* (Howse, 1997) and online papers by Dolan (2023) provide an excellent overview of how the Observatory, now known as the Royal Greenwich Observatory, was founded in 1675 for the improvement of navigation at sea and how chronometers came to be developed in parallel with astronomical methods for determination of longitude.

Almost by accident, I had been investigating time signals in Scotland and New Zealand that were almost diametrically opposite on the Earth's surface. Both countries have maintained outstanding archives, making it possible to understand how local politics and economic development, as well as available technology, had influenced time signal provision. An overview of how time signals for mariners evolved worldwide was missing.

The Cape of Good Hope had a critical strategic position on the sea route between Europe to India, Indonesia, the Far East and Australasia in the days of sail, before the advent of steam propulsion and the opening of the Suez Canal in 1869. Cape Town has remained an important port ever since. Many of the earliest time signal developments had originated in South Africa and much had been done by distinguished astronomers at the Cape Observatory to explore the development of time signals there. But there were gaps and it became apparent that some information in official lists and archives of correspondence by the Astronomers Royal at Greenwich did not tally with records at the Cape. A chapter in Part II shows how some of those contradictions have been resolved. As often, I received considerable help from overseas contacts, many introduced to me by Wayne Orchiston, who were invariably supportive.

A time signal, described as a ball, had become operational in Mauritius before the famous time ball at Greenwich, but who had designed and erected it and how did it work? That became another challenge. St Helena and India also had important early time balls, but there was still much to be discovered about their design and history. I was able to build on pioneering work by Bartky and Dick in the United States, carried out in the 1980s, concerning worldwide time balls and the distribution of time from the United States Naval Observatory (Dick, 2003). Since then, improved access to archives worldwide through the growth of the internet has made it possible to uncover historical material that throws new light on early developments.

My interests broadened again when I was invited to participate in a mini-conference organised by Wayne Orchiston about the history of astronomy in Southeast Asia. It was held in Chiang Mai, Thailand in February 2020. My contribution was an oral paper "Time signals for mariners in SE Asia during the pre-wireless era: time balls, discs, lights, guns and bells". It was just before the Covid-19 pandemic made travel and work on physical archives almost impossible for an extended period. There were some fascinating signals in the region, with an extraordinary range in Indonesia and Singapore (Kinns, 2021). Was it possible to create a similar picture for the whole world? That led to a sequence of papers covering the Atlantic Islands, Africa, India and Mauritius, which form the basis for chapters in Part II. I had to accept that I could not expect to access archives everywhere, partly because of language constraints and national restrictions and partly because many might have been destroyed during times of war.

3.1 Origin of the Time Ball

Another stimulus for this book was the history of time signals at Portsmouth, England. This was the location of the first time ball trial in 1829, but newly uncovered documents have exposed incorrect assumptions about the arrangement and about later time balls at Portsmouth (Kinns et al., 2021). Paul Fuller has long been working on the general history of British time signals. Douglas Bateman was responsible for many changes to the arrangements at Greenwich in modern times (Bateman, 2013). Both have contributed guest chapters to this book.

3.2 Evolution of Time Signals for Mariners

The National Library of Scotland in Edinburgh has a particularly fine collection of Admiralty lists of time signals worldwide. As soon as social distancing allowed during the Covid-19 pandemic, I could work through volumes for different geographical regions and years, building a picture of their physical locations, appearance to mariners, accuracy, and evolution from one signal type to another. By the end of the nineteenth century the number of time signals worldwide far exceeded the number in the then British Empire. It was a rewarding task that appears not to have been attempted previously. It led to Part II of this book, which has evolved from my review paper in the *Journal of Astronomical History and Heritage* (Kinns, 2022).

Different countries had different approaches to the signalling of exact time. A principal aim of Part II is to explore the broad evolution of visual time signals for mariners, focusing on the period after 1880 when comprehensive lists were published by the British Admiralty. Electric time lights first appeared in the early twentieth century and replaced or complemented many time balls. They had been almost completely ignored in published books and papers about time signals, not least because they were never used as official signals in the British Isles or the United States (Kinns, 2023). Time guns were operated widely, many in parallel with time balls. Wireless signals became widely available after 1911 and were then listed separately. They are excluded from the present study, although some time balls were operated using wireless signals. The number of time balls and other visual signals peaked in the period after the First World War and declined markedly after the Second World War.

4 Content of Subsequent Chapters

An outline history of how chronometers became available and how methods using astronomical observations were developed to allow determination of longitude at sea is included in Chap. 2. This led to ideas for visual time signals that could be used to rate chronometers at principal ports. These are described in Chap. 3, with a particular focus on the work of Robert Wauchope, who was the originator of precise time signals under control by astronomical observations and is often described, rather loosely, as the inventor of the time ball.

The history of the time ball at Greenwich from its first operation in 1833 to the present day is described in Chap. 4 by Douglas Bateman, followed in Chap. 5 by a description of time ball apparatus made by Maudslay, Sons and Field that set the standard for systems supplied by other manufacturers. Admiralty lists of time signals are introduced in Chap. 6.

Time signals for navigation might have existed independently of other public signals, but expansion of the railway and telegraph systems and the increasing mobility of populations, soon led to the introduction of time zones and wider regulation of human activities. Chapter 7 by Paul Fuller, shows the extent to which time signals for general public use proliferated while the number listed by the Admiralty remained limited.

Part II begins with an overview of the types of signals that appeared in Admiralty lists from 1880 to 1947, showing how they evolved from one type to another and then declined with the advent of wireless signals. Succeeding chapters include maps and tables for individual countries and groups of countries. Details of individual time signals are accompanied by photographs and images of them, providing a reference catalogue for future extension.

Acknowledgements Many of the time ball illustrations throughout this book have come from the collection of postcards formed by Klaus Hülse in Germany. He has published them on his own website (Hülse, 2023) and has been most generous in allowing me to use them freely in my research. He has often provided additional clarification. Wayne Orchiston has never failed to offer encouragement over 15 years.

Many friends, family members and colleagues working in different fields have given freely of their time in reading draft text and in helping to bring this book to its final form. These include Richard Maudslay, Lesley Abell, Anne Robertson, Dr. Michael Kinns, Sir Roger Tomkys and my guest authors Paul Fuller and Douglas Bateman. I shall always be grateful for the historical information I received from correspondents in Britain and many countries overseas. They will be acknowledged in the individual chapters. If there are any residual errors in such a wealth of data, I take full responsibility for them. Finally, I very much appreciate all the support I have had from the team at Springer Nature.

References

Bateman, D. (2013). The time ball at Greenwich and the evolving methods of control – Part I. *Antiquarian Horology, 34*(2), 198–218; Part II. *Antiquarian Horology, 34*(3), 332–346; Part III. *Antiquarian Horology, 34*(4), 471–488.

Clarke, D., & Kinns, R. (2011). Some new insights into the history of the Glasgow time ball and time guns. *Journal of Astronomical History and Heritage, 15*(1), 59–67.

Dick, S. J. (2003). *Sky and ocean joined: The U.S. Naval Observatory 1830–2000*. Cambridge University Press.

Dolan, G. (2023). *The Royal Observatory, Greenwich*. http://www.royalobservatorygreenwich.org/articles.php?article=0

Foster Petree, J. (1949). *Henry Maudslay, 1771–1831: And Maudslay, Sons & Field, Ltd.* Commemorative brochure to mark the establishment of the Maudslay Scholarship, published by the Maudslay Society in 1949 and reprinted with amendments in 1956; now available on CD from the Kew Bridge Steam Museum. [Foster Petree was Secretary of the Maudslay Society.]

Howse, D. (1997). *Greenwich time and the longitude*. Philip Wilson Publishers Ltd.

Hülse, K. (2023). *Themen -Seite Zeitsignale: Zeitbälle oder Zeitkanonen*. http://www.leuchtturm-welt.net/HTML/TIMEBPK/TIMEBALL.HTM

Kinns, R. (2009). Time keeping in the antipodes: A critical comparison of the Sydney and Lyttelton time balls. *Journal of Astronomical History and Heritage, 12*(2), 97–107.

Kinns, R. (2010). Time balls, time guns and Glasgow's quest for a reliable local time service. *Journal of Astronomical History and Heritage, 13*(3), 194–206.

Kinns, R. (2011). The early history of the Edinburgh time ball and time gun. *International Journal for the History of Engineering & Technology, 81*(2), 264–290.

Kinns, R. (2014). Did the Edinburgh time ball really weigh 15 cwt? *International Journal for the History of Engineering & Technology, 84*(2), 160–174.

Kinns, R. (2021). Chapter 12: Time signals for mariners in SE Asia: Time balls, discs, bells, guns and lights. In W. Orchiston & M. N. Vahia (Eds.), *Exploring the history of Southeast Asian astronomy: A review of current projects and future prospects and possibilities* (pp. 411–458). Springer.

Kinns, R. (2022). Visual time signals for mariners between their introduction and 1947: A new perspective. *Journal of Astronomical History and Heritage, 25*(4), 601–713.

Kinns, R. (2023, January). The emergence of time light signals for calibration of marine chronometers. *International Journal for the History of Engineering & Technology, 93*(1), 16–46.

Kinns, R., & Abell, L. (2009, January). The contribution of Maudslay, Sons & Field to the development of time balls in Australia. *International Journal for the History of Engineering & Technology, 79*(1), 59–90.

Kinns, R., Fuller, P., & Bateman, D. (2021). Exploring the Portsmouth time balls. *Journal of Astronomical History and Heritage, 24*(3), 751–769.

Chapter 2
Determination of Longitude and Chronometer Rating

Roger Kinns

Abstract This chapter provides a simplified view of developments that led to visual time signals. Astronomers of many different nationalities provided the essential foundation. The search for a method to determine longitude at sea lasted for centuries. Recognition of methods that would allow determination of longitude preceded development of accurate instruments and provision of nautical almanacs.

It was recognised that longitude might be determined using a timekeeper, later known as a chronometer, that could display accurate time at a chosen meridian on the earth's surface. Astronomical and chronometer solutions became available at almost the same time in the second half of the eighteenth century. The challenge then was to check and maintain the accuracy of time measurement during long sea voyages. Time signal observations on successive days could be used to rate chronometers, complemented by understanding of factors that caused rate variation at sea. That was the justification for visual time signals around the world.

1 Introduction

The international search for a method to determine longitude at sea lasted for centuries and has been described by many authors, including Howse (1997). It is well-told by Dolan (2023) on *The Royal Observatory, Greenwich* website. Recognition of astronomical methods that would allow longitude to be determined by navigators preceded development of accurate instruments. Invention of the sextant allowed measurements at sea of the required accuracy. Various methods were developed and used, but the ultimately preferred approach was to use the method of lunar distances or "lunars". This exploited the predicted position of the Moon relative to the stellar background, which follows a cycle of 18 years 11.3 days (Howse, 1997, 66). Careful measurements over that period were needed to predict future positions. Sufficient accuracy in tabular presentations was achieved at last by Tobias Mayer, a practical

R. Kinns (✉)
The Maudslay Society, Pembroke College, Cambridge, England

R. Kinns (ed.), *The World of Visual Time Signals for Mariners*, Historical & Cultural Astronomy, https://doi.org/10.1007/978-3-031-57334-7_2

astronomer in Göttingen, who sent his results to the British Admiralty in 1755. It was a remarkable product of international collaboration that could be used to generate nautical almanacs. The accuracy was improved further over the years, but the principles remained the same.

It was also recognised that longitude might be determined using a timekeeper, later known as a chronometer, that could display accurate time at a chosen meridian on the earth's surface. For centuries, that had seemed an impossible requirement. Astronomical and chronometer solutions became available at almost the same time in the second half of the eighteenth century. The challenge then was to design and manufacture reliable chronometers at a price that could be afforded by ship-owners and navies. The challenge after that was to check and maintain the accuracy of time measurement during long sea voyages. That led to visual time signals around the world, the subject of this book.

This chapter provides a simplified view of developments that led to visual time signals. Astronomers of many different nationalities provided the essential foundation, but Greenwich and the British Admiralty played a central role in facilitating accurate world-wide navigation. The Greenwich time ball is still in service as the best-known visual time signal.

2 Sidereal and Mean Solar Days

Only a tiny proportion of the population who were not astronomers or navigators would have recognised the significance of the sidereal day, about 4 minutes shorter than the mean solar day, whose length corresponds to one complete rotation of the Earth relative to the remote stellar background. Transit telescopes, which were used to regulate astronomical clocks and time signals, observed the sidereal day. These telescopes were designed and erected to observe the daily passage of selected "clock stars" across the NS meridian at their known elevations. The mean solar day of 24 hours, familiar to the population at large, is longer because it includes the average time needed for the Earth to rotate further towards the centre of the Sun during a complete orbit.

3 Greenwich and the Board of Longitude

The observatory at Greenwich had been founded in 1675 specifically to improve accuracy in navigation via provision of precise astronomical observations (Howse, 1997). The Observatory work was directed by the Astronomer Royal. The first to have this title was John Flamsteed (1646–1719), who was appointed by King Charles II, then the British monarch. Flamsteed served between 1675 and 1719 and demonstrated that the Earth rotated at a constant rate, allowing longitude to be expressed in terms of either angular distance or time difference between two meridians. It was to be over two centuries before tiny variations in the rate of rotation could be identified.

The Board of Longitude was established by Act of Parliament in 1714 with the Astronomer Royal as an ex-officio member, to encourage solutions to the problem of longitude measurement at sea. This was paralleled by similar encouragement in other countries (see Howse, 1997; De Grijs, 2020). Initial progress was slow and erratic but accelerated rapidly in the 1740s. Flamsteed was followed by Edmond Halley, James Bradley and Nathanial Bliss. The fifth Astronomer Royal was Nevil Maskelyne (1732–1811) who served from 1765 to 1811. He was in post when it first became possible to determine longitude at sea with high accuracy using either a timekeeper or the method of lunar distances.

3.1 Chronometer Invention

In recent times, Maskelyne has been presented as an opponent of John Harrison (1693–1766) who was the first to build a timekeeper that would remain accurate over a long period at sea (Sobel, 1996). Maskelyne favoured the use of astronomical measurements to estimate longitude but did not seek to profit financially from that preference (Howse, 1993). Harrison had been awarded the prestigious Copley Medal of the Royal Society in 1749, for his promising early development of timekeepers H1–H3 (King, 1993). Harrison was eventually granted a large award in 1765, usually described as the Longitude Prize, for his H4 timekeeper. However, it still fell short of being an affordable solution. Many others made significant contributions to the development of marine chronometers that could be supplied in quantity at a reasonable price (Howse, 1997). They too received awards for their work.

3.2 Determination of Longitude from Astronomical Observations

Maskelyne had published a plan for an annual nautical almanac before his appointment as Astronomer Royal (Maskelyne, 1763). This would include predictions of the lunar position relative to the stellar background during the year after publication (Howse, 1997). Maskelyne persisted with the plan and published the first British nautical almanac after he became Astronomer Royal (Maskelyne, 1766).

One of the last important acts of the Board of Longitude before its dissolution in 1828 was to initiate development of an observatory in the southern hemisphere at the Cape of Good Hope and appoint the Reverend Fearon Fallows (1788–1831) as its first director in October 1820 (Gill, 1913). He introduced the first time light.

The sixth Astronomer Royal between 1811 and 1835 was John Pond (1767–1836), who was in post at the time of the Portsmouth time ball trial in 1829–30 and presided over installation of the Greenwich time ball in 1833.

George Airy (1801–1892) was the seventh Astronomer Royal at Greenwich from 1835 to 1881 and was at the heart of time signal development during that time. Airy was a meticulous correspondent who was often approached about provision of time signals worldwide. Many of those exchanges feature in this book. He was succeeded by William Christie (1845–1922), who served from 1881 to 1910. Their archived correspondence is a valuable reference concerning the development of visual time signals in the pre-wireless era. Christie was succeeded by Frank Dyson (1868–1939), who introduced the Greenwich time pips in wireless signals during his service from 1910 to 1933. Later Astronomers Royal had ongoing involvement in time determination and transmission as technology evolved towards the era of space travel.

4 Requirements for Chronometer Accuracy

Time at the prime meridian had to be known accurately: an error of 1 second in time corresponds to an angular rotation error of 15 arc-seconds, equivalent to 0.46 km at the Equator or half that at Latitude 60°. That placed heavy demands on the reliability and accuracy of chronometers. It is worth noting than an error of 1 second per day corresponds to time measurement accuracy that is close to 0.001%.

4.1 Chronometer Rating

A chronometer would be "rated" by its manufacturer and ideally checked independently by a qualified authority before use at sea, so that its user could apply a first-order correction to the indicated time. It was not unusual for a chronometer to gain or lose 5 seconds per day, while the rate might depend on factors such as temperature and the local magnetic field.

A chronometer was not adjusted until return to its supplier or a qualified authority. Many ships carried several chronometers on long voyages, partly to guard against chronometer failure and partly in the expectation that an average reading, after rate correction, would be more accurate than a single measurement.

Repeat rating of chronometers at different ports was viewed as essential for accurate navigation using chronometers of average quality, as well as checks using the method of lunar distances. This is well-illustrated in the following letter concerning a voyage to Madras, which also mentioned a longitude error in the stated location of the Madras observatory (Toynbee, 1859).

> December 31st, 1858.—Last evening we arrived in Madras, and by the flash of the gun I find that on applying the errors and rates to the chronometers supplied by the maker before leaving England, the best chronometer was 56½ miles too far East, and the worst one 175

> miles, while the lunar errors were right to one mile, and the lunar rates to four-tenths of a second. I may add that the Nautical Almanac gives the longitude of the Madras Observatory as 5h. 21m. 4s., but my friend Mr. W. H. Bayley tells me that some years ago Mr. Taylor, the Astronomer here, detected a mistake in this, and that it really is, 5h. 20m. 57.3s. Major Worster, now in charge of the Observatory, also says that this is the correct longitude.

In this letter, longitude is specified as time in advance of Greenwich, 1 hour corresponding to 15° angular rotation. The flash of a gun served as the Madras time signal in 1858 (Kinns, 2020).

4.2 *Effects of Magnetism on Chronometer Rates*

Chronometer rates might be affected by many different environmental factors, including seaway motion, temperature, lubrication and wear. There was particular concern about the effects of the local magnetic field. Any structure made of iron onboard would tend to modify the Earth's magnetic field. The deflection of a compass needle showed the influence of the ship's structure on the local field direction. Moving a magnet close to a chronometer could also be shown to affect its rate. To many, that was sufficient evidence to introduce time signals that could be observed without moving a chronometer from its onboard location. Others were more sceptical about the importance of the magnetic field.

Early trials at sea to identify influences on chronometer rates were inconclusive (Akkermans, 2020), largely because it was difficult to separate the effects of seaway motion and temperature variation, for example, from magnetic field effects. Any possibility of magnetic field influence was reduced by improved chromometer design, while care was taken to position chronometers away from any fittings that might change the magnetic field. It was a pragmatic approach. The general understanding of electricity and magnetism was to improve dramatically with the work of James Clerk Maxwell (1831–1879). The possible effects of magnetic fields on chronometer rates emerged again when ships used extensive electrical machinery. A special study was commissioned to determine what field strength would influence a chronometer rate by 1 second per day (Chapman & Lewis, 1912). It was a complex study, but the broad conclusion was that the field strength would have to be considerably greater than that of the Earth's field at Greenwich.

4.3 *Effects of Temperature on Chronometer Rates*

It was shown by John Chapman Hartnup (1806–1885), Director of Liverpool Observatory, that the main reason for the difference between rates provided by the supplier and rates measured at sea was temperature variation. It then became possible to improve longitude accuracy by measuring the rate of a chronometer at different temperatures in a laboratory facility and correcting readings at sea by monitoring temperature during a voyage (Obituary, 1886).

Hartnup demonstrated key relationships between chronometer rates and temperature that made it possible to correct chronometer readings during lengthy sea voyages, provided that rates had been pre-determined under laboratory conditions. He evaluated the effect of temperature on the rates of 100 chronometers in 1854 and was later able to deduce formulae for the general behaviour of chronometers (Obituary, 1886). The following extract shows how his work was applied in practice.

> In his Report for 1872 to the Marine Committee of the Mersey Docks and Harbour Board, Mr. Hartnup gives a formula for calculating the correction due to thermal error, for any temperature of a chronometer whose rate has been determined by observation in the three definite temperatures of 55, 70 and 85 degrees Fahrenheit. An example showing the advantage of such corrections to the rates of two chronometers during a voyage from Liverpool to Calcutta is given in No. 2, vol. xxxv of the Society's Monthly Notices.

It then became possible to check chronometers using land-based signals without the same urgency to measure the rate using repeat measurements over several days at different ports. Many owners of chronometers, including the Pacific Steam Navigation Company from 1877, sent them to Hartnup's observatory in Liverpool for the necessary calibration. Similar calibration could be carried out at other observatories worldwide. Hartnup's work was admired internationally (The Japanese Navy, 1878, for example).

5 Requirements for Time Signal Accuracy

Precise measurement of longitude was needed to determine signal times. Once the longitude of the signal had been established, a transit telescope could be used to ensure that the signal was made at the same GMT and local mean time each day. These transit observations regulated astronomical clocks which often incorporated devices to drop a time ball or fire a gun automatically. Longitude could be determined initially using the method of lunar distances or observations of eclipses and occultations.

Longitude accuracy was improved in the early years by shipping a large number of carefully calibrated chronometers between observatories and averaging results. In 1845, over 60 chronometers were sent 16 times backwards and forwards between Altona near Hamburg and Pulkowa near St Petersburg while 40 chronometers were used in 1846 in a similar manner to establish the longitude difference between Altona and Greenwich (Howse, 1997, 117).

Telegraph links, including long-distance underwater cables, were later used extensively for accurate determination of longitude, following pioneering work in the United States (Dick, 2003). Allowances had to be made for the signal propagation speed along telegraph wires, which could be as low as 3000 km/s (Thomson, 1856).

The accuracy of the measured longitude between Washington and Paris improved from about 1 second using chronometers before 1850, to a few tenths of a second using telegraph cables in the 1860s, and to a few hundredths of a second after the introduction of wireless signals in the twentieth century (Dick, 2003, 73).

5.1 *Random and Bias Errors*

There were two types of error in the timing of land-based signals. The first was a bias error, resulting primarily from inexact determination of longitude. Sometimes, it had to be recognised after careful deliberation that there was an error in the stated longitude. One example is the longitude of Wellington in New Zealand, which was corrected in April 1874 by 1′ 33″, corresponding to a timing error of 6.15 seconds (Longitude of Wellington, 1874). Another example was an error in Portugal, found using telegraphic determination by the US Navy in 1878–79: "… the longitude of the observatory at Lisbon has up to the present time been in error more than two miles" (Green et al., 1880, 11).

Stated longitudes and signal times were subject to adjustment in official lists when more accurate determinations became available. A bias error could also result from factors such as a consistent delay in time ball release after an electric signal had been received. Sometimes, this type of error was minimised by arranging for the release signal to be sent slightly early. This approach was used in Singapore, for example (Kinns, 2021). This would not affect determination of chronometer rates and would be announced in official notices.

The second type of error was random, resulting from factors such as imprecise transit observations or extended reliance on an astronomical clock when bad weather prevented astronomical observations. Random errors were dependent on the skill of local operators and the quality of available equipment.

Ideally, daily signals should be accurate to better than ±1 second for chronometer rating. The best-equipped observatories, such as those at Greenwich and The Cape of Good Hope, aimed for an accuracy of ±0.2 seconds in visual signals. Measurements over 5 days would then give a rate accuracy of better than ±0.1 seconds per day. Annual reports from individual observatories sometimes included statistics concerning time signal accuracy, such as *rms* errors and the proportion of signals outside stated error bands.

References

Akkermans, E. J. (2020). *Chronometers and chronometry on British Voyages of Exploration, 1819–1836* (PhD thesis). University of Edinburgh.

Chapman, S., & Lewis, T. (1912). The effect of magnetism on the rates of chronometers and watches (communicated by the astronomer royal). *Monthly Notices of the Royal Astronomical Society, 72*(7), 583–608.

De Grijs, R. (2020). A (not so) brief history of lunar distances: Lunar longitude determination at sea before the chronometer. *Journal of Astronomical History and Heritage, 23*(3), 495–522.
Dick, S. J. (2003). *Sky and ocean joined: The U.S. Naval Observatory 1830–2000*. Cambridge University Press.
Dolan, G. (2023). *The Royal Observatory, Greenwich*. http://www.royalobservatorygreenwich.org/articles.php?article=933
Gill, D. (1913). *A history and description of the Royal Observatory, Cape of Good Hope*. London: H. M. Stationery Office.
Green, F. M., Davis, C. H., & Norris, J. A. (1880). *Telegraphic determinations of longitudes of the east coast of South America: Lisbon ... , with the latitudes of several stations in 1878 and 1879*. Washington Government Printing Office.
Howse, D. (1993). The Astronomers Royal and the problem of longitude. *Antiquarian Horology, 21*(1), 43–51.
Howse, D. (1997). *Greenwich time and the longitude*. Philip Wilson Publishers Ltd.
King, A. (1993). The Astronomers Royal and the problem of longitude. *Antiquarian Horology, 21*(1), 52–59.
Kinns, R. (2020). Time signals for mariners in India, Burma and Ceylon. *Journal of Astronomical History and Heritage, 23*(3), 523–552.
Kinns, R. (2021). Chapter 12: Time signals for mariners in SE Asia: Time balls, discs, bells, guns and lights. In W. Orchiston & M. N. Vahia (Eds.), *Exploring the history of southeast Asian astronomy: A review of current projects and future prospects and possibilities* (pp. 411–458). Springer.
Longitude of Wellington. (1874, April 1). *Wellington Independent*, 3.
Maskelyne, N. (1763). *The British mariner's guide*. Printed for the Author.
Maskelyne, N. (1766). *The nautical almanac and astronomical ephemeris for the year 1767*.
Obituary, Hartnup, J. (1886). *Monthly Notices of the Royal Astronomical Society, 46*, 188–191.
Sobel, D. (1996). *Longitude: The true story of a lone genius who solved the greatest scientific problem of his time*. Fourth Estate.
The Japanese Navy. (1878 September 4). Article in the *Hampshire Telegraph and Sussex Chronicle*, 2.
Thomson, W., later Lord Kelvin. (1856). On the theory of the electric telegraph. *Proceedings of the Royal Society of London, 7*, 382–399. https://royalsocietypublishing.org/doi/pdf/10.1098/rspl.1854.0093
Toynbee, C. (1859). The westerly route for crossing the line. Letter to the Editor of the *Nautical Magazine*, 170–177.

Chapter 3
Ideas for Visual Time Signals

Roger Kinns

Abstract The first ideas for land-based visual time signals controlled by astronomical observations that could be used to rate marine chronometers have been credited to Robert Wauchope. Time guns had often been used previously but they were not then precision signals. The principal aim of this chapter is to demonstrate how the dropping time ball was introduced as the preferred time signal, following Wauchope's first suggestion of a shutter signal in 1818. He later proposed the hauling down of a flag at noon. The Admiralty changed this signal to a dropping ball at 1 pm which was trialled at Portsmouth in 1829 and favoured by Wauchope thereafter. He continued to press the Admiralty and other authorities worldwide to introduce time balls, finally succeeding with the time ball at Greenwich. Preceding independent solutions included a shuttered oil lamp and a shuttered time ball or disc. The expansion of time balls and other signals for mariners is also discussed.

1 Contributions by Robert Wauchope

The first ideas for land-based visual time signals controlled by astronomical observations that could be used to rate marine chronometers have been credited to Robert Wauchope. Time guns had often been used previously but they were not then precision signals.

The principal aim of this chapter is to demonstrate how the dropping time ball was introduced as the preferred time signal in 1829, following Wauchope's first suggestion of a shutter signal in 1818. The later expansion of time balls and other signals for mariners is also discussed.

R. Kinns (✉)
The Maudslay Society, Pembroke College, Cambridge, England

R. Kinns (ed.), *The World of Visual Time Signals for Mariners*, Historical & Cultural Astronomy, https://doi.org/10.1007/978-3-031-57334-7_3

1.1 The Career of Robert Wauchope

Robert Wauchope (1788–1862) was a talented officer in the Royal Navy from an aristocratic family (Patterson, 1858). He had demonstrated his navigation skills, determination and personal bravery in various actions during the Napoleonic Wars and was promoted to Post-Captain in 1814 (Dictionary, 1849). He was in command of HMS *Eurydice* at St Helena from 1816 to 1819 and recorded his initial ideas for time signals in his "remark book" (Wauchope, 1818). He was, however, a complex man whose religious fervour after his conversion to evangelical Christianity in 1819 led him into conflict with many senior officers and naval practices of the day. He was on half-pay without a ship from 1819 to 1834, as were many other officers when the Royal Navy reduced in size. Initially offered and then denied command of HMS *Thalia* because of his declared opposition to allowing prostitutes on ships in harbour, and reduced again to half-pay, he won an appeal through his family connections and took command of HMS *Thalia* from 1834 to 1838. His main task was to intercept slave ships that were then plying between West Africa and the Americas. It was his last seagoing appointment. Subsequent promotions through increasing seniority in the Navy List took him to flag rank in 1849. Initially Rear-Admiral of the Blue, he became a full Admiral in 1861, not long before his death (Obituary, 1862). Figure 3.1 is a photograph of him, taken towards the end of his life.

Wauchope believed that land-based time signals visible from ships in harbour were essential to avoid having to remove a chronometer from its shipboard location

Fig. 3.1 Robert Wauchope towards the end of his life (https://wornandwound.com/time-balls-horological-origins-new-years-ball-drop/robert-wauchope/)

for rating ashore. This overcame concerns that the rate might be influenced by modification of the Earth's magnetic field due to onboard equipment made of iron, as well as by any shock experienced during transfer. It also reduced the risk of chronometer loss or damage. Wauchope worked hard to persuade others to recognise the benefits of land-based time signals regulated by astronomical observations.

Wauchope has been recorded as the inventor of the time ball for almost two centuries (see, for example, Dictionary, 1849), although careful reading of Wauchope's own written statements indicate that the time ball was actually introduced by the Admiralty as a modification to his proposed flag signal. The dropping ball was certainly favoured with enthusiasm by Wauchope after its successful trial in 1829, although he was always careful to claim invention of the land-based visual signal, often referred to as "my signal", rather than the time ball *per se*. His written statements and reactions to them are reviewed in the following sections. He and others believed that the inventor of a concept should have higher status than the engineers and entrepreneurs who brought that concept to practical fruition. Invention of the time ball has to be seen in that light.

1.2 Wauchope's 1818 Proposal for a Telegraph Signal

Wauchope's first idea for visual signals used a mechanical telegraph to signal time to ships offshore. This may have been rooted in the mechanical telegraph system that was then under development for communication between the Admiralty, Portsmouth and Chatham (Holmes, 1983). That telegraph system was implemented by Henry Maudslay (1771–1831) whose precision engineering company, later named Maudslay, Sons & Field, was to provide the time ball apparatus for Greenwich in 1833.

Wauchope's entry in the remark book for HMS *Eurydice* is reproduced below (Wauchope, 1818).

> As time pieces are now become so common I beg leave to insert a plan of a Telegraph for taking observations for ascertaining the Rates of Time Pieces by an Artificial Horizon without moving the Time Piece out of the Ship, which is always attended with danger particularly abroad, where frequently there is no place of safety to leave the Chronometer onshore during the time the Ship is at Anchor, rendering it necessary to take it backwards and forwards for each day's observations.

The diagram is shown in Fig. 3.2. Wauchope's description of the design and its operation is included in Kinns et al. (2021). He was suggesting a land-based signal that used a combination of sound and a simultaneous visual signal to communicate respectively with the astronomical observer and the ship. The flap or shutter could be lowered as a preparatory signal. The shutter would eventually migrate to a time ball and preparatory signals including flags and the raising of a ball would be introduced.

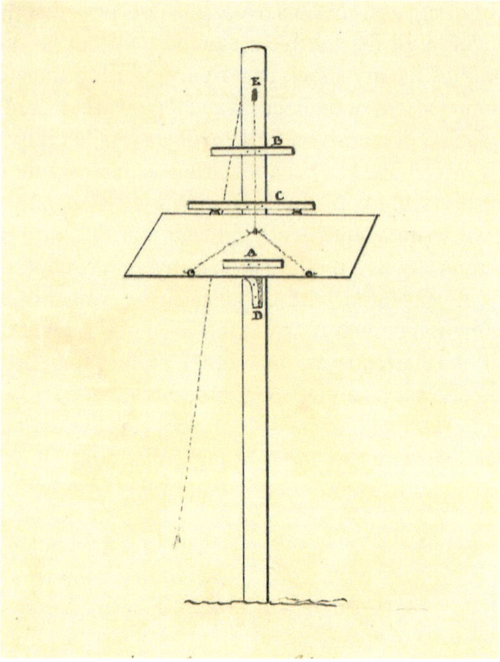

Fig. 3.2 Wauchope's telegraph signal proposal in 1818 (UK Hydrographic Office, File MP 84-34)

2 Other Early Ideas for Visual Signals

2.1 Fallows' 1823 Lamp Signal

Many of the early developments, including the first operational time light under astronomical control in about 1823, were at the Cape of Good Hope (Editorial, 1835; Bartky & Dick, 1981). This light had been introduced by the Reverend Fearon Fallows, the first astronomer at the Cape who had been appointed by the Board of Longitude in 1820 and arrived at the Cape in 1821. Fallows had to endure budget cuts and poor support from the Board that forced him to use portable instruments that he had brought from England in 1821 for almost the whole of his tenure. He operated them from a temporary hut in unfavourable working conditions but was still able to contribute valuable astronomical observations. He died in 1831, not long after permanent observatory buildings had been erected at his preferred site (Gill, 1913).

Fallow's time light post-dated Wauchope's 1818 proposal but no evidence has been found of any communication between Wauchope and Fallows. A shutter was opened to expose an oil lamp as a preparatory signal. Closure of the shutter, which made the light invisible to ships, was the signal. This type of land-based signal re-emerged in the form of electric time lights during the twentieth century.

2.2 *Owen's Rocket Signals*

Rocket signals at the Cape had been proposed in 1826 by Captain William Owen (Editorial, 1835). Their explosion times were to be determined by the Observatory and required subsequent notification to ships, so they would have been inferior signals. They were also rejected on grounds of expense, although they appear to have been used and discarded in favour of time balls at St Helena (Findlay, 1867; see Kinns, 2022). Owen had been involved in earlier experiments using rockets (Akkermans, 2020, 129–130).

3 Wauchope's Proposal for a Flag Signal

Wauchope continued to consider land-based time signals and again pressed the Admiralty to introduce them in December 1824. He received a polite acknowledgement from the Admiralty (Barrow, 1824):

> With reference to your letter of the 20th instant, submitting a plan for the regulation of chronometers, I am commanded by my Lords Commissioners of the Admiralty to convey to you their lordships' thanks for your communication.

Wauchope's 1824 proposal was not published. A submission by Wauchope dated June 1827 was located by Adrian Webb (2010) during his research on the history of hydrographic data management and is reproduced below (Wauchope, 1827). It may have been a resubmission of his 1824 proposal without further alteration. He was then favouring a flag signal at noon. He emphasised the effect of the magnetic field as being "very obvious", although later research showed that it was usually less significant than other effects such as temperature variation.

The text of his proposal was transcribed by Webb and is reproduced below.

> The chief reason why merchant ships do not carry Chronometers so frequently as they ought, is not so much owing to the expence of the Instrument, as to the difficulty of ascertaining a good rate, arising from the inability of masters of ships to do so; as the operation requires considerable time, and great accuracy.
>
> And likewise on board of H.M. ships, the custom is, to send their chronometers on shore to their great risque, and after all it is frequently found that the shore rate, and sea rate does not agree; the cause of such disagreement being very obvious, arising from the same source of error to which compasses are subject – viz the magnetic influence which the mass of iron in the ship has upon the Instrument. Many advantages would therefore accrue. Could an accurate rate be found, without moving the Chronometer out of the positions, or place it is intended to remain in when at sea. The plan I propose to obviate all these difficulties is this.
>
> Suppose at Portsmouth for instance, it was notified to all the ships at Spithead, to merchant ships, as well as to Men of War, that five minutes before noon at Greenwich, a particular flag was to be hoisted at the Platform; and that the instant it was noon at Greenwich it would be hauled down; this would give every ship within sight an opportunity of comparing their Chronometers with Greenwich time: and they would get a rate by this means, far more accurately than could be obtained by the best sextant and Artificial-horizon, as the rate of

> the watch which regulated the hauling down of the Flag would be ascertained by the Transit Instrument at Portsmouth Observatory.
>
> Should this plan be approved of: the Master Attendant at the different Dock-yards might be entrusted with the Flag and the Transit-instrument which might be put up at every year (*yard?*), the expence would be very trifling, as the window of a convenient storehouse would answer perfectly well for fixing the Transit-Instrument in the Meridian of the place, without any building being erected for the purpose. And the men of war at the different ports would always be a check against any carelessness or inattention on the part of the Master-Attendant.

Wauchope's proposal was presented to the Lords of the Admiralty on 1 September 1827. He had obtained a supporting statement from James Inman, Professor of Nautical Mathematics at the Royal Naval College, which was included with his proposal (Inman, 1827). Most of the ideas in the paper were eventually adopted, but there was still no mention of a time ball. His expectation appears to have been that if his plan was approved by the Admiralty then dockyard staff would design and implement a detailed arrangement.

4 Emergence of the Time Ball

Wauchope published two papers in the 1830 issue of the *Edinburgh New Philosophical Journal* (Wauchope, 1830a, b) which have often been interpreted as evidence that he invented the time ball. In reality, they indicate that Wauchope's recommendation of a flag signal was discarded in favour of a ball signal by those who had been instructed to prepare a trial of Wauchope's plan. Also, his earlier preference for a noon signal was changed to a signal at 1 pm. Wauchope recognised immediately that these were significant improvements and promoted time balls thereafter, injecting his own design ideas.

4.1 The First Time Ball Trial

Documents located after the pioneering work by Bartky and Dick (1981) showed that the first time ball at Portsmouth in 1829 did not use the arrangement published by Wauchope after the trial had started (Wauchope, 1830b). Also, the location was not as previously assumed (Kinns et al., 2021). Figure 3.3 shows the announcement of a time ball trial in October 1829, recording that the time signal was the moment of release and that the ball was located on the "High Tower" at Portsmouth. The drop time was 1 pm.

Wauchope, then on half-pay, did not appear to have had any direct involvement in the trial and wrote to the Admiralty in June 1830 enquiring whether it had been viewed as a success (Wauchope, 1830c).

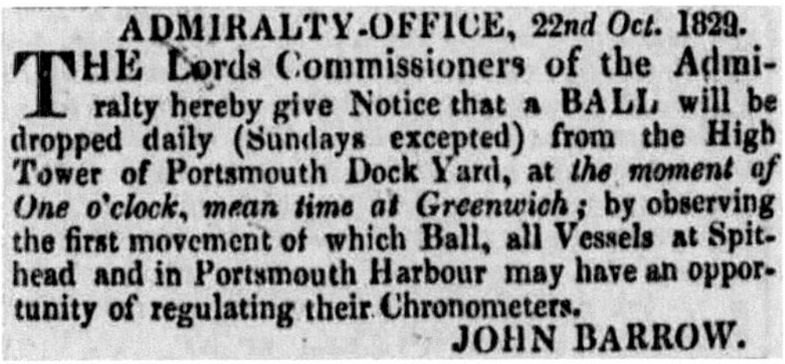

ADMIRALTY-OFFICE, 22nd Oct. 1829.
THE Lords Commissioners of the Admiralty hereby give Notice that a BALL will be dropped daily (Sundays excepted) from the High Tower of Portsmouth Dock Yard, at *the moment of One o'clock, mean time at Greenwich;* by observing the first movement of which Ball, all Vessels at Spithead and in Portsmouth Harbour may have an opportunity of regulating their Chronometers.
JOHN BARROW.

Fig. 3.3 Time ball announcement on 22 October 1829

> May I request that you will be pleased to lay before their lordships the anxiety I feel to ascertain whether the plan for finding the rates of chronometer, which I had the honour of laying before their lordships, and which is at present on trial at Portsmouth, has met with their lordships approbation; and if so, whether I may be permitted to endeavour to get the signal established upon the coasts of France and America, as the chief utility of the plan must depend on its being generally adopted.
>
> Should their lordships favour me with their approbation, I can desire no higher honour for myself that the expression of that approbation.

Wauchope received a prompt and positive reply (Barrow, 1830).

> In return to your letter of 22d ultimo, I am commanded by my Lords Commissioners of the Admiralty to acquaint you that the plan adopted at Portsmouth for finding the rates of chronometers after your suggestion, appears, by a Report which their Lordships have received, to have succeeded; but with regard to your proposal, of establishing signal stations on the continents of France and America, their Lordships cannot interfere.

The reply mentions a report on the trials which has not been located. No mention of the first time ball has been found in contemporary local newspapers, other than the initial announcement in Fig. 3.2. Wauchope later acknowledged that "This plan was adopted at Portsmouth, and by a letter to me appeared to have succeeded, but, from not having an Observatory there, it appears to have fallen to the ground" (Wauchope, 1852).

It remains unclear whether the 1829 apparatus was simply a trial arrangement to prove the concept of a shore-based signal or whether it was used for some years afterwards. The Portsmouth time ball was not mentioned in discussions after 1837 that led to the 1845 Liverpool time ball, the only references to time ball design being to the arrangement at Greenwich. No previous time ball at Portsmouth was mentioned when a new time ball there was under discussion in 1856 (Wood, 1856). The biographical entry for Wauchope implied, however, that it was an operational signal (Dictionary, 1849).

4.2 *Wauchope's Papers in 1829–30*

Wauchope's first paper in the *Edinburgh New Philosophical Journal* was published in 1830, and was dated 23 November 1829 (Wauchope, 1830a). The second part, which contained his time ball design proposal, was dated 1 February 1830 (Wauchope, 1830b). Parts of the papers are reproduced in this section, with some additional commentary. The first paper began with an enthusiastic report of initial success of the trial at Portsmouth. "The accompanying plan" in the first paragraph is a reference to his appended communication to Professor Jameson (editor of the journal). The reported conversation with an admiral preceded erection of the trial signal, so it is a comment on Wauchope's plan for a visual signal, not on the time ball.

> In sending you the accompanying plan for regulating chronometers by an instantaneous signal, I may mention that the subject has occupied my attention since 1818, when I first proposed a telegraph for the purpose to the Lords of the Admiralty and, in 1824, I again wrote on the same subject, and received a letter of thanks from their Lordships.
>
> I now feel particularly happy in stating that their Lordships have directed a trial of the above plan to be made at Portsmouth; and, in a letter of the 12th of this month, from an officer in an official situation there, who has taken a great interest in and superintendence of the signal, he says, – "The Admiral and myself were at the King's Stairs, when a boat from a line-of-battle ship landed with her chronometers, &c. not only to set them by the clock at the Observatory, but they had brought their artificial horizon and sextant to take sights. It was blowing very fresh, and the ship I believe was waiting for this inconvenient, and after all unsatisfactory process. I then repeated to the admiral (which I had several times mentioned before) your plan, marking this as a case in point. From that moment he pursued its adoption with energy, and it is now, although in an infant operation, quite sufficiently established to give proof of *complete success*."

4.3 The Transition from a Flag Signal to a Time Ball

The last part of Wauchope's first paper, which followed presentation of his proposal for an Edinburgh time ball, was his communication to Professor Jameson. This was similar to the letter that he had submitted to the Admiralty in 1827, transcribed above (Wauchope, 1827). The significant change was the addition of two final paragraphs, whose key content is reproduced below.

> … Perhaps one hour after noon might be better for the signal, to allow an observation being made to ascertain the time of day a short time before the signal is made.
>
> The accompanying signal by means of balls, is that which has been adopted at Portsmouth, and which appears to have answered perfectly.

The penultimate paragraph reflects the Admiralty decision to make the signal at 1 pm in the first time ball trial (Fig. 3.3), in preference to a signal at noon GMT which had been proposed by Wauchope. This practice was widely used subsequently because it allowed astronomers to concentrate on solar observations when the sun was near its zenith.

The final paragraph suggests that the Admiralty decided to replace the hauling down of the flag with a signal "by means of balls" in the trial. It was the first mention of a ball signal in any of Wauchope's known correspondence. Wauchope might justifiably have seen the time ball as a refinement of his proposed flag signal and he certainly promoted time ball signals thereafter with great persistence and enthusiasm. One advantage of a ball over a flag is that it always appears as a circular object, while the visibility of a flag signal depends on wind strength and direction. Another is that the moment of release is a better-defined time signal. Nevertheless, the hauling down of a flag on receipt of an observatory signal was still being used as a reduced-cost alternative in the twentieth century, so it was not abandoned entirely.

4.4 Wauchope's Recommendation for Edinburgh

The other paragraph in Wauchope's first paper was a recommendation that a time ball be introduced in Edinburgh. Its position in the paper before his appended letter to Jameson tends to obscure the progression from his proposals between 1818 and 1827 to the time ball trial in 1829.

> Before concluding, I cannot help expressing a hope that the Edinburgh Astronomical Institution may take the plan into their consideration, as the Observatory on Calton Hill is so admirably adapted to the purpose. A flag-staff of a very moderate height, and a ball of four feet diameter, would be perfectly well seen by shipping in the Leith Roads; besides the advantage which would accrue to every watch-maker in Edinburgh and Leith, by giving them the power of comparing their time-pieces with true time. I may mention that the very ingenious and very excellent watch-maker, Mr. Whitelaw, No.16, Prince's Street, has stated to me that he could with perfect ease, adapt the Observatory clock, or any other time-piece, so as to disengage the ball at a given hour, without injury to the time-piece, which would certainly be a very great convenience.

Those were significant ideas, ahead of their time. There was no operational time ball in Edinburgh until 1854. By then, it had been decided to install it on the nearby Nelson's Monument, rather than at the Observatory. As at Greenwich, the apparatus was supplied by Maudslay, Sons & Field (Kinns, 2011).

4.5 Wauchope's 1830 Time Ball Design

Wauchope sought unsuccessfully to promote a particular time ball arrangement in his second paper (Wauchope, 1830b). This is shown in Fig. 3.4.

The signal was the time of drop by one ball diameter. The design was submitted on 1 February 1830, over 2 months after his first paper. It is likely to have been the plan that was sent to Admiral Charles Adam in February 1830 which elicited the following response (Adam, 1830; see also Wauchope, 1852):

> I am much obliged to you for sending me your plan for instantaneous signals.
>
> It appears to me that the scheme you have suggested is admirably adapted for the purpose, and that so correct and public a method of making known to ships the exact *Greenwich time* would be attended with the greatest advantage to navigation in general. I hope therefore that it will be adopted on the most extensive scale.

Charles Adam had been Wauchope's captain during his initial service on HMS *Resistance* in 1805.

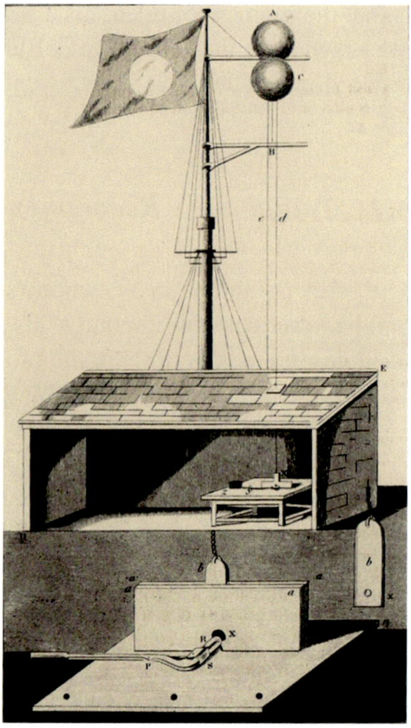

Fig. 3.4 Wauchope's proposed arrangement (Wauchope, 1830b)

Wauchope's concept design was never used for operational time balls, not least because the time to drop by a specified distance might be influenced by weather conditions and sources of friction. He commented that "The ball C will take about four-tenths of a second to fall its own diameter, five feet, which separation of the two balls constitutes the signal. The halliards, should, therefore, be disengaged four-tenths of a second before the true time." An object in free-fall under gravity takes 0.55 seconds to fall 5 feet, a time that will be increased by any friction. It was a surprising error. Wauchope focused particularly on the trigger arrangement. He did not address the challenge of bringing the dropping ball to rest without risk of damage, which was a problem with many early time ball designs.

Wauchope suggested a preparatory flag signal, whereby the flag would be raised 5 minutes before the signal and hauled down 1 minute before. He also suggested that the ball signal should be dropped three times at 1 minute intervals "for the convenience of observers". Several time balls in France after 1870 were dropped at 10 and 10.02 am, so that the basic idea was adopted occasionally. Raising of the ball, sometimes in stages, became the preferred preparatory signal. However, a preparatory flag signal was used in Calcutta, now Kolkata, for many decades (Kinns, 2020). It was hoisted 20 minutes before the time signal at 1 pm, while three separate time balls were each raised 5 minutes before 1 pm. The balls were dropped and the flag hauled down as the time signal. Preparatory flags were also used in conjunction with time guns at a small number of locations.

4.6 *Wauchope's Recommendation for Greenwich*

Wauchope pressed the Admiralty to introduce a time ball at Greenwich (Wauchope, 1833).

> Their Lordships having been pleased to approve of, and to adopt, my plan for finding the rates of chronometers by instantaneous signal; a copy of their Lordship's letter to me on the subject I have the honour to enclose. The importance of having a similar signal established at Greenwich Observatory having been suggested to me by Sir Jahl. Brunton, I beg to state that I have examined for this purpose, the East and West India exportation docks and canal, when nearly low water, in all of which situations the signal could be most distinctly seen; in many of the ships it could be seen from the cabin windows. ...
>
> ...
>
> If their Lordships still approve of my plan for ascertaining the rates of chronometers by signal, I need not further insist upon the importance of Greenwich Observatory as a signal-station, from the immense number of shipping that will be benefited, as well as the facilities which will attend its erection there. I have to request you will be pleased to lay the above before their Lordships for their consideration.

Wauchope made his points with clarity and the response was immediate (Barrow, 1833a, b).

> Having laid before My Lords Commissioners of the Admiralty your letter of the 17th instant, I am commanded by their Lordships to acquaint you, that your plan for the (*sic*) ascertaining the rates of chronometers by an instantaneous signal has been referred to the Astronomer Royal.

The Greenwich time ball was in operation a few months later. It represented the essential transition from a concept design to a properly engineered solution. The Greenwich design by Maudslay, Sons & Field set the standard for time ball machinery and is described in more detail in Chap. 4. It was erected, installed and tested between June and October 1833. The time was very short for a sophisticated new design, so preparatory design work may have been carried out before funds were made available for its construction and before Wauchope's final stimulus (Wauchope, 1833). It was extraordinarily successful and the same apparatus remained in use until the First World War with only minor modifications and repairs. Most of the apparatus was then replaced, as described in Chap. 4 (see Spencer Jones, 1952; Bateman, 2013).

4.7 *Announcement of the Greenwich Time Ball*

The official notice of the Greenwich time ball on 28 October 1833 is transcribed below (Barrow, 1833b).

> The Lords Commissioners of the Admiralty hereby give notice, that a ball will henceforward be dropped every day, from the top of a pole on the eastern turret of the Royal Observatory, Greenwich, at the moment of one o'clock. P. M., mean solar time. By observing the first instant of its downward movement, all vessels in the adjacent reaches in the

> river, as well as in most of the docks, will thereby have an opportunity of regulating and rating their chronometers. The ball will be hoisted half-way up the pole at five minutes before one o'clock, as a preparatory signal; and close up at two minutes before one.

It is similar in most respects to the 1829 Portsmouth notice (Fig. 3.3), with the addition of preparatory signals. The raising of the ball in two stages at stated times before the exact signal was often used for later time balls.

5 Selection of Time Ball Parameters

5.1 Visibility Trials

The operational signal at Greenwich was preceded by visibility trials during August and September 1833, using balls of different diameters fixed to a pole (Greenwich Observatory, 1833). There was some public confusion as to whether these were operational time balls and whether the intended drop time was at noon or 1 pm, to the extent that members of the public waited patiently to observe the dropping of the fixed ball. There was consternation when one of the balls fell to the ground in a gale on 31 August (Editorial, 1833). The tested ball diameters may have been as large as 8 feet, but the final choice was a ball of 5 feet diameter, falling 10 feet.

5.2 Weight Requirements

An important feature of time ball design was weight control. Time balls had to be heavy enough to fall smoothly in variable weather conditions, while being within the weight range that could be handled by the braking apparatus without unacceptable risk of damage to either the ball or supporting arrangement.

A spherical time ball with a diameter of 5 feet (1.5 m) has a surface area of about 7 m^2. Airy commented in correspondence concerning a possible time ball at Portsmouth that time balls of this size weighed about 90 kg (Wood, 1856). This included the mass of the interior wooden frame and attachments, but not the shaft and piston that were rigidly connected to the ball and entered a vertical cylinder to provide braking using an air spring.

Many time balls supplied between 1850 and 1880 with a diameter of 1.5 m used a zinc skin nailed to a wooden frame. Zinc has a density of 7130 kg/m^3, so a skin with a thickness of only 1.06 mm (1/24 inch) would have a mass of about 50 kg. Many other time ball designs were used, with large variations in diameter and mass according to local preferences. Some used wickerwork construction, while others used a metal frame structure without an external skin. There were large variations in the braking arrangement, ranging from the air spring that was first used at Greenwich, through retarding weights and springs to friction brakes.

The first time ball at St Helena, although only in use for short period, was small and light (Editorial, 1835). Most time balls in the United States had small diameters and masses, but there was a notably heavy copper time ball at Boston, introduced in 1878 with a diameter of 1.2 m and mass of 180 kg (Bartky, 1983). Some later time balls used aluminium to reduce weight, while retaining large diameters. One example is the design of time balls for Singapore which had a diameter of 1.8 m, a mass of only 35 kg, and used guide wires to control the drop (The Singapore Free Press, 1892).

6 Wauchope's Struggle for Recognition

Throughout his life, Wauchope struggled for recognition that he was the originator of land-based visual time signals. Coming from an aristocratic family with naval connections at senior level, he was able to elicit written endorsement of his statements that he was the first to introduce the idea in 1818 and that he had presented a plan to the Admiralty in 1824 which led to the Portsmouth trial in 1829. He never claimed explicitly to have introduced a time ball in place of a flag but felt that too much credit was given to engineers and contractors who developed arrangements that were essentially implementations of his plan.

6.1 1830s Correspondence

There was extensive correspondence in the *Nautical Magazine* in the 1830s, following announcement of a time ball at St Helena in 1834 with sceptical comments by the editor (Editorial, 1835). Wauchope felt that it failed to give him sufficient credit for originating the idea of land-based visual time signals (Wauchope, 1836):

> Having seen an account of the time-ball at St. Helena in your number for November, 1835, containing also some allusions to myself, I have taken the liberty of enclosing to you my correspondence with my Lords Commissioners of the Admiralty upon the subject.
>
> I had pointed out the advantages to be derived from the plan for communicating time by means of telegraphs so far back as 1818, in my remark-book transmitted to the Admiralty when in command of the Eurydice, at that time on the Cape and St. Helena station. Sir Jahl. Brenton was then naval commissioner at the Cape, and an extract from a letter of his, dated 15th November, 1833, to Mrs. Wauchope, will, I think, establish my prior claim to the invention, before the Rev. Mr. Fallows, (who, it appears, was astronomer at the Cape in 1820,) or Captain William Owen, R.N. In reference to the chronometer signal, Sir Jahl. Brenton says, "I regret very much that it has not been called by an appropriate name. I feel quite convinced, at the same time, that the originator is well known, and will become more and more so, as the utility of the measure is fully understood and appreciated. I beg that Captain Wauchope will not hesitate to use my name whenever he thinks it may be of use; he can state positively, that he did mention the idea of such a signal to me at the Cape many years before its adoption; and I believe he can also say, that it was adopted in consequence

of an interview with Sir Edward Owen, when a lord of the Admiralty, to whom, I think, I mentioned him by letter as the originator of the plan in question."

I not only mentioned my views at that time to Sir Jahl. Brenton, but spoke freely upon the subject to all my brother officers, pointing out the advantage which shipping in general would derive from the chronometer signal being generally adopted; but I had no conception at that time how many difficulties I should have to contend with, before I could obtain a public trial of the plan in question. The Eurydice was paid off in 1819, and many things after this occurring to occupy me, it was not till 1824, when, finding that nothing had been done by any one that I knew of upon the subject, I again turned my mind to the importance of getting the chronometer signal generally established. You will find enclosed my letter to the Admiralty, and their lordships' letter of thanks to me, signed by Mr. Barrow, in which he says, "I am commanded by my Lords Commissioners of the Admiralty to convey to you their lordships' thanks for your communication."

With regard to the mode of displaying the ball, much must depend upon the situation in which it may be placed; and as to the various kinds of machinery which may be employed for hoisting and dropping it, this may be varied ad infinitum. What I contend for is, having originated the system of giving the time by means of a signal for ascertaining the rates of chronometers; and having perseveringly (from conviction of its utility) urged it forward, ever since 1818 to its general adoption.

In 1830, as soon as the experiment had been satisfactorily made at Portsmouth, I laid the plan before the French and American ambassadors in London, and urged its adoption by their respective governments. I have by me the American ambassador's letter in reply, which is highly satisfactory. In Edinburgh and Liverpool I have also done my best to get it established, both for the use of shipping, and also for regulating the time in the towns, which is often of so much consequence in mercantile concerns, particularly as to the starting of coaches, &c. You will observe by the printed correspondence I enclose, that it was my letter to their lordships which was laid before the astronomer royal, that induced the Admiralty to cause the time-ball to be placed at Greenwich Observatory.

Having now made you acquainted with my claims, may I beg that you will in justice do me the favour to publish them in the next number of your valuable magazine, as the article in No. 45 makes it appear that I only took up the plan at third hand.

Wauchope enclosed letters of endorsement with his response, which were published in the *Nautical Magazine* (Barrow, 1830, 1833a; Adam, 1830; Wauchope, 1833; Brenton, 1836). His 1824 proposal to the Admiralty was not, however, published with the letters. The 1836 endorsement from Rear-Admiral Sir Jahleel Brenton, writing from Greenwich Hospital, is shown below. It confirms, again, that his 1818 remark book entry was considered to be the origin of subsequent developments.

I beg leave to certify, that the statement in Captain Wauchope's letter of the 2d April 1836, respecting his being the originator of the instantaneous signal for the regulation of chronometers, is perfectly correct; that he did communicate his plan to me at the Cape of Good Hope in 1818, which I believe to be the first suggestion of the kind.

An important feature of Wauchope' published letters was the absence of a specific claim to time ball invention. He was then given due, but limited, credit by the editor of the *Nautical Magazine* (Editorial, 1836). This is transcribed below.

We have a great horror of detracting from the merits of anyone, and having laid the foregoing documents before our readers, we need scarcely add that we find the proposal alluded to by Captain Wauchope in his remarks for the Eurydice. The plan there proposed of giving the instant of time, is by the observer with the sextant and artificial horizon onshore, causing at the instant of observation the dropping of a shutter on its hinges, the instant of its fall being noted on board as shown by the time-keeper. We need not stay to offer any remarks on this method, our purpose now is with the regard to priority of invention. On this point, sufficient has been advanced by Captain Wauchope, not only to establish the fact of his plan having been proposed at the time he states, but to be the same in principle as the system adopted at Portsmouth and Greenwich, although at the latter place the means of carrying the method into effect are most superior, and, as we have remarked, "combine elegance with durability, on truly scientific principles." The very ingenious contrivance at Greenwich, described in our last volume, p. 584, is the work of other hands; but Captain Wauchope's claim appears to be, the right of being considered the originator of the system at present adopted, – a claim to which he is no doubt entitled, and, in addition thereto, the credit due to the perseverance with which he pursued his method until it was established; although we would not be supposed to assert, that the communication of time by signal, for a similar purpose, was never before adopted.

6.2 1852 Correspondence

As late as 1852, when he was a retired Rear-Admiral living in Penrith, he wrote to the press to complain that his name was not being mentioned when a new time ball was being considered for Deal (Wauchope, 1852). He appended the letters that had been published in the 1836 *Nautical Magazine*. The supporting editorial included the following statements, which added comment on the new time ball for The Strand (Editorial, 1852).

It is rather hurtful to those unpatronised inventors who devote their time and experience with anxious and unremitting attention towards professional improvements, that a comparatively new class has crept into society, which not only carries off all the profit, but would seize upon the honour also. ... Those who suggest and those who elaborate are each worthy of a share of public consideration; but surely he who invents, and who also applies that invention, has infinitely superior claims to both of them.

We have been led into this reflection by the pompous announcement the other day of the time-ball in the Strand, which is to drop with that at the Royal Observatory at Greenwich, and thence to be extended in all directions. ... By a convincing little pamphlet now before us, we see that Rear-Admiral Wauchope has long laboured for ascertaining the rates of chronometers by an instantaneous signal, and that a time-ball for this most vitally useful object employed his sagacity nearly five-and-thirty years ago.

The final comment is a reference to Wauchope's 1818 proposal for a telegraph signal, illustrating how invention of the time ball had become confused with the invention of a visual time signal.

7 Early Time Signals Worldwide

7.1 The Shuttered Time Ball at Port Louis, Mauritius

A novel implementation, which used a stationary black ball or disc behind a shutter in a white painted building, was operational at Port Louis, Mauritius in April 1833. Its development appears to have been independent of the Admiralty and it preceded the Greenwich time ball by 6 months. It will be described in detail in a later chapter.

7.2 Time Balls on Principal Trade Routes

The British East India Company recognised the merits of Wauchope's proposals and erected early time signals for chronometer rating in the 1830s (Howse, 1997). These were located on the trade routes between Europe and the Far East and implemented using local resources. The first of these was a dropping ball at St Helena in January 1834, which was noted by Wauchope when he was in command of *HMS Thalia*. Another time ball was erected in Calcutta, now Kolkata, in 1835 (Phillimore, 1958). There are references to an early time ball at Madras, now Chennai, but a plan to operate a regular time ball service there may have been undermined by a reluctance to allocate resources to it. It appears that a gun was used instead until the 1850s (Kinns, 2020). There was an early time ball at the Cape Coast Castle near Accra, now in Ghana (Maclean, 1839). The first time ball outside the British Empire was erected in 1839 by the Dutch at Batavia, now Jakarta, in Indonesia (Melvill & Smits, 1850).

7.3 Time Balls in the USA

An alternative approach, favoured by the United States Navy for several decades after time balls and other land-based signals had been first introduced elsewhere, was to purchase only high-quality chronometers and to rate them carefully before issue to the fleet (Dick, 2003). This was at least in part due to the high cost of providing observatories and time signals along extended coastlines in the era before long-distance telegraph links from a central observatory had become feasible and reliable. The earliest public time signal in the USA was introduced in 1845, when a time ball was dropped at the Naval Observatory in Washington, DC (Dick, 2003). It was not then intended as a signal for rating chronometers, which was undertaken by Observatory staff. Bartky (1983) explored the subsequent history of time balls in the United States, noting the variety of time balls that were eventually arranged around the Atlantic and Pacific Coasts, the Gulf of Mexico and the Great Lakes. The North American signals will be discussed later in more detail.

7.4 *Use of Time Guns*

Time guns were widely used as approximate public time signals but were only of value for rating chronometers when they were under observatory control. A flash pistol was introduced at the Cape of Good Hope in 1833 (Warner, 1979). A time gun in St Helena was fired in 1834 when the time ball was seen to drop. The flash of the gun was the time signal, not sound which propagates slowly at about 330 m/s.

A common early practice was for a gun to be fired by an army contingent at an approximate stated time. The exact time of the flash would be noted by astronomers who would make this time available to ships via a notice on the following day. An example of this practice, and criticisms of it, was published in the *Nautical Magazine* (Madras, 1841; see Kinns, 2020):

> For ascertaining the error of chronometers, you must note the time of the flash of the 8 p.m. gun fired at the fort, and the corresponding Madras mean time, as noted at the observatory, will be sent off the next morning from the master attendant's office; but as the flash is not always distinctly seen at the observatory, too much reliance is not to be placed on this mode of ascertaining the error.

7.5 *Introduction of the Electric Telegraph*

Development of the electric telegraph, often described as the introduction of galvanism, allowed time balls to be dropped and guns to be fired automatically using signals from either a local or a remote observatory. It was introduced at Greenwich in 1852 as a modification to the 1833 arrangement. There was an important remote implementation at Deal, on the east coast of Kent in England, about 100 km from Greenwich, which used the railway telegraph system and entered service in 1855. A signal from Greenwich caused the ball to drop automatically. A return signal to confirm that the ball had dropped was included in the original arrangement. A detailed description of its development up to 1884 has been presented by Beresford and Combridge (1990).

Although it might have revolutionised time ball operation, the use of telegraph lines to control a time ball directly made the process vulnerable to line breakages and false signals due to electric storms for example. There was a declared process for signalling an inaccurate time ball drop, but there were ongoing concerns about reliability of the telegraph signals. A more conservative approach was to regulate an astronomical clock using telegraph signals and to control the time ball drop or firing of a gun from the clock using an additional mechanism.

Charles Piazzi Smyth, the second Astronomer Royal for Scotland, introduced a time gun on the battlements of Edinburgh Castle in 1861 to complement the time ball on the Nelson Monument in Edinburgh which had become operational in 1854. It was fired automatically using a gun clock that was regulated by telegraph from the Monument.

7.6 Airy's Lists of Time Signals in 1861

George Airy (1801–1892) was Astronomer Royal at Greenwich from 1835 to 1881 and was at the heart of time signal development during that time. He had a large impact on the provision of time signals world-wide and compiled lists of time balls that he either knew to exist in 1861 or believed to have been proposed but not implemented (Airy, 1861a, b).

The first list is transcribed below (Airy, 1861a).

> Greenwich; Deal; London (E and I Telegraph Co), City Observatory; Liverpool (E and I Telegraph Co), Victoria Tower; Portsmouth; Edinburgh; Glasgow (a ball was known to exist in 1859 but no particulars relating to it have been found); Cape of Good Hope (Simon's Town); Madras; Calcutta; Sydney; Quebec; Williamstown, Vic, Aus; Washington, US.

The list includes 9 in Great Britain, but not all were visible to mariners and some were intended for domestic use. Two of these were in London: in the Strand (The Electric Time Ball, 1852; Clark, 1852) and at the "City Observatory" at Cornhill, the home of many chronometer makers in London. Only 7 overseas were listed, of which two were in India and two in Australia. The first time ball in the USA in Washington DC (see later), had been established in 1845 for domestic use. The list is clearly not complete, because it did not include St Helena or Ascension, for example, nor did it include Bombay. This may be because many of the early signals were established without involvement by the Astronomer Royal. Early time balls in the Netherlands and Indonesia had been replaced by time discs in the 1850s (see Chaps. 13 and 21).

The single entry for the Cape of Good Hope is surprising. The first time ball at the Cape Observatory was erected in 1836 and supplemented by a repeater time ball to improve visibility to ships in Table Bay (Maclear, 1852, 1853; see Kinns, 2022). The first time ball at Simon's Town was erected in 1857 and operated electrically from November 1861 (Cape Town, 1861; see Kinns, 2022).

The second list of projected time balls is transcribed below (Airy, 1861b).

> Hamburgh (*sic*), 1857; Copenhagen, 1857; River Tyne (Shields), 1859, Gravesend, 1860; Mauritius, 1860.

Again, the list is very short, with only two in Great Britain and three overseas. The first time signal in Mauritius (Lloyd, 1833) had been discontinued and a new time ball there was not operational until 1866 (see Kinns, 2022).

7.7 Semaphore Signals

Correspondence in 1860 concerning a possible time ball at Gravesend illustrates how the Astronomer Royal was asked for advice and how he responded. This is reproduced below. The approach was from Stephen Leach at the Thames Conservancy (Leach, 1860a). His letter to Airy includes the remark "Mr Main, upon whom I called with reference to the subject, in your absence, informed me that Messrs. Maudslay & Field supplied the mechanical part of the apparatus & in order to save you trouble I have written to them for particulars". The Reverend Robert Main was then the First Assistant at Greenwich. Airy replied promptly (Airy, 1860).

> In regard to the Time Signal, if you decide on having a Ball, you cannot do better than consult Messrs. Maudslays but it has long since appeared to me probable that an efficient construction might be made on the Semaphore principle, at a much smaller expense. …

Leach wrote to Airy again a fortnight later (Leach, 1860b).

> I have made a drawing of a Semaphore signal which I think would answer the purpose & which I shall be happy to submit to you. I find however a strong preference on the part of the Conservancy for a Ball, as being the commonly recognised mode of exhibiting the signal.

A few semaphore signals were used in South Africa and at a small number of other locations but most were replaced by time balls. The only semaphore time signals in Admiralty lists from 1898 onwards were at Havana in Cuba, Colombo in Ceylon (now Sri Lanka) and Madras in India. Those at Colombo and Madras had been replaced by time balls by 1919 (Kinns, 2022). A semaphore signal at Moro Castle in Havana was listed in 1911 but had also been replaced by a time ball before 1915. A few other signals, some described as discs at the end of a lever arm, might have been described as semaphores but were listed as discs. A late example was the multiple arm signal on the Fullerton building in Singapore (New Time Signal, 1930).

7.8 *Disc Signals*

Discs rotating about a horizontal axis to simulate a disappearing ball were favoured by the Netherlands in preference to dropping time balls. Time balls on guardships at Hellevoetsluis and Nieuwe Diep had been introduced there in 1846, but shore-based time discs were introduced at Vlissingen in 1854 and replaced the guardship time balls soon afterwards (see Chap. 13). They were used subsequently at various locations in the Netherlands, Belgium, France and Indonesia.

7.9 *Other Alternative Signals*

Other alternatives to a dropping spherical ball were used at some locations. These included cones and drums, a few of which could be collapsed as the signal. A collapsible cone was introduced at Devonport in England in June 1861 (Washington, 1861). The notice included the statement "The time signal is not made as usual by the dropping of a ball, but by the collapse of a Cone, which, when not in use, hangs in a closed state on the Flagstaff. …" The hauling down of a flag was also used as a lower-cost but less accurate alternative.

7.10 *Provision of Time Signals for Mariners Between 1860 and 1880*

Time signals for mariners continued to be introduced worldwide during the period from 1860 to 1880, when the first Admiralty list was published. Seven time balls in Germany were introduced in the 1870s, while France introduced six time balls, a rotating disc at Cherbourg and a flag at Brest. The first time balls at Cadiz in Spain, Lisbon in Portugal and at Trieste, Fiume and Pola in Austria, now in Italy and Croatia, were established. Many more were established in the then British Empire, notably in South Africa, India, Australia, New Zealand, Canada and in island colonies on principal trade routes. Time guns were introduced at many locations, often in parallel with time balls or other visual signals. A weakness of time guns used alone was the lack of preparatory signals, such as the raising of a ball in two stages to warn observers that a time signal was imminent.

Additional time discs were introduced in the Dutch East Indies (now Indonesia), a falling drum was introduced at Rio de Janeiro in Brazil, while flag signals were introduced at Curaçao in the Netherlands West Indies and at Paramaribo in Suriname. Other signals included collapsible objects.

7.11 *Time Signal Proliferation*

The earliest time balls and other time signals were developed for rating marine chronometers but time signals were soon adopted inland, inspired particularly by the development of railway networks and the desire for operation to exact timetables. Most in the British Isles were operated using electric telegraph signals from Greenwich or Edinburgh. Those signals were accurate, but the mechanisms used to operate the visual signal required regular checking and maintenance. A time ball on a prominent public building or the firing of a time gun was often regarded as a statement of civic prestige, but many authorities came to recognise that the costs of erection and particularly maintenance were incompatible with available budgets. It was all too common for an inaccurate time signal to become an object of derision and many time balls and time guns had short lives. The wide range of time signals for public use in the British Isles is described in Chap. 7, showing how the total number of signals was much greater than those listed by the Admiralty for rating of marine chronometers.

References

Adam, C. (1830). *Letter to Captain Robert Wauchope*, dated 13 February 1830. [Annexed to Wauchope, 1836 and Wauchope, 1852].

Airy, G. B. (1860). *Letter to S. W. Leach*, 4 August 1860. RGO 6 614, Leaf 30, Papers of George Airy, Cambridge University Library.

Airy, G. B. (1861a). *List of time balls in actual operation,* May 1861. RGO 6 615, Item 2, Papers of George Airy, Cambridge University Library.
Airy, G. B. (1861b). *List of places at which time balls have been projected, the erection of none of which is known to have been proceeded with.* RGO 6 615, Item 3, Papers of George Airy, Cambridge University Library.
Akkermans, E. J. (2020). *Chronometers and chronometry on British Voyages of Exploration, 1819–1836* (PhD thesis). University of Edinburgh.
Barrow, J. (1824). *Letter to Captain Robert Wauchope*, dated 24 December 1824. [Annexed to Wauchope, 1836].
Barrow, J. (1830). *Letter to Captain Robert Wauchope*, dated 6 July 1830. [Annexed to Wauchope, 1836 and Wauchope, 1852].
Barrow, J. (1833a). *Letter to Captain Robert Wauchope*, dated 20 June 1833. [Annexed to Wauchope, 1836 and Wauchope, 1852].
Barrow, J. (1833b). Announcement of the Greenwich time ball, 28 October 1833. *True Sun*, 29 October 1833, 3.
Bartky, I. R. (1983). Time dissemination, sky with ocean joined. In S. J. Dick & L. E. Doggett (Eds.), *Proceedings of the sesquicentennial symposia of the U.S. Naval Observatory, December 5 and 8 1980.* US Naval Observatory, Washington DC.
Bartky, I. R., & Dick, S. J. (1981). The first time balls. *Journal for the History of Astronomy, 12*, 155–164.
Bateman, D. (2013). The time ball at Greenwich and the evolving methods of control – Part I. *Antiquarian Horology, 34*(2), 198–218; Part II. *Antiquarian Horology, 34*(3), 332–346; Part III. *Antiquarian Horology, 34*(4), 471–488.
Beresford, C. F. C., & Combridge, J. H. (1990). The Deal time ball. *Antiquarian Horology, 19*(1), 33–43. [The paper includes a description by Beresford that was published in *The Military Telegraph Bulletin* in September 1884, with additional explanatory material by Combridge].
Brenton, J. (1836). *Letter to Captain Robert Wauchope*, dated 18 June 1836. [Annexed to Wauchope, 1836].
Cape Town. (1861). Notice headed "Cape Town, Cape of Good Hope, Nov. 21, 1861". *The Times*, 7 January 1862.
Clark, E. (1852). Letter to the Editor, *The Times*, 8 September 1852.
Dictionary of Naval Biography. (1849). *Entry for Robert Wauchope, Rear-Admiral of the Blue*, 1261.
Dick, S. J. (2003). *Sky and Ocean Joined: The U.S. Naval Observatory 1830–2000.* Cambridge University Press.
Editorial. (1833). Article about time ball visibility trials. *John Bull*, 15 September 1833, 7.
Editorial. (1835). The time-ball of St. Helena. *The Nautical Magazine, 4*, 658–660.
Editorial. (1836). Published in *The Nautical Magazine, 5*, 463–464. [Response to Wauchope, 1836].
Editorial. (1852). *The Witness*, 20 October 1852 [with appended letter and testimonials submitted by Wauchope (1852)].
Findlay, A. G. (1867). *Sailing guide for the Ethiopic or South Atlantic Ocean* (5th ed.). Holmes Laurie.
Gill, D. (1913). *A history and description of the Royal Observatory, Cape of Good Hope.* H. M. Stationery Office.
Greenwich Observatory. (1833). Article about time ball visibility trials. *The Standard*, 2 September 1833, 1.
Holmes, T. W. (1983). *The Semaphore: The story of the Admiralty-to-Portsmouth shutter telegraph and semaphore lines 1796–1847.* Arthur H Stockwell Ltd, Ilfracombe.
Howse, D. (1997). *Greenwich Time and the Longitude*. London, Philip Wilson Publishers Ltd.
Inman, J. (1827). *Included in papers marked "Admty 1 September 1827" relating to the Admiralty visitation of the Dock Yards in 1827*, on which occasion the annexed statement is supposed to have been obtained from Professor Inman, UK Hydrographics Office TNA, ADM1/3466.
Kinns, R. (2011). The early history of the Edinburgh time ball and time gun. *International Journal for the history of Engineering & Technology, 81*(2), 264–290.

Kinns, R. (2020). Time signals for mariners in India, Burma and Ceylon. *Journal of Astronomical History and Heritage, 23*(3), 523–552.
Kinns, R. (2022). Visual time signals for mariners between their introduction and 1947: A new perspective. *Journal of Astronomical History and Heritage, 25*(4), 601–713.
Kinns, R., Fuller, P., & Bateman, D. (2021). Exploring the Portsmouth time balls. *Journal of Astronomical History and Heritage, 24*(3), 751–769.
Leach, S. W. (1860a). *Letter to G.B. Airy*, 1 August 1860. RGO 6 614, Item 29, Papers of George Airy, Cambridge University Library.
Leach, S. W. (1860b). *Letter to G.B. Airy*, 18 August 1860. RGO 6 614, Item 33, Papers of George Airy, Cambridge University Library.
Lloyd, J. A. (1833). Rating Chronometers in the Mauritius, 19 April 1833. Notice published in the *Nautical Magazine, 5*(1835), 136.
Maclean, G. (1839). Notice to mariners, dated 27 July 1839, published in *The Naval Chronicle*, 1840, 128.
Maclear, T. (1852). The Observatory Time Ball – Cape of Good Hope, dated 7 July 1852. Notice in the *Nautical Magazine and Naval Chronicle, 21*, 611–612.
Maclear, T. (1853). *Establishment of an Additional Time Ball at the Cape of Good Hope*. Notice to Mariners [No. 153]. Issued by the Hydrographic Office, 13 December 1853. Reproduced in the 1854 *Nautical Magazine and Naval Chronicle, 23*, 54.
Madras. (1841). Entry in the *Nautical Magazine and Naval Chronicle for 1841*, 363.
Melvill, P. (Baron Melvill van Carnbee and Lieutenant, DRN), & Smits, H. D. A. (Lieutenant DRN). (1850). *The Seaman's guide*, published by the Hydrographic Office, Admiralty, London. Republished as a digitised facsimile edition by the British Library, printed in Great Britain by Amazon.
New Time Signal. (1930). Article, new time signal on post office: Electric contact with Mount Faber. *The Singapore Free Press and Mercantile Advertiser*, 4 February 1930, 17.
Obituary, Wauchope, R. (1862). *Naval and Military Gazette,* 21 June 1862, 386–387.
Patterson, J. (1858). *History and genealogy of the family of Wauchope of Niddrie-Merschell*, Edinburgh, printed for private circulation. https://deriv.nls.uk/dcn23/9538/95384318.23.pdf
Phillimore, R. H. (1958). *Historical Records of the Survey of India* (Vol. IV). Survey of India.
Spencer Jones, H. (1952). The Greenwich time ball. Letter to the editor of *Engineering*, 13 November 1952. [Harold Spencer Jones was Astronomer Royal 1933–1955].
The Electric Time Ball at the Strand. (1852). *The Times,* 19 August 1852.
The Proposed Time Balls. (1892). Article published in *The Singapore Free Press and Mercantile Advertiser*, 21 March 1892, 3.
Warner, B. (1979). *Astronomers at the Royal Observatory, Cape of Good Hope*. Balkema.
Washington, J. (1861). *Time signal for Plymouth sound, notice to mariners [No. 21]; England – South Coast,* 18 June 1861. Hydrographic Office, Admiralty, London. [Washington was the Hydrographer of the Navy].
Wauchope, R. (1818). *Remark book of HMS Eurydice*. File MP 84-34. UK Hydrographic Office, Taunton.
Wauchope, R. (1827). *Letter to John Wilson Croker*, dated 21 June 1827, annotated by the Admiralty on 1 September 1827, included in papers marked "*Admty 1 September 1827*" relating to the Admiralty visitation of the Dock Yards in 1827, UK Hydrographics Office TNA, ADM1/3466.
Wauchope, R. (1830a). Plan for ascertaining the rates of chronometers by signal. *Edinburgh New Philosophical Journal, 8*, 160–162. [This first part was submitted on 23 November 1829].
Wauchope, R. (1830b). Plan for ascertaining the rates of chronometers by signal. *Edinburgh New Philosophical Journal, 8*, 289–291. [This second part, which included the diagram, was submitted on 1 February 1830].
Wauchope, R. (1830c). *Letter to the Admiralty*, dated June 1830.
Wauchope, R. (1833). *Letter to Captain the Hon. George Elliot, Admiralty*, dated 17 June 1833. [Annexed to Wauchope, 1836 and Wauchope, 1852].

Wauchope, R. (1836). *Letter to the Editor of The Nautical Magazine*, 2 April 1836, with supporting testimonials concerning time ball invention, published in *The Nautical Magazine, 5*, 460–463.

Wauchope, R. (1852). *Letter to the Editor of The Naval and Military Gazette*, 13 September 1852, with supporting testimonials concerning time ball invention, previously published in *The Nautical Magazine, 5*, 460–464. Republished with a supporting editorial in *The Witness*, 20 October 1852.

Webb, A. (2010). *The expansion of British Naval Hydrographic Administration, 1808–1829* (unpublished PhD thesis). University of Exeter.

Wood, H. (1856). *Time ball on semaphore at H.M. Dock Yard Portsmouth: Answers to questions submitted by professor Airy*, dated 4 September 1856. Royal Greenwich Observatory Archives RGO 6 613, Papers of George Airy, Cambridge University Library.

Chapter 4
The Greenwich Time Ball and Its Evolution

Douglas Bateman

Abstract The Greenwich time ball is surely the best known time ball, given its success and prominent location, both on, and adjacent to, buildings of architectural and historical importance. It has operated almost continuously since 1833, with only a break during the Second World War and up to 1952. It was designed and erected between June and October 1833 and remained in service with only minor modifications for over 80 years, setting a remarkable standard for later time ball arrangements.

This chapter describes the early history and the rebuild in 1914-1919, leading to automation in the 1960s. For the latter the source of time was a pendulum clock controlled by the six-pips from the BBC, eventually being superseded by electronic control. In this case the time was derived from the low frequency radio time service, linked to the time standards in the National Physical Laboratory, and then, in 2017, by time signals from the GPS satellites.

1 Introduction

The Greenwich time ball is surely the most well-known, given its success and prominent location, both on, and adjacent to, buildings of architectural and historical importance. Figure 4.1 shows its modern appearance. It has operated almost continuously since 1833, with only a break during the Second World War and up to 1952. Apart from renovations to the Observatory in 1958, this "premier" time ball has had a very long period as an accurate public timekeeper.

This chapter describes the early history, the rebuild in 1914–1919, leading to automation in the 1960s. For the latter the source of time was a pendulum clock controlled by the six-pips from the BBC, eventually being superseded by electronic control. In this case the time was derived from the low frequency radio time service, linked to the time standards in the National Physical Laboratory, and then, in 2017, by time signals from the GPS satellites.

D. Bateman (✉)
Independent Researcher, Crowthorne, UK

R. Kinns (ed.), *The World of Visual Time Signals for Mariners*, Historical & Cultural Astronomy, https://doi.org/10.1007/978-3-031-57334-7_4

Fig. 4.1 A view of the observatory on high ground, some 160 feet (about 50 metres) above the Thames

2 The Early Days

As stated earlier Captain Robert Wauchope had suggested methods of giving an 'instantaneous' signal of observatory time so that sailors could check their chronometers. After several years of lobbying and a trial system at Portsmouth, a letter and "proof" of the concept finally persuaded the Admiralty to proceed. A letter from his London address, dated 17 June 1833, states "Their Lordships having been please to approve of, and to adapt my plan for finding the rates of chronometers by instantaneous Signal, a copy of which, containing their Lordships letter to me on the subject, I have the honor to inclose" (The National Archives). This letter is date stamped within an inked octagon JU. 21 1833. The very same day, with the same date stamp, (see Fig. 4.2) a memo was sent to the 6th Astronomer Royal, John Pond, FRS, as follows:

> 18 June [1833]
>
> "Send the enclosed to Mr Pond & direct him to state whether he sees any way by action to a measure of this kind being adopted at the Royal Observatory for the information of ships in the river by which they may ascertain the rating of their Chronometers.
>
> Initials".

Someone, perhaps not understanding the concept, had added in pencil "Pray what?".

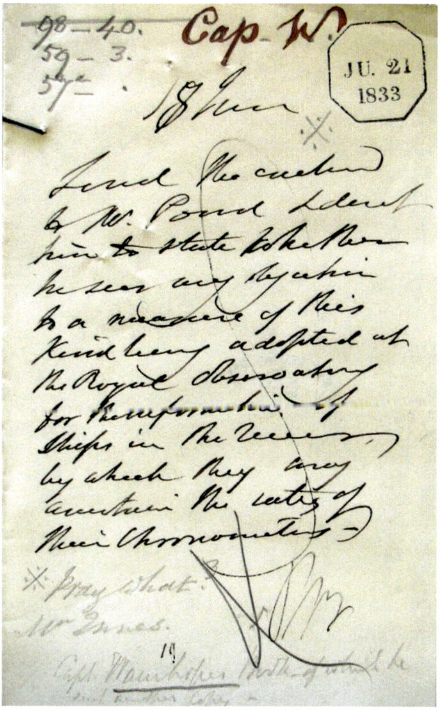
Cap. W.
JU. 21 1833
18 June
Send the circular to Mr. Pond & desire him to state whether he sees any objection to a measure of this kind being adopted at the Royal Observatory for the information of Ships in the River by which they may ascertain the rates of their Chronometers –

Fig. 4.2 The memo in the National Archives with the implicit authorisation to construct the time ball

Pond acted more or less immediately, and within a period of 3–4 months the time ball was constructed by Maudslay, Sons and Field of Lambeth, a company that was founded by Henry Maudslay and later made time balls for other locations (Maudslay & Field; see Chap. 5). The cost was £180, which is equivalent to about £18,000 now.

There are four factors that contributed to the success of the installation. The first is the presence of two ornamental turrets at the east and west corners of Flamsteed House, the original observatory. Perhaps these are the architectural features that Sir Christopher Wren, in his design allowed "a little for pompe". The turrets are easily seen from the river Thames, and little modification was required to the roof timbers to give strong support to the ball mast. Secondly, the mast rising from the centre of the turret conveniently allowed the mechanism to be reasonably close to the structural walls, as shown in Fig. 4.3. The third factor is the choice of the competent and very experienced engineers, Maudslay, Sons and Field. Fourthly, and by no means the least, the time ball was managed by the very astronomers who were responsible for maintaining mean time for the nation.

To set the scene further, Fig. 4.4 is a familiar view of Flamsteed House where we can imagine where the mechanism is located. Figure 4.5 is the Octagon room where the mechanism is hidden in a tiny room called the time ball cupboard.

Soon after installation a number of engravings appeared in various journals, and the first comprehensive description, Fig. 4.6, appeared in the *Nautical Magazine* in 1835, with a full-page illustration (Nautical Magazine, 1835). In 1845 we come to

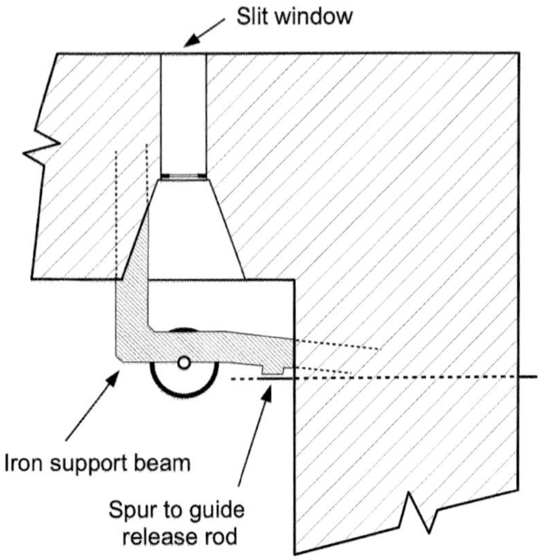

Fig. 4.3 A scale drawing of the location of the piston and cylinder. The iron support beam is original and the narrow slit window is just detectable to the left of the left tall window, as in Fig. 4.1

the next publication, The *Illustrated London Almanack* (1845) which contains the much more widely reproduced drawing, Fig. 4.7. The *Almanack* also added details of component parts such as the base of the mast, see Figs. 4.8 and 4.9 (Illustrated London Almanack).

Of historical importance is the illustration from the Scenes of the observatory, from *The Graphic* of 8 August 1855, as in Fig. 4.10 (The Graphic, 1855).

3 Reliability

Each year the Astronomer Royal was obliged to submit a formal report to the Board of Visitors, which was comprised of representatives of the Admiralty, Nautical Almanac office, the Royal Astronomical Society, and leading academics (Dolan, 2024). The primary purpose was to describe the astronomical achievements with sections on the telescopes and any improvements. Further sections covered the buildings, their condition, alterations or additions. Staff matters were given, initially in broad terms, then in some detail. For example, the report of 1916 stated that "Of the 26 computers who were employed at the commencement of the war, 17 volunteered for military service."

Bad weather was the main cause of the ball not dropping although only from 1870 onwards was this mentioned in the reports. Most years have the stock sentence "On (typically 3 to 10) days the ball was not raised on account of the violence of the wind". Although the Observatory was recording wind speeds there is no mention of a limiting speed for raising the ball, and one must presume some notional value or judgement. Snow and ice also caused infrequent failures. Occasionally the "clips"

Fig. 4.4 The top of the mast carries the N, E, S, W cross tree with a weather vane above, the visible part above the ball extending 7 feet. Below the roof of the turret the mast extends to be level with the roof of Flamsteed House (at the level of the cornice). The lift mechanism and air-brake cylinder fits all the way from the roof line to the floor of the Octagon room, a short distance below the tall windows. A crank is on the wall inside the roof space of the sloping slate roof, originally extending down into the lobby for the release lever and clock

that held the raised ball would fail. If the drop did fail due to mechanical (the clips), electrical or (very rare) operator error, the ball was dropped at 2 pm precisely.

Apart from keeping a factual record of successes and failures, there will have been some sensitivity to wind speed amongst the staff caused by the collapse of the mast on 6 December 1855. The Observatory has a watercolour of this event, as painted by one of Airy's sons. However, there is no mention of this in the annual reports!

The mast was made of soft wood and more disposed to rotting than oak or teak. There is, however, something of a design flaw that would encourage the wood to decay where the ball comes to rest. Rainwater would be expected to run down the mast and into the drain set into the metal mast support inside the turret.

On 2 April 1895 the chain broke as the ball was being raised, and the ball was not raised again until 2 May, after a new chain had been obtained and the cylinder which acts as an air cushion had been cleaned. This was found to be very rusty, and it was discovered that the upper portion of the wooden framework of the ball was rotten,

Fig. 4.5 The Octagon Room with the time ball mechanism in a cupboard to the left of the door and behind the replica clocks. The doorway leads to the staircase to the lobby below

which allowed the rain to run down the mast into the cylinder. A rusty spring in the cylinder was found in 2011.

4 Improvements, 1914–1919

The time ball and mechanism operated without too much trouble for many years although the annual reports do mention failures of the "clips" which held the time ball when raised. No doubt wear on the clips and piston edge contributed to slippage. Of course, the ball was not raised in very severe weather, with total failures amounting to a few days a year. The decision was made to renew the whole structure and the 1914 Annual Report states:

> The renewal of the Time-Ball and the general overhaul and reconstruction of its mechanism has been sanctioned. A new teak mast is being constructed for the Observatory by the Works Department of the Admiralty. Messrs. E. Dent & Co. are supplying a new aluminium ball, and a new winch for winding, and are repairing and adapting parts of the existing machinery. The winch will be removed from its present position near the Octagon Room to the Ball Lobby.

The Observatory may have chosen Dent & Co because of their proven ability in making precision pendulum clocks, and their supply of the Westminster clock, popularly known as Big Ben. Also, they had modified the Hardy regulator that gave the eye-and-ear check for the astronomers at the Airy Transit Circle (the primary

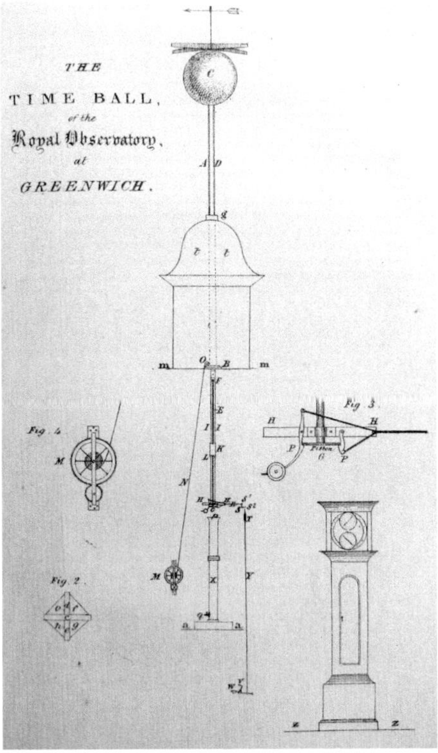

Fig. 4.6 Note that the cylinder is in two parts and the piston is raised clear of the cylinder. The clock shown is a rather questionable representation of 'Graham no 3'

telescope). Dent had also made the chronograph and rotary pendulum in 1855. Nevertheless, Dent tackled a job closer to heavy engineering! In March, 1914, they submitted an estimate as follows (Dent, 1914).

> New 5 foot Aluminium Ball with hollow square centre and lignum vitae bushes, weather ring and aluminium plate for raising rods. All painted. £45
>
> Phosphor bronze and steel tubes for raising ball. Steel tube piston rod and plunger, cast iron brackets and remaking discharging escapement with tempered steel hexagon locking plate. £35
>
> New winch for raising ball, including lifting plate lines, pulleys etc. £15
>
> Erecting, fixing etc. £20

(As an aside, note the use of the phrase "discharging escapement" by a company that specialises in clocks, whereas the then current term was "clips" and nowadays they are called latches.) The ball was reported as ready for delivery in October 1914, but as we have seen, the war intervened. The next letter in March 1919 is to Dent to say that their work may proceed concurrently with that of the Civil Engineers Department of the Royal Naval College. This is for the new mast, which was to be fitted with a lightning conductor by the College. In parallel, enquiries were made about electro-plating the ball, but this was not pursued. The work proceeded, but not without some argument about final costs because extra requirements were added.

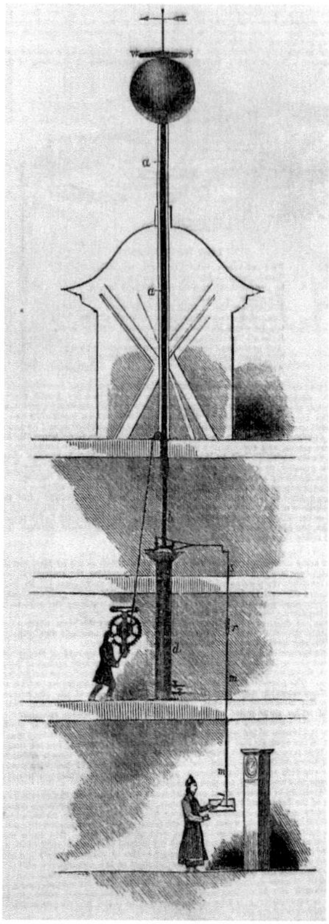

Fig. 4.7 The popularly reproduced engraving of 1845

Then in May the firm submitted an invoice for additional work not included in the original estimate as follows:

> New locking lever and additions to winch; heavy iron shoe for mast and making pattern for same; 2 large counterpoise weights with framed pulleys; 1 pulley to lead up to winch; 1 plunger; 1 large spiral spring with flange; 1 heavy brass plate for electric apparatus and fitting up; removing bruises from Ball, paint etc. £96 6s

The final total was recorded as £226 6s and after some stern letters by both parties, the bill was reduced to £206 by Dent, claiming that this left them "absolutely without profit on the transaction". There had been late deliveries and complications due to the Observatory asking for extras such as the counterpoise weights, and Dent had failed to list all the planned items in their estimates.

Taking the structural changes, the teak mast is an extremely well made and very sturdy structure and although not very obvious, it had been shortened. Originally it gave a 10 foot drop. The 1919 shortening will have been based on three considerations: a) the amount of the drop is not critical for the time indication, b) it will reduce the risk of wind damage when the ball is raised, and more importantly, c) to incorporate the mechanical design changes due to not having the latches engage directly on the piston. The drop is currently 2.19 m (7 feet 2 inches), and adding the

Fig. 4.8 The base of the mast with a rim to collect any rainwater that may come down the mast

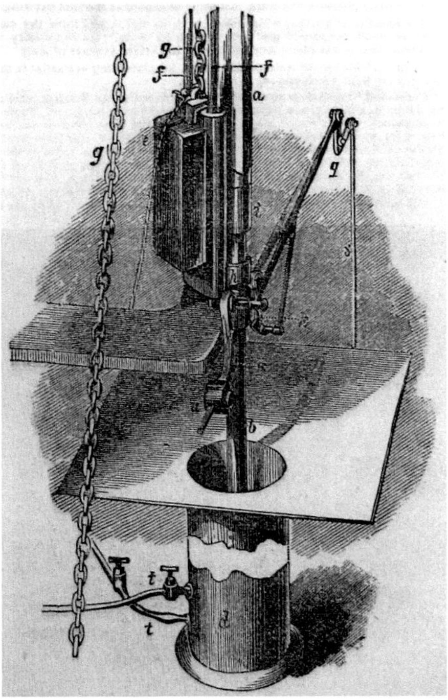

Fig. 4.9 The release mechanism with counterweight for the lift chain. The 'clips' and elbow crank are shown although the engraving omits the extent of L-shaped support beam. The clips engaged directly on the piston when raised. Note the taps at the base of the cylinder

Fig. 4.10 The individual caption in the Graphic of 8 August 1855, clockwise, are: The Octagon room (Flamsteed's observatory) from the courtyard; The chronograph – reading off the observations of the previous night; The mean solar standard – the timekeeper of the Country; The public clock at the entrance to the observatory. Note the hand resetting the time ball release levers

diameter of the ball and about 0.15 m "clearance" at the top and bottom of the mast, the external mast length is about 3.91 m (12 feet 10 inches).

The other external feature is the ball itself, a cleverly made sphere from sheet aluminium. A detail was the addition of a rainwater deflector skirt—the weather ring in the estimate. This ensures that the surface of the ball itself is no longer a conduit for water to the base of the mast. The size is generally quoted as 5 feet in diameter, Indeed this was confirmed from some work on the time ball (with scaffolding) in 2013 as being 5 feet 1 inch in diameter (1.55 m) (Bateman, 2013). The ball will have had a smooth finish, rather than the somewhat dented appearance which was due to careless workmanship during major renovations to Flamsteed House in the period 1957–1960. It was speculated that the ball had been dented by the builders, in fact the diligent researcher, Graham Dolan, who has interest in

Fig. 4.11 The ball and mast in 2011

Greenwich and the history of the Royal Observatory found a film clip showing the building work (Dolan, 2022). The clip has a release date of 24 October 1958 by British Pathé, as at http://tinyurl.com/timeballdent (Dolan, 2022). The film runs for a total of 65 seconds and shows building work on Flamsteed House and about 3 seconds of the builders rolling the ball. Much of the film clip shows the untidy nature of the building site. The extract from the film shows that the ball is being rolled, probably from an out-of-the-way corner. The ball is clearly about to suffer more damage but the film clip shows that the workmen are not being deliberately careless—just a case of not respecting the historical artefacts as at the present time.

Figures 4.11 and 4.12 show the ball and details. Although these photographs were taken in 2011 they almost certainly represent the structure as in 1919. Inside the turret there is the mast and the base plate, see Figs. 4.13 and 4.14.

5 The Revised Lift Mechanism, 1914–1919

Instead of a single chain hoist acting on a single wooden triangular "rod" to lift the ball, the new arrangement consisted of a pair of rods to the ball, a pair of lift tubes and a pair of wire ropes attached to a lift plate, with the pair of ropes combining to a single rope to the new winch (and in 1960, to an electric motor). The other major mechanical change was the separation of the piston face for the latches to an

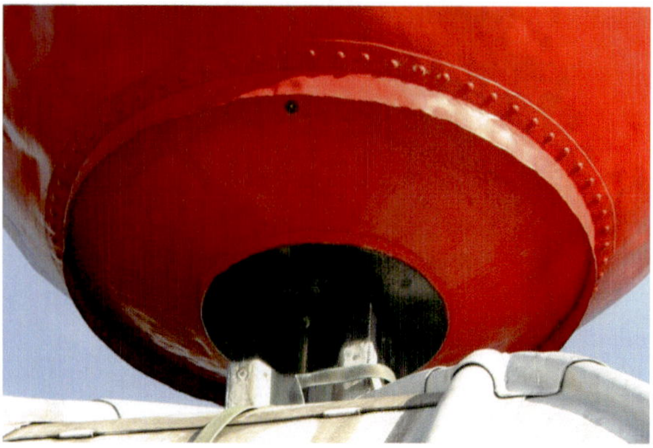

Fig. 4.12 The rainwater deflecting skirt. The lead flashing fits tightly around the mast and the lightning conductor threads its way inside the ball up the side of the mast. The small black disc (actually a 2 pence coin) covers an aperture where an endoscope was used to inspect the inside of the ball. The west-side phosphor-bronze support rod is also visible

Fig. 4.13 The upper part of the mast inside the turret and the counterweights fitted in 1919

independent latch bar on the piston rod. The latter inevitably led to a shortening of the travel of the piston in the cylinder, and hence the (beneficial) shortening of the mast. Figure 4.15 shows the general arrangement of the main components.

An interesting detail is that the latch bar on the piston rod pivots about an axis through the diameter of the piston rod so that the weight of the assembly, when raised, is resting evenly on each latch. The latch bar has short hard steel faces that engage with the latches. Note that this assembly takes the entire weight of the ball, rods and piston when fully raised whereas the weight is taken by a strong yoke at ceiling level *b* when the pulleys, *c* are raising the ball. Another detail is the piston in two parts as discovered during an overhaul, and although not detailed in the Dent rebuild estimates, it was probably part of the changes in 1919 (Watkinson, 2011). The overhaul also unexpectedly revealed a large coil spring buffer in the bottom of

Fig. 4.14 The replacement base plate with a drain hole (yellow arrow) and a crack in the casting (red arrow). The buffer springs compress by about 6 mm when the ball is at rest. The counterweight steel wires are linked to each collar on either side of the mast

the cylinder—see below. It is likely that the only original components from 1833 are the thick iron latch support beam set in the walls, and the cylinder, and the latch release relay crank the other side of the wall.

It has to be presumed that the single wire rope from the curved yoke *f* to the winch followed some pulley arrangement to the winch in the time ball lobby, but there are no obvious apertures in the wall or remnants of fixings to give a clue as to the route taken by the rope. At present, the wire rope passes over no fewer than seven pulleys to reach the motorised winch, as described below.

6 Automation in 1960 and Improvements in 1982

There are no clear records and photographs of the work on the time ball itself during the period 1958 to 1960, although the clock, motor and pulley system almost certainly date from this time. This was a period of transition, the Old Royal Observatory being incorporated into the National Maritime Museum, and building work undertaken by the Ministry of Works (Littlewood and Butler, 1988). At one point a press release stated that the dropping of the time ball would be "discontinued shortly for a few months" and that "It has now become necessary to carry out an extensive refit of the mechanism … ". Among the few documents found in the National Maritime Museum files is another in the form of a press release, signed by Lt Cdr D W Waters, Curator of Navigation and Astronomy, dated 26 October 1960 (National Maritime Museum, 2012). The document, headed "Flamsteed House, Time Ball and Electric Clocks", states:

> The Clock at the gate of Flamsteed House, the Electric "Tompson" (sic) Clock in the Octagon Room, and the Time Ball are all worked off a Master Electric Clock in the Basement of Flamsteed House. The Accuracy of the Master Clock is checked weekly by the

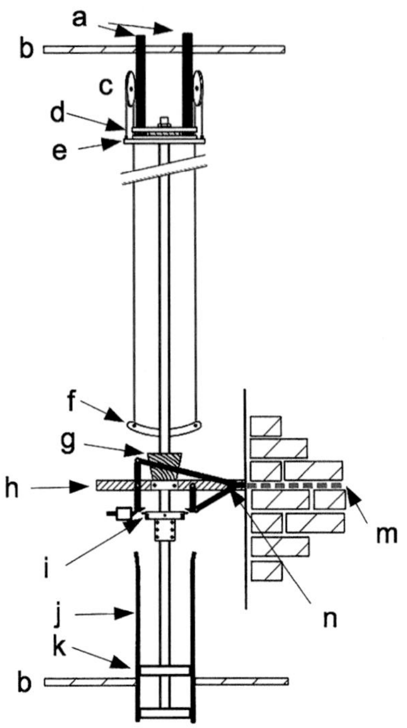

Fig. 4.15 Schematic view of the revised lifting mechanism from 1919 about to engage the latches. Note that the lift rods and piston rod are actually tubes, but that the word rod, as in push rod, is a more convenient term. (a) Lift rods extending to the mast. (b) Floor levels. (c) Lift pulleys. (d) Plate connecting piston and the lift rods. (e) Lifting plate raised by the wire ropes. (f) Yoke between twin lift ropes and rope to the winch. (g) Wooden block to support the at-rest lifting plate. (h) Latch support beam, 1833. (i) Latch bar about to engage latches. (j) Cylinder extension. (k) Dual piston. (m) Rod to release relay crank. (n) Peg to limit travel of latch levers

> Ministry of Works, which is the Department for its maintenance, and it keeps time to within about 1 second of Greenwich Mean Time.

A Synchronome clock system (described in more detail in the next section) was the time reference with attendant sequencer, relays and electronics for the three actions: stage 1, raise the ball halfway; stage 2, raise the ball to the top of the mast; stage 3, release. The clock is of standard design with impulses every half-minute, which is of no consequence for the impulse on the precise second for release, and similarly to finish the raising at 2 minutes before 1 pm. However, for the pulley system and safety check, stage 1 is started 5½ minutes before 1 pm so that the ball starts to rise at 5 minutes before the hour. This starting sequence was called the 'lost motion' by the engineers in 1981, see below.

The heart of any automatic system is the motor and winch as shown in Fig. 4.16. Electrically, the motor is reversible to raise the ball and then lower the lift plate prior to the release, and has a brake that locks the motor shaft when the supply is removed. This is necessary to "hold" the ball when half raised. To raise or lower the ball completely by hand, a crank handle fits on the squared shaft and the brake panel is pressed

Fig. 4.16 The motor and drum without the chain safety guard. From left to right the drum, 4:1 chain reduction, 40:1 gear box, reversible motor, electromagnetic brake and squared shaft for a handle

down to release the brake. The name plate gives the rating and output speed of the motor as 26 rpm. The on-line *Grace's Guide to British Industrial History* shows that Normand Electrical were in business on the north side of Clapham Common from 1937 and the website has an advertisement for a motor of 26–2800 rpm (Graces Guide).

Although the mechanism of a cable-raised system is fundamentally the same as in 1919, changes were made to suit the motorised method using a safety mechanism known as a "lost motion". For access to the cables, pulleys and the latch mechanism, it is necessary to climb the iron ladder to the next level—a floor level halfway up the height of the Octagon Room. The first thing to be seen is the rather puzzling complex of rods, wire ropes, pulleys and switches. These form the "lost motion" and safety mechanism. Because the sequence starts at 5½ minutes before the hour, 30 seconds has to be "lost" before the ball starts to be raised. The method of losing the time is for a carriage to slowly climb the rods. Most importantly, it carries springs that compress about 30 mm against a pair of collars on the rods. If the ball is jammed, the springs compress further until a switch is operated and the motor stops, thus acting as a safety mechanism. Figure 4.17 shows the general arrangement. Figures 4.18, 4.19, and 4.20 show the main components of the mechanism as in 2012.

7 The Solenoid That Is Not a Solenoid

Historically the time ball was released by a simple long rod, assumed with sufficient weight, that pulled the latches via a crank the other side of the east wall of the Octagon Room, with the rod descending to the time ball lobby by the entrance. Initially worked by hand, the "trigger" was released electrically from 1852. For the automation in 1960 a "mechanism" was installed in the tiny loft space above the time ball lobby and close to the crank. This is referred to as the "solenoid/thruster" on the notes and drawings by the Property Services Agency for the planned

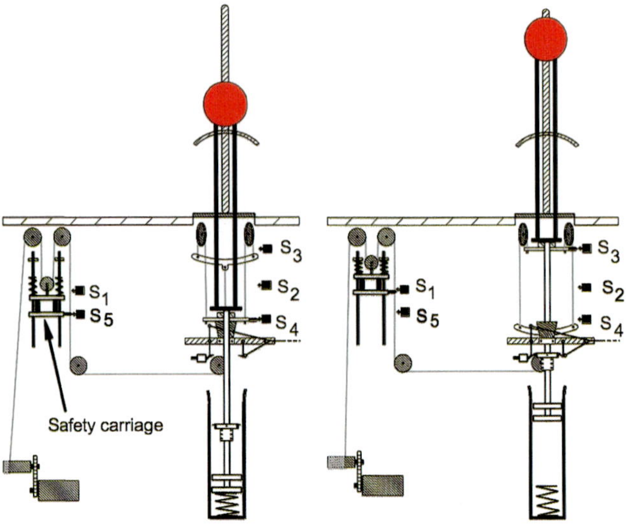

Fig. 4.17 The complete sequence starts with either the manual push buttons or electronic control with stage 1 that starts the motor so that the 'lost motion' safety carriage rises to press against the collars: this process takes 30 seconds. If the ball is jammed through ice or snow or other reason, the carriage compresses the springs until the switch S1 activates and the sequence is cancelled and all power is cut. In normal operation the motor continues and raises the ball until S2 is activated and the motor stops. After the time delay the motor starts and the lift plate continues until S3 is activated when a timed relay is set running. At around 12:59 pm the relay switches the motor to reverse, the latches engage and the lift-plate descends to S4 and the safety carriage descends to S5, leaving 10 seconds before 1 o'clock during which a warning bell rings. At 1 pm the latches are withdrawn and the ball drops. (not to scale and pulley layout simplified)

Fig. 4.18 A view of the main lift and latch assembly. From the top, the piston rod and lift rods (actually pipes) to the mast and plate, the lifting plate with wire ropes in its lowest rest position, the 1833 iron beam, parts of the latch pull rods (with counter-balance), and a cover over the cylinder, and just below, part of the pulley beam

Fig. 4.19 The 'lost motion' mechanism with the ball raised half way. The figure shows, far left, the wire rope from the winch, cradle and partially compressed springs, wire rope descending to the pulleys for the lift, lift platform, lift rods to the mast and piston rod

Fig. 4.20 The ball raised and resting on the latches. The latches follow the 1919 redesign and were renewed in 1998 by Peter Watkinson. Note that the lift plate has returned and is resting on the shaped wooden block

improvements in 1981. Figure 4.21 shows all that can be seen with the aid of a torch through the small hatch into the loft.

It was reasonable to call it a solenoid thruster from this cursory inspection because the cylindrical shape is consistent with a coil pulling an armature. Enough of a crank can be seen to show that there is an upward push on a rocker to pull the bell crank down to exert a pull on the release latches the other side of the wall. Further investigation was required and the Curator of Horology, Rory McEvoy, wriggled through the hatch with the camera to take some close-ups. Visitors ascending the stairs to the Octagon Room were amused to see a pair of legs dangling from the hatch. Figure 4.22 shows the mystery machine, clearly not a solenoid but some form of motor with a pair of thrust rods pushing a rocker bar that in turn pulls the crank, thus mimicking the original downward pull to the release trigger below. The crank is almost certainly one of the few original items dating from 1833.

The fact that it is other than a solenoid had obviously not been passed on to Observatory staff during the 1960s, and the term solenoid was used until the recent

Fig. 4.21 What little can be seen of the 'solenoid/thruster' through the small hatch into the loft space above the time ball lobby

investigations. It was "discovered" to be otherwise by Peter Watkinson in 1985, who regularly serviced the time ball machinery. In response to a query he replied:

> Well, I suppose I expected a solenoid as that is what I had been led to believe, but I remember thinking prior to viewing it that its action was atypical of a solenoid – unless damped in some way and the sound was wrong also (Watkinson, 2011).

The thruster was removed during 2010 by Watkinson for a service. The label shows it to be a powerful mechanism, trade name "Thrustor", with a 180 watt motor and rods delivering a thrust of 75 lb; the label shows it to have been made by Associated Electrical Industries, a company with a high reputation (Associated Electrical Industries). Watkinson describes the machine as a centrifugal pump that circulates hydraulic oil against a large diameter annular piston to which the two external rods are attached. It thus delivers a smooth linear thrust, and the power and speed of response is such that there is virtually no delay between the signal to the motor and the release of the ball.

8 The Mystery Spring, Overall Weight and 'Pneumatics'

Just as the solenoid was found to be an electro-hydraulic mechanism, another surprise was finding a very large spring in the base of the cylinder. During 2011 some unusual noises had been heard at the end of each descent of the ball – both in the

Fig. 4.22 Not a solenoid but a motor driven device with push rods that push a rocker that pulls on the crank that connects through the wall to the latches

cylinder and a thud transmitted throughout the rods and piston, presumably due to the buffer springs at the base of the mast "bottoming". With some ingenuity, Watkinson unbolted the cylinder from its base flange and jacked the whole cylinder up enough to reveal a large spring in two pieces. The dirt on the spring is certainly reminiscent of oil and water contamination. Watkinson reported that the old spring was 250 mm free-height, 100 mm at max compression and 150 mm at loaded rest. A new spring was made close to these measurements (Watkinson, 2011).

Given that Dent quoted for a spring "1 large spiral spring with flange" it is assumed that the spring will have been there since 1919 and somehow some water or moisture will have entered the cylinder.

In order for the spring maker to fabricate the spring it was necessary to give a load, or rather, the weight of the time ball assembly. Watkinson assembled a lever that could lift the whole time ball system by raising the main lifting plate. With a spring balance and a lever ratio of a little over 4:1, he 'weighed' the system at about 133 kg. This compares with DB's calculated estimate of about 90 kg. In both cases the weight of the counterweights has been deducted. For these weights (see Fig. 4.13) the annual report for the 1914–1919 rebuild gave a figure of 1 cwt (51 kg) whereas both Watkinson and myself, from the dimensions and density of cast iron, estimate 60 kg. The calculated estimate is based on the total of all the lift components: phosphor-bronze rods, steel tubes, piston tube, pistons, lift plate and latch blocks, and the ball itself. Because the dimensions of all the lift components are known with some accuracy, their total, by calculation, is 96.2 ± 2 kg. For the ball, the estimate is based on a thickness of aluminium alloy of (imperial 16 SWG) 1.6 mm for a sphere diameter of 1.55 m giving 41 kg. There is an aluminium base for the ball of say 10 kg, and perhaps 10 kg for the internal wooden frame, giving a total of about 60 kg. The combined estimates are 156.2 kg with 60 kg deducted to give the weight on the latches of about 96 kg, rounded to 100 kg. Comparing Watkinson's weighing of 133 kg with the calculated estimate of 100 kg, there is a discrepancy of 33 kg. Given that the component parts can be estimated with reasonable accuracy, the ball must account for the difference, suggesting a weight of 103 kg. In other words, the two methods give a ball weight ranging from 60–93 kg. If the aluminium is the imperial 14 SWG, 2.03 mm, the calculated estimate would raise the mass by about 10 kg. In 2006 the interior of the ball was inspected with an endoscope to check for any obvious weakness or rotting of the internal wooden structure. None was found, but neither were the observations (made with some difficulty) sufficient to give much idea of the overall mass of the wooden structure.

Concerning the pneumatics, the rate of descent is controlled by the "spring of air" and all the early accounts and illustrations refer to the rate of descent being controlled by a tap at the base of the cylinder. Figure 4.23 is the general view inside the time ball cupboard, at the floor level of the Octagon Room (see also Fig. 4.5) and there is certainly a tap near the base of the cylinder, but from the covering of dust on and around it, does not appear to have been turned for many years.

Close inspection of the cylinder shows that a series of holes have been drilled in a helix, at about 1 foot intervals and then blocked with tapered plugs. This suggests some experimentation to control the rate of descent. One of the holes, open for many years, is responsible for the large oil stain, caused no doubt by the rush of air when the piston descends. Referring back to "the spring of air" it is obvious to all spectators that when the ball descends, a little more after halfway down it halts and *rebounds* a few centimetres. This happens when the piston passes the hole and cuts off the escape of air, and descent continues more slowly, relying on the leakage of air around the piston. If we take the side hole aperture of 8 mm and assume the gap between the piston and cylinder wall to be of the same area, then the clearance around the piston is about 0.08 mm, or 0.003 in. This clearance may sound generous but whilst the term cylinder has been used, in reality it is a simple iron *pipe*. Note that if the piston was a tight fit it would fall a little and then stop!

The tap may have been more important in the early days, such as before the current ball with a rain deflecting "skirt" to drain away *rain water*. In fact, the Annual Reports to the Board of Visitors for 1895 stated: "[the cylinder] ... was found to be very rusty, and on further examination it was discovered that the upper portion of the wooden framework of the ball was rotten, and allowed the rain to run down the mast into the cylinder." The cylinder itself is a relatively straightforward affair, in three sections of 5 ft 8 in, 12 in and 3 ft 8 in to give a total length of 10 ft 4 in (3.15 m).

Fig. 4.23 Inside the time ball cupboard, from left to right, the steel ladder to the next sub-floor, an interface relay box (1992), the control relays box (1982), the reversible motor, the (brown) push buttons for manual control, a dirty oil stain on the wall, and the cylinder

9 The Synchronome Clock—The Timepiece for the Automation

When the motor driven update took place in 1960 the source of time was a Synchronome clock with a series of cams to initiate the movement at 5½ minutes before 1 pm, raise fully at 2 minutes and the drop at precisely 1 pm. The cam system, however, was not able to give the Daylight Saving changes. (Synchronome clocks had a long productive period from the 1920s to the 1960s providing "master" clocks operating slave dials in distributed time systems in factories, hospitals, offices, colleges and schools. The slave dials would all advance together at half minute intervals (Miles, 2019).

Figure 4.24 shows the clock in a cabinet in the basement, which at one time will have housed other time references for the Observatory.

The clock was kept to time by means of the BBC six-pips with a detector that received the pips, and if the clock was slow it was advanced (or retarded) by a

Fig. 4.24 The Synchronome clock for the time ball release in 1990 in the basement of Flamsteed House (photographed in 1990). The pendulum is on the left with relays for each step: stage 1, raise halfway; stage 2, raise to top; stage 3, drop. on the top of the cabinet is an audio amplifier for the six-pips from the BBC to the galleries. Below is a battery charger and the wooden box on the right contains a speculum mirror by Herschel

magnetic corrector. Although the clock gave its time every half second, each "tick" was to time within 0.1 second. This particular clock worked steadily from 1960 until 1992 when it was superseded.

The panel behind the clock was a tangle of wiring with adjacent low and high voltages and well below current electrical wiring regulations. In addition some of the time units showed signs of overheating. In 1990 one of us (DB) was invited to the Observatory to see if the system could be controlled electronically with the time derived from the low frequency radio service managed by the National Physical Laboratory. The signals came from a radio station near Rugby, set up in 1927 for the very low frequency transmissions on 16 kHz, and a series of standard frequencies were transmitted from 1951 for scientific reference purposes. Eventually in 1972 the low frequency standard of 60 MHz (old call sign MSF) remained with the addition of a rapid binary time code, all within part of the extended 'pip' at the minute mark. This fast code was discontinued after a number of years. In 1977 a "slow" code was added which, within each minute of the seconds markers, designated markers were elongated or not as the case may be, to give a 'code' for the year, month, day of the month, day of the week, hour, minute, British Summer Time and another detail to give DUT1, which is relevant to navigation. This dissemination of time has become very widespread for the now ubiquitous radio controlled clocks and watches. In 2007 the Rugby transmitter was closed and the antenna at Anthorn in the Solway Peninsula took over. A similar service is provided in Germany, with the transmitter, DCF77, near Frankfurt.

The new requirement, therefore, was to automate the dropping of the ball, drive the Shepherd Gate clock, and give pulses to two Tompion replica dials then in the apertures in the Octagon Room. In case of the failure of MSF, six-pips by land line from the BBC were to give synchronisation to the internal quartz crystal oscillator.

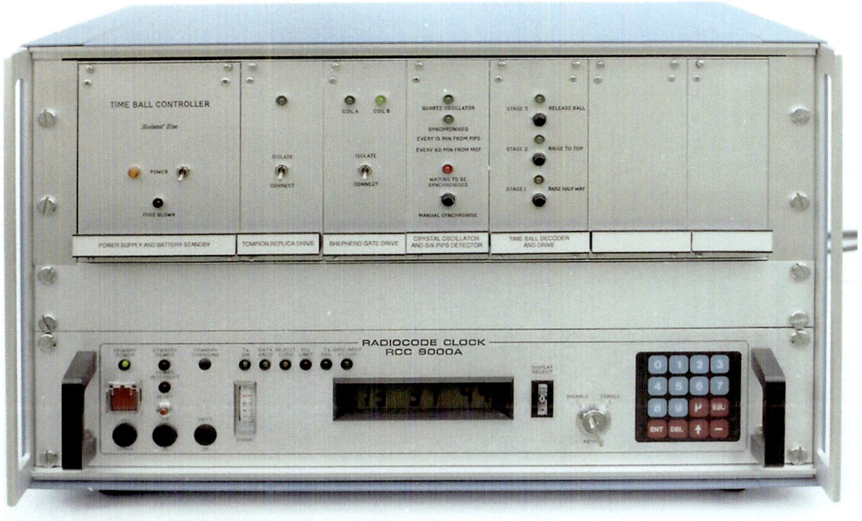

Fig. 4.25 The controller that took over the timing for the ball in 1992. At the upper level, power supply and battery standby, Tompion regulator 'ticks', Shepherd gate clock, crystal oscillator and six-pips detector, the three stages, raise half way, top and release together with manual operation buttons

The MSF signal contains a code for Daylight Saving. A specification was drawn up resulting in a commission for DB to construct the necessary controller, as in Fig. 4.25. The radio receiver was a commercial type that found use in radio and TV studios, power stations, or similar applications where events could be programmed into a control panel. The controller was put into use in 1992 on a shelf in the time ball cupboard, pending a move to the front entrance as part of a major revision of the entrance for the public.

The system continued in use until 18 August 1997 when the Observatory suffered a major lightning strike. Extensive damage to alarm and security systems was caused, and as can be imagined with control wiring not far from the struck area and lightning rods, the controller and MSF receiver were damaged. Some of the electrical damage could have been repaired although the more sophisticated radio receiver needed to be returned to the manufacturer. Pending a new method of controlling the release, the ball was dropped manually via the three large buttons in the time ball cupboard. At this time, the National Maritime Museum had better internal technical resources and the decision was made to use a simpler MSF receiver by Galleon and a personal computer (PC) to carry out the decoding of the relevant moments in time, Fig. 4.26.

These venerable computers continued in use until they were replaced in 2017 by a micro-processor and GPS time source.

Fig. 4.26 The 1997 electronic control system with small MSF receiver with two old style personal computers, the upper for controlling the Shepherd gate clock. The larger black box contains the time ball relays

10 Colour and Appearance of the Time Ball

Black is undoubtedly the best colour for an object to be seen at a distance, particularly in poor visibility. The Portsmouth ball may be inferred to have been black from the early illustrations and if not, it would probably have been mentioned in the written descriptions. However, in the twentieth century the Greenwich ball changed colour and there is lack of agreement over dates from black to red, and then it may have been black and red again! An early mention of the Greenwich ball being black is in the *Illustrated London Almanack* of 1845 and a few years later in Dickens' *Household Words* for 1850. A watercolour by Hubert Airy after the ball was blown down in 1855 also shows the ball as being a very dark colour, almost certainly black. In 1857 Airy, in correspondence with the Copenhagen Observatory, declares that "As far we can judge, black is the best colour" (Airy, 1857). One of the preeminent historians of the Observatory, the late Commander Derek Howse, states that the ball was originally black and repainted red some time in the 1920s (Howse, 1975). Howse may be quoting anecdotal information, but this is reinforced by correspondence relating to the new aluminium ball installed by Dent in the rebuild of 1914–1919 (see later). The quotation by Dent for the new ball gives the detail "All painted" (Dent, 1914). A later letter addressed to "The Aluminium Plating Company" enquiring about nickel plating the ball quotes that "it is coated on all exposed surfaces with a preservative paint" (Astronomer Royal, 1919a, b). Tantalisingly, the colour is not given, but is highly likely to have been the recognised preservative paint, red lead primer, containing the red oxide of lead. The plating process was not pursued. It could well be that the Observatory was relaxed about the red ball because it would still be quite visible to the ships in the Thames, and which by then could receive time checks by wireless. In addition, other time balls were red. Fortunately, strong evidence is given by the prolific and eminent painter of nautical scenes and the river Thames, W. L. Wyllie, member of the Royal Academy. The relevant painting, dated 1924, is hanging in the Greenwich Heritage Centre, a mere 3 miles from the Observatory. The close up reveals, with some accuracy, details of the Observatory and a ball that is coloured with a fair amount of red. Further confirmation arises from a colour film made in 1925/26 by Claude Friese-Greene called *The Open Road*, and a relevant extract can be seen on YouTube (YouTube). The attendant text by the British Film Institute (BFI) takes pride in the fact that the film was a great technical advance and that Britain could be seen in colour for the first time.

Although the ball was definitely red in 1926, Admiralty lists for 1919, 1922, 1928, 1931, 1934, and 1937 all record a black colour. This discrepancy may be explained by the fact that Hydrographic Office, who produce the lists, is 160 miles from Greenwich in Taunton, Somerset, and did not check for local knowledge. There may have been some repetitive reporting of "black". However, frequent changes of colour are not at all likely given the difficulty of climbing around the turret to paint the ball. The ball may well have got very dirty in the smoky and polluted atmosphere in London. The Observatory will have contributed to smoky and sooty air with its own coal fires, both in the buildings and porter's lodge. Figure 4.27 shows a very dirty ball. In some of the Admiralty reports there may well have been an element of expectation! It is also worth noting that the Observatory annual reports only mention failures due to strong winds, and make no mention of scaffolding or other work to paint the ball.

The horological curator, Jonathan Betts believed that the ball was painted a dark shade of red in 1939, and a dark brown in the late 1950s, and that the Museum's

Fig. 4.27 A rare photograph of the time ball taken in 1946. Wartime allotments in the park can be seen, with Greenwich power station in the background. The ball is spherical and very dirty after years of pollution. We have been unable to trace the copyright holder. (From a postcard supplied by G. Dolan, 2012)

establishment Officer, Reg Lowen, decided then to "smarten it up" by having it painted the present bright red (Betts, 1996).

In 2006 scaffolding was erected around Flamsteed House for redecoration and some renovation. The scaffolding was extended to the top of the time ball mast to grease the squeaking pivot of the weathervane and re-gild the letters. The opportunity was taken to paint the ball with a fresh coat of letter-box-red.

The rather battered appearance of the ball excites occasional comment, and there is documentary evidence that the ball was damaged in transit to the Observatory after it was made in 1914, and "Arrangements have been made to put it into store" (Astronomer Royal, 1914). Work on reconstruction was delayed due to the war and did not start again until 1919, and when in another letter relating "to the estimated cost of restoring the New Time Ball to its true shape, I beg to inform you that as the cost is so high and the result to be achieved is one of appearance only, it has been decided to leave the ball in its present condition" (Astronomer Royal, 1919a, b). Then, as stated earlier, during major renovations to Flamsteed House, the ball was moved around by builders causing the dents. The film clip mentioned gives the date of rolling the ball over obstacles is dated 24 October 1958. Removing the dents may have been considered but there was a risk that the now age-hardened thin aluminium could well have split or cracked by bending back into shape. The bright red paint is now part of the history of the time ball, but with the minor disadvantage that a gloss finish emphasises any defects!

In conclusion, the Greenwich time ball was the first ever accurate public time check, used by mariners for nearly 100 years. It then became a popular tourist attraction and symbol of Britain's maritime history, now approaching 200 years of use overall. Further extensive reading may be obtained in the Journal of Antiquarian Horology (Bateman, 2013).

References

Airy, G. B. (1857). *Letter to HCFC Schjellerap*, 10 February 1857, RGO Archives, RGO 6/145, folios 90–93.

Associated Electrical Industries (AEI) was a British holding company formed in 1928 through the merger of the British Thomson-Houston Company (BTH) and Metropolitan-Vickers electrical engineering companies. In 1967 AEI was acquired by the General Electric Company (GEC), to create the UK's largest electrical group at the time.

Astronomer Royal. (1914). *Letter from Astronomer Royal to E. Dent & Co*, 14 October 1914, RGO Archives, RGO 8/40.

Astronomer Royal. (1919a). *Letter from Astronomer Royal to The Aluminium Plant and Vessel Co.*, Wandsworth, 4 July 1919, RGO Archives, RGO 8/40.

Astronomer Royal. (1919b). *Letter to the Aluminium Plating Company*, dated 2 May 1919, RGO Archives, RGO 8/40.

Bateman, D. (2013). The time ball at Greenwich and the evolving methods of control. *Antiquarian Horology, 34*: Part 1 (June 2013), 198–218; Part 2 (September 2013), 332–346; Part 3 (December 2013), 471–488.

Bateman, D. (2014). Greenwich time ball. *Antiquarian Horology, 35*, 1014.

Betts, J. (1996). The planet watchers of Greenwich. *Antiquarian Horology, 22*(Spring 1996), 436–437. See also Littlewood, K., & Butler, B. (1988). *Of ships and Stars*. Athlone Press, p. 178.
Dolan, G. (2012). Private communication, The time ball and wartime allotments postcard.
Dolan, G. (2022). The film clip may also be located on the website www.britishpathe.com, search for Royal Observatory and then look through the clips for UK: Time Passes On.
Dolan, G. (2024). *Graham Dolan maintains the extremely informative website*. The website in turn gives links to all the Annual Reports to Visitors. https://www.royalobservatorygreenwich.org
E. Dent & Co. Ltd. (1914). *Estimate, time ball*, dated 11 March 1914, RGO Archives, RGO 8/40.
Graces Guide. https://www.gracesguide.co.uk/Normand_Electrical_Co (Internet search 2024).
Howse, D. (1975). *Greenwich observatory, volume 3: The buildings and instruments* (p. 136). Taylor and Francis.
Illustrated London Almanack. (1845). pp. 28–29.
Littlewood, K., & Butler, B. (1988). Chapter 8. In *Of ships and stars* (pp. 176–179). Athlone Press.
Maudslay & Field. Henry Maudslay (1771–1831) was a pioneer builder of all-metal machines, and is considered a founding father of machine tool technology, including screw cutting lathes and establishing standards. He began work aged just twelve at the Royal Arsenal, Woolwich. He excelled as a blacksmith, and quickly became London's most sought-after mechanic. An early project involved building machines for Joseph Bramah to construct his patent unpickable lock. Later, Maudslay worked with Marc Brunel to construct the world's first production line, making pulley blocks for the Royal Navy at Portsmouth. The company made many marine steam engines and at Kew Bridge Steam Museum there is a Maudslay pumping engine of 1838. He had his main works in Lambeth, south London, and Lambeth North underground station now covers this site. There is a commemorative window of Henry Maudslay in Greenwich Town Hall.
Miles, R. H. A. (2019). *Synchronome, masters of electrical timekeeping* (Rev. ed.). Antiquarian Horological Society.
National Maritime Museum. File: NMM 10: C41/21 A, Flamsteed House (entry located by Geraldine Charles, Museum Archivist & DRO, Royal Museums Greenwich, 2012).
Nautical Magazine. (1835). Neptune, The time ball at Greenwich. *Nautical Magazine, IV*(1835), 584–586.
The Graphic. (1855). *The Graphic* of 8 August 1855.
The National Archives, Kew, reference ADM 1/2729, folio 56, and memo to Pond, folio 59/3.
Watkinson, P. (2011). *Turret clock maker, private communication.*
YouTube link located by G Dolan, 2022. www.youtube.com/watch?v=9m1KEH0Enpc

Chapter 5
Time Ball Supply by Maudslay, Sons & Field After 1852

Roger Kinns

Abstract Although the time ball apparatus built by Maudslay, Sons & Field was of exceptional quality, only four time balls were ever supplied by the company after Greenwich in 1833. These were built for Deal and Edinburgh in 1853, for Sydney, NSW in 1855 and finally for Siemens Brothers in 1873, who added electrical and other equipment before shipping the complete system to Lyttelton, NZ in 1873. Like Greenwich, all used an air spring arrangement. This was widely adopted by other suppliers until the 1880s.

A first design with a rack and pinion hoist was used for Deal and Edinburgh. A second design with modifications was used for Sydney and replicated for Lyttelton. The emphasis in this chapter is on the design of the mechanical apparatus, including the structure of the ball itself. The photographic and other evidence that showed the origin of the design for Lyttelton is described in detail.

1 Introduction

Maudslay, Sons & Field (abbreviated to Maudslays in this chapter) was a wide-ranging general engineering business (Foster Petree, 1934, 1949/1956). It had evolved from the one-man business founded in 1797 by Henry Maudslay (1771–1831) who is regarded as the pioneer of precision engineering. Company records were almost entirely destroyed after liquidation of the firm at the end of the nineteenth century (Foster Petree, 1949/1956). However, the 12 notebooks of Charles Sells, who was Chief Draughtsman at Maudslays, have survived (Sells, 1842–1883). They are concerned mainly with the performance of marine steam engines which were the main business of the company, but they included a single note about time balls which stimulated the work in this chapter.

This note is a list in Sells' handwriting on the inside back cover of his last notebook. It is shown in Fig. 5.1. The dates are of order placement, rather than first

R. Kinns (✉)
The Maudslay Society, Pembroke College, Cambridge, England

R. Kinns (ed.), *The World of Visual Time Signals for Mariners*, Historical & Cultural Astronomy, https://doi.org/10.1007/978-3-031-57334-7_5

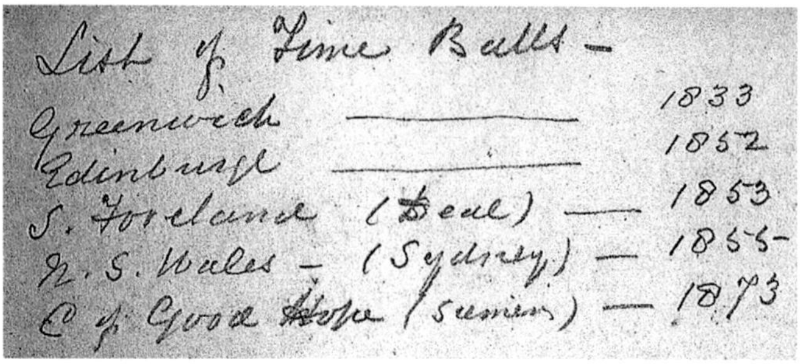
List of Time Balls –
Greenwich ——— 1833
Edinburgh ——— 1852
S. Foreland (Deal) — 1853
N. S. Wales – (Sydney) — 1855
C of Good Hope (Siemens) — 1873

Fig. 5.1 The list of time balls in the notebook of Charles Sells

operation. The 1833 apparatus for Greenwich has been described separately in Chap. 4. The final entry is the only indication that there was a relationship with Siemens Brothers for time ball supply in 1873. It did not, however, result in a time ball at the Cape of Good Hope. Detailed investigation of design features has shown that the apparatus was sent instead to Lyttelton, New Zealand (Kinns, 2009).

Although the apparatus built by Maudslays was of exceptional quality and was often recommended by the Astronomer Royal at Greenwich in his correspondence with third parties, only five time balls were ever supplied by the company. All used an air spring arrangement, whereby a piston connected to the ball entered a cast iron cylinder after the ball had dropped a short distance in free fall. It compressed air in the cylinder to bring the ball to rest. This design, which was then more reliable than other designs using pulleys, weights and springs to arrest the ball's descent, is attributed to Joshua Field (Smyth, 1853). The air spring arrangement was widely used by other suppliers until the 1880s when the alternatives of a friction brake and other arrangements had been developed to offer sufficient reliability at reduced cost. Initially, the time ball at Greenwich used a chain hoist, subsequently replaced by stranded wire ropes (see Chap. 4). The later four used a rack and pinion hoist.

A first rack and pinion design was used for Deal and Edinburgh. A second design with modifications was used for Sydney and replicated for Lyttelton. The external appearance of the time ball and tower structures, and their overall histories, are described in Chaps. 10, 22, and 23 relating to their geographic locations. The emphasis in this chapter is on the design of the mechanical apparatus, including the structure of the ball itself.

2 Time Ball Design

The time balls themselves are all believed to have been made to the same pattern after 1852. The ball at Deal is now a fibreglass replica. The ball at Sydney was changed to one made from Muntz metal in the 1870s. Only the Lyttelton ball remained in its original form, restored after earthquake damage in 2011.

2.1 Time Ball Construction

Each original ball had a diameter of 5 feet (1.5 m). It had a mahogany frame and a thin zinc skin whose original thickness was about 0.8 mm. It was nailed to 16 ribs. The internal structure included bracing and a wooden box that was a close fit to the mast. Airy indicated that the ball mass was about 200 lbs (90 kg) when he was proposing a similar arrangement for Portsmouth (Wood, 1856).

The ball was attached to a long shaft and piston which entered a cylinder to compress air and bring the dropping ball to rest before it reached the bottom of the cylinder. The shaft and its attachments probably weighed about 50 kg (Kinns, 2014). The diameter and length of the cylinder were also key parameters.

The design balance was chosen with care but careful maintenance was needed to ensure only light contact with the supporting tower after the ball descent. Damage to the ball did occur when air bled out too quickly. It was tempting to reinforce the ball structure and increase the thickness of its skin, notwithstanding the increased load on the air spring.

The ball supplied for Lyttelton remained unaltered after delivery, apart from changes to its colour. The lower part of the ball was badly damaged after it fell to the ground in the 2011 aftershock that caused the tower to collapse. Figure 5.2 shows the upper interior of the ball, before its restoration. The restored frame is shown in Fig. 5.3.

Fig. 5.2 Upper interior of Lyttleton time ball before restoration. (Author's photograph, 2016)

Fig. 5.3 Lyttleton time ball frame after restoration. (Courtesy Starks Engineering, Christchurch, 2017)

Fig. 5.4 Edinburgh ball frame, pre-restoration in 2009. (Courtesy Alan Marshall)

The Edinburgh time ball, supplied in 1853, is still usable, but it has been subject to repairs on many occasions and its weight increased over the years. The extent of reinforcement in Edinburgh is shown in Fig. 5.4. The number of ribs and spacers is the same as at Lyttelton. It appears that the vertical ribs have been reinforced by adding additional layers of wood, so that the ball mass increased considerably. The restored ball, based on its final working state, weighs almost 200 kg (Kinns, 2014). The original air spring arrangement to arrest the ball's descent is still in use but requires careful maintenance to avoid unacceptable impact loads on Nelson's Monument.

2.2 *Time Ball Mass*

The need for a heavy ball and development of appropriate arrangements for controlled deceleration, in order to ensure reliability and availability in all but extreme weather, were described in a paper concerning the Edinburgh time ball (Smyth, 1853). The requirements and their solution were well-described qualitatively, but apparently unnoticed numerical errors created a myth that has proved hard to suppress.

2.2.1 The Origin of the Mythical Edinburgh Mass

Statements that the time ball weighed an implausible three-quarters of a ton were still being perpetuated in twenty-first century guidebooks (Kinns, 2011, 2014). Their origin was an uncorrected error in a presentation by Charles Piazzi Smyth to the *Royal Society of Arts* that "the ball is made very heavy, say 15 cwt." (Smyth, 1853). This weight was repeated later in the same paper, when Smyth discussed the forces needed to operate the release catches:

> … But if, as is found, 1/300 of 15 cwt., or 10 lb, (*sic*), will pull the trigger of, and drop, the big ball; then 1/300 of 10 lb., or ½ oz, will be enough to drop a 10 lb. weight inside the apparatus, so as to fall on, and efficiently to pull, the great trigger. …

It was a strange sequence of errors. The real mass of the ball is likely to have been over eight times smaller than 15 cwt (768 kg), while the shaft and other attachments would have added about 50 kg.

2.2.2 Changes in Mass Due to Repairs and Alterations

If ball damage did occur, the temptation was to strengthen the ball, which increased its mass and made a future crash more likely. That could easily lead to a vicious circle, which is believed to have occurred in Edinburgh, for example. The restored ball, based on the reinforced construction present in 2009, weighs about 200 kg (Kinns, 2014). The diameter of the air cylinder used to brake the ball has not been changed.

Modifications to the Sydney apparatus in the 1870s led to substantial increases in the mass of the ball and of the attached shaft. The diameter of the air cylinder was increased to suit, possibly later than Russell's modifications that are described below.

2.3 *Time Ball Colour*

The balls at Edinburgh and Deal were originally painted black, as were most time balls that would be sighted against the sky by ships in harbour. The Edinburgh ball colour changed to pale grey before restoration but it is now black again. The Sydney ball was originally black. It is now painted yellow. The colour of the Lyttelton ball changed several times. An early painting showed it as red with a white band, while some photographs indicated a uniform dark colour. It had been black with a red band after its restoration in the 1970s, but the newly restored ball has returned to red with a white band.

3 Principal Design Features of the Rack and Pinion Hoist

The bronze rack was made in sections, bolted to a hardwood shaft. The choice of wood is uncertain but it may have been greenheart or teak, chosen to resist rot. The slotted external pine mast had overall dimensions of 8 inches (203 mm) square and allowed a drop height of 10 feet (3 m). It was slotted to allow the ball to be pushed up using an iron rod, so that the ball and its shaft could be supported by catches at the top of their travel. The gear mechanism allowed the rack to be raised by windlass (also described as a capstan) using a 10-tooth pinion in the final drive which could be moved along its shaft to disengage from the rack once the ball was supported by catches. A ratchet was used to prevent counter-rotation of the windlass during winding. The catches were released using a link mechanism that required only small forces for its release. The rack had a width of 1½ inches (38 mm) and a pitch of slightly less than 1 inch (24–25 mm).

The lower end of the shaft was connected to a piston that entered a cast-iron cylinder after the ball had dropped a short distance. The cylinder had an internal diameter of 8 inches (203 mm), sized to provide an air spring that arrested the descent of the ball and its shaft attachments within the drop height. The air in the cylinder first underwent adiabatic compression. A bleed valve was used to limit rebound and to allow a slow final descent as the air cooled and exhausted from the cylinder. Design of an effective seal between the piston and cylinder was a significant challenge at the time of construction. Various devices were used to prevent unacceptable collision with the tower structure, including an external cushion and mechanical spring arrangements at the bottom of the cylinder. These varied between installations and are likely to have changed during the operational life of the apparatus.

The detailed differences between the four implementations will be illustrated using photographs. Original engineering drawings of the apparatus have not been found, but Siemens Brothers sent a copy to New Zealand in 1874, with added operational notes, as part of their supply for Lyttelton. It has been retained in Wellington, NZ and is included in Chap. 23.

4 The First Design for Deal and Edinburgh

The time ball apparatus for Deal and Edinburgh were both constructed in 1853 and most mechanical design features were almost identical. They were the first to use a rack and pinion arrangement for hoisting the ball. The dated manufacturer's plate is still in place at both locations, as shown in Fig. 5.5. The Deal time ball was operated by electric telegraph from Greenwich, over 100 km away. Extra time was needed for its development, so it did not enter service until January 1855, over 9 months after the start of the Edinburgh service which used an adjacent observatory.

The apparatus for Deal was ordered by the Admiralty and included the royal coat of arms. The apparatus for Edinburgh was ordered by Charles Piazzi Smyth, the second Astronomer Royal for Scotland, in consultation with George Airy. It included clock design features by Ritchie of Edinburgh that allowed automatic ball release (Kinns, 2011).

The casings for the first design were made of wrought iron, with corner pieces and bolted plates. The bolted plates can be seen in Figs. 5.6 and 5.7, together with the external gearing and windlass arrangements. The Deal apparatus included additional apparatus for release by remote telegraph which can be seen in Fig. 5.6. The casing had a square cross section measuring 251 × 251 mm at Edinburgh.

A key feature of the first design was the use of a guide wheel on the back of the shaft, opposite the pinion that engaged with the rack. These guide wheels can be seen in Figs. 5.6 and 5.7. Another key feature was the arrangement for simultaneous catch release on either side of the shaft. The Edinburgh and Deal gear segments for this purpose are shown in Fig. 5.8.

Edinburgh Deal

Fig. 5.5 The Maudslay, Sons & Field plates in 2009. (Photographs by the Author, Edinburgh: Michael Kinns, Deal)

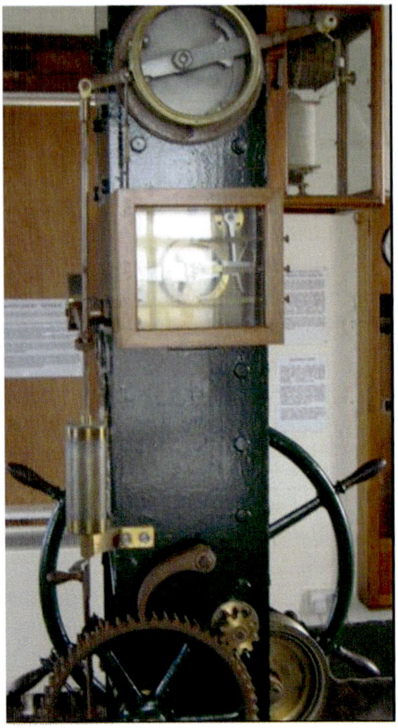

Fig. 5.6 The Deal arrangement. (Photograph by Michael Kinns, 2009)

Fig. 5.7 The pre-restoration Edinburgh arrangement. (Courtesy Alan Marshall)

Edinburgh (above) Deal (below)

Fig. 5.8 The gear segments, balance weights and inspection holes at Edinburgh and Deal. (Photographs by the Author, Edinburgh: Michael Kinns, Deal)

The principal features of the gear trains at Deal and Edinburgh are similar, but the windlass is on the opposite side of the casing to the external gears at Deal, with differences in the handles.

4.1 Edinburgh Restoration in 2009

The Edinburgh apparatus was carefully restored in 2009 and can still be operated (Marshall, 2009). Figure 5.9 shows many of the components awaiting restoration. The original electric trigger at Edinburgh has been substituted by a manual rope pull; smoke from the Edinburgh Castle gun having replaced the time ball as the primary signal. Figure 5.10 shows the restored rack and pinion in 2011.

Fig. 5.9 Edinburgh components awaiting restoration. (Author's photograph, 2009)

5 The Second Design for Sydney

The 1855 design for Sydney included various refinements of the design for Deal and Edinburgh. The casings in the second design were cast as integral structures, presumably to give increased rigidity. A single guide wheel opposite the pinion was used at Deal and Edinburgh. This was changed to a design that used two pairs of guide wheels, engaging with the sides of the shaft. Another important change was in the arrangement for release of catches on either side of the rack. Gear segments and catches at Deal and Edinburgh were replaced by lever-linked roller catches.

Maudslays built the time ball apparatus for Sydney, NSW in 1855. It became operational on 5 June 1858, following completion of Sydney Observatory buildings (Pickett & Lomb, 2001). It will be shown that a replica of this second design was made for Siemens Brothers in 1873.

5.1 Sydney Modifications by Henry Russell

An 1899 letter from Henry Russell, the NSW Government Astronomer, to Sir Charles Todd in Adelaide was thought to be the only surviving evidence in Australia of Maudslays supply (Russell, 1899). Confirmation was found in the Todd

Fig. 5.10 The restored Edinburgh mechanism. (Author's photograph, 2011)

correspondence, archived in Adelaide (Todd, 1899; see Kinns & Abell, 2009). Henry Russell, the NSW Government Astronomer, modified the Sydney apparatus to a limited extent during the 1870s, but most principal features, including the rack and pinion mechanism and the casing, were retained (Russell, 1899).

The 1899 letter from Russell to Todd included criticisms of the original apparatus in Sydney and a brief description of the principal modifications that Russell made to it during the 1870s. The complete correspondence was transcribed by Kinns and Abell (2009). Extracts from the Russell letter are shown below.

> Some of the machinery which it is necessary to get at sometimes was enclosed in a cast iron post that made it extremely difficult to do what was necessary.
>
> The electrical dropping gear was complicated and clumsy in the extreme. So much so that I removed it and put in a direct acting trigger. …
>
> When Maudsly (sic) sent it out the ball had a wooden frame covered with zinc nailed to the wood, of course the rain got in and it required frequent repairs. I had it removed and had a Muntz ball made without wooden ribs and it has never given us any trouble.
>
> The original electric trigger took 0.6s–0.7s to act. I altered it so that the action is instantaneous.
>
> The original shaft that carries the rack for winding it up was of wood and caused us much trouble until I replaced it with iron. …

In his reply (Todd, 1899), Todd made the following observations:

> … It may interest you to know that I very carefully inspected the Time Ball at Maudsleys (sic) in 1855 before I came out to Australia. It was then considered a very fine piece of work; but I quite recognised the objections you mention from its being too much closed in.
> The Time Ball at Deal, which was under my care when at Greenwich Observatory in 1855, was also made by Maudsley (*sic*), and upon almost identical lines with those of your Time Ball at the Sydney Observatory.

The present Sydney mechanism retains the features mentioned in Russell's letter. There have been various repairs and additions since the letter was written more than a century ago. For example, the pinion has been replaced on more than one occasion, because the teeth were stripped when it was not withdrawn correctly before release. Also, the cast iron cylinder that is now used to provide the air spring has a larger diameter than the original, in order to accommodate the increased weight of the Muntz metal time ball and iron shaft. The ball can now be raised using a modern electric motor, but hand operation using the windlass is still possible.

The original drawings for Sydney have been lost, so the original design has to be inferred from knowledge of later alterations and inspection of the Lyttelton apparatus.

5.2 Other Time Balls in Australia with Maudslay Connections

Two other time balls in Australia are known to have been influenced by experience with designs by Maudslays (Kinns & Abell, 2009).

The apparatus at Semaphore was designed by Charles (later Sir Charles) Todd, who had been responsible for operation of the time balls at Greenwich and Deal in 1855 before his appointment to Adelaide (Todd, 1899). It was built locally and is likely to have exploited his experience at Greenwich (Abell & Kinns, 2010; Bateman 2013). There is no evidence of direct involvement by Maudslays in the design for Semaphore, although Henry Maudslay's daughter Isobel was resident in Adelaide with her family at the time of its construction.

The Sydney design provided the basis for the apparatus that was installed in the new customs house at Newcastle, NSW, which became operational on 21 February 1878. It incorporated Russell's Sydney modifications and other improvements, which included an open structure for the mechanism casing, and was manufactured by Potter & Sons of Sydney during 1877.

Russell remarked (Russell, 1899):

> The one we have in Newcastle was made in Sydney and is much better than the original in every way. The cast iron bore was replaced by four wrought iron corner pieces which leave the machinery all open and reduces the cost. It has been nearly 20 years in its place and has never cost anything for repairs.

5.3 Potential Supply for Other Locations in England

Airy wanted to use an arrangement with remote operation from Greenwich at various naval dockyards and at the Start Point in Devon, exploiting the refinements for Sydney. He obtained a quotation from Maudslays for supply of an apparatus for Portsmouth and communicated with dockyard staff about its detailed location (Wood, 1856; see Kinns et al., 2021). He was unable to secure the necessary funds and Maudslays never received further orders for locations in England.

6 The South African Mystery

The Sells notebook entry for 1873 (Fig. 5.1) was interpreted by Foster Petree (1934) to imply that Siemens might have supplied the electrical components for Maudslays mechanical apparatus, which was thought to have been sent to South Africa. He had no reason to doubt the stated destination.

6.1 1873 Time Ball at Alfred Docks

A time ball was installed on a North Quay warehouse at Alfred Docks in Cape Town during 1873 (Spencer Jones, 1993). This is the only time ball that is known to have been installed in Cape Colony at that time. There do not appear to be any surviving records of the 1873 arrangement or of its supplier. It was, however, the principal time ball for Table Bay in the first edition of the Admiralty list of time signals, published in 1880. The ball diameter was not recorded, but the stated drop height of only 6 feet (1.8 m) was much smaller than the 10 feet (3 m) favoured by Maudslays. No evidence has been found in South Africa that Maudslays ever supplied a time ball apparatus for the Cape.

6.2 1894 Time Ball at Alfred Docks

The Alfred Docks site was redeveloped during the 1890s and "In 1894 a new time ball was erected in a much more conspicuous position near the Resident Engineer's Office of the Cape Town Docks" (Gill, 1913). The 1894 apparatus was supplied by Gillett and Johnston (1893). It is known from 1894 drawings that its diameter and drop height were then 1.5 and 3 m. It used a rack and pinion mechanism, but its design details were not those favoured by Maudslays. Its elevation was increased in 1904, by adding an extra storey to the tower. Unfortunately, the first Admiralty list entry for 1880 was not changed in editions after 1894 to reflect changes in the installation, adding further confusion.

7 Re-use of the Second Design for Lyttelton, New Zealand

The time ball at Lyttelton, New Zealand was supplied by Siemens Brothers in London and it had been assumed before 2009 that the mechanical apparatus was to their design. It has now been shown that it was actually a replica of the design for Sydney and had been built for Siemens by Maudslay, Sons and Field (Kinns, 2009). Siemens provided the electrical apparatus, and related equipment such as the astronomical clock, and acted as the systems integrator. It was the only heavy-duty time ball apparatus to be supplied by Siemens and the last to be made by Maudslays.

Siemens Brothers shipped the apparatus for Lyttelton from London in July 1874, following an order in March 1873. Siemens had become a principal supplier of telegraphic equipment, with heavy commitments to supply and installation of telegraph cables at the time. As at Sydney, time ball erection had to await completion of the necessary tower. The apparatus was unpacked in April 1876 and installed in the new tower at Lyttelton in September that year. It finally became operational on 23 December 1876. It was the third time ball in New Zealand, those at Wellington and Dunedin (Port Chalmers) having become operational in 1864 and 1867.

7.1 Time Ball Supply by Siemens Brothers

The principal biographies of Sir William Siemens and Siemens Brothers do not include any mention of time balls (Pole, 1888; Scott, 1958). Time ball supply would have been a minor business activity in relation to design, supply and installation of submarine telegraph cables. The main preoccupations of Siemens Brothers during 1873 and 1874 were the manufacture and laying of the ill-fated Platino-Brasiliera cable, manufacture of the Direct United States cable and commissioning of the cable-laying ship *Faraday* (Scott, 1958). These carried huge financial risk, with the fate of the company depending on a successful outcome. It would not be surprising in these circumstances if Siemens decided to subcontract provision of a time ball apparatus, electing to act as systems integrator for their client. Siemens does not have any record of time ball supply by either the British or German companies, other than for Lyttelton (Kinns, 2009).

The following letter to the Astronomer Royal suggests that Siemens Brothers first became aware in 1871 of the time ball designs supplied by Maudslays (Siemens, 1871).

> An Australian correspondent of mine wishes to establish Time Balls in the Colony (*sic*) and asks me for some details of the arrangements adopted by yourself.
>
> Would you kindly allow my Assistant Mr Naglo to examine the Apparatus with a view of satisfying my correspondent.

Clearly, Siemens did not have previous experience of time ball design. Siemens' enquiry may relate to the time ball at Semaphore, South Australia, or the replacement apparatus for Newcastle, NSW which came to fruition in 1875 and 1877 respectively.

An enquiry from a jeweller in Sheffield elicited a response mentioning Siemens Brothers as a potential supplier of apparatus for a small time ball in a shop (Christie, 1877).

> In reply to your inquiry the Astronomer Royal requests me to say that if you want an external Time Ball (of large size), Maudslay, Sons & Field, engineers, Westminster Bridge Road S.E. would probably be the best firm to apply to, as they have made several (at Greenwich, Deal & Edinburgh). If however you only wish a small one for your shop window, Messrs. Siemens Bros. 120 Queen Anne's Gate S.W. might perhaps undertake it, as they are Telegraph Instrument Manufacturers, or you might apply to some Clock manufacturer as E Dent & Co. 61 Strand.

As in other correspondence, the letter mentions Maudslay, Sons & Field as the favoured supplier of heavy-duty apparatus. Christie, who signed it, later succeeded Airy as Astronomer Royal. They appeared to be unaware in 1877 either that the Lyttelton time ball system had been supplied by Siemens Brothers, or that the mechanical apparatus was made for Siemens by Maudslay, Sons & Field.

7.2 Restoration of the Lyttelton Time Ball Apparatus

The Lyttelton time ball service was discontinued in 1934. The building and apparatus fell into disrepair, but a group of local enthusiasts was formed in 1969 to restore the famous landmark. They suffered an early setback when vandals damaged the mechanism, smashing the manual gearwheels and stealing the electromagnet. The electromagnet was never recovered, but the Siemens Brothers nameplate was found in nearby scrub; it has since disappeared again. A replica of the electromagnet was made using a photograph of the original. Some new parts for the mechanism were made during restoration by local engineering firms and it was possible to raise the ball again in December 1969. Restoration of the tower was a difficult and ultimately expensive challenge. The project was finally completed in 1978 (The Lyttelton Time-Ball Station, 1979).

8 Comparisons of the Sydney and Lyttelton Apparatus

Table 5.1 shows a comparison of some principal mechanism components at Sydney and Lyttelton. It includes dimensions of the rack and pinion; the upper and lower guide wheels; and the casings. Dimensions are all to the nearest mm.

The rack at Sydney was re-backed with an iron post by Henry Russell, replacing the original wooden post which is still extant at Lyttelton. The rack dimensions are the same to 1 mm accuracy.

Each cast-iron casing has a square section, with the same breadth and depth of 10 $\frac{5}{8}$ inches (270 mm). The arched apertures for access to the rack and pinion also have similar dimensions, although the shape changed slightly between Sydney and Lyttelton. The aperture at Lyttelton is now covered by a modern perspex window, which can be removed when necessary. The sliders and yokes for the rack pinion, which act as simple clutches at Sydney and Lyttelton, appear to be identical. This applies also to the brass plates on the back and front of the casing that support the trigger levers and roller catches for the rack.

The upper and lower guide wheels at Sydney and Lyttelton are identical, given small tolerances in manufacture and measurement. The inner width of the guide wheels is 3 $\frac{5}{16}$ inches (84 mm).

Pairs of photographs allow comparisons of particular design features. Figure 5.11 shows the layout of the gears and lower guide wheels. Figure 5.12 shows the trigger mechanisms and Fig. 5.13 shows the upper guide wheels.

The rack and pinion arrangements are the same, but the outer gears and capstan differ in detail. The outer gearwheel and pinion have 80 and 15 teeth at Sydney, but only 62 and 10 teeth at Lyttelton. The Lyttelton gears have the same number of teeth as at Edinburgh and Deal, so Sydney is the exception. The 6-spoked capstans differ in detail, that at Sydney having radial handles as well as a handle on one of the spokes. The original Lyttelton drawing shows a handle at the capstan periphery, now positioned on one of the spokes. The capstan at Edinburgh does not have radial

Table 5.1 Principal mechanism dimensions at Sydney and Lyttelton

Feature		Sydney	Lyttelton
Pinion on Rack	No of teeth	10	10
	Width	40 mm	
	Diameter	80 mm	
Rack	Width	39 mm	39 mm
	Pitch	24 mm	25 mm
Upper guide wheels	Outer width	99 mm	103 mm
	Inner width	84 mm	84 mm
	Diameter	332 mm	332 mm
Lower guide wheels	Outer width	100 mm	97 mm
	Inner width	84 mm	84 mm
	Diameter	150 mm	149 mm
Main casing	Width	270 mm	270 mm
	Depth	270 mm	270 mm

Sydney Lyttelton

Fig. 5.11 The gears, clutch and aperture. (Author's photographs, 2007–8)

Sydney Lyttelton

Fig. 5.12 The trigger mechanisms. (Author's photographs, 2007–8)

handles, while that at Deal is similar to Sydney. Details like this are easily changed to suit the operator and may have been modified during the working lives of the apparatus.

Sydney (above) Lyttelton (below)

Fig. 5.13 The upper guide wheels. (Author's photographs, 2007–8)

The trigger mechanisms are significantly different. The Lyttelton mechanism is more complicated, with an extended series of levers between the electromagnets and the ball release catches. This is likely to be the original arrangement at Sydney that was criticised by Henry Russell as being "complicated and clumsy in the extreme". His modifications to the Sydney apparatus gave a much more direct link between the electromagnets and the release catches. An upper link to the electromagnet contacts is still present at Lyttelton, but the link has been disconnected at Sydney and the right hand lever is now redundant. In effect, the original trigger arrangement for Sydney could be seen at Lyttelton.

Fig. 5.14 Salvaged casing of Lyttelton time ball. (Author's photograph, 2016)

The present slotted mast at Sydney is made of metal and has a circular section, while the slotted mast at Lyttelton is made of wood with a square cross-section. Russell modified the Sydney mast at the time the ball was changed; early photographs show a mast with a square cross-section. The mast at Lyttelton was similar to that at Edinburgh.

The Lyttelton apparatus was severely damaged during 2011 and may not be repairable, but as much as possible was rescued from the rubble of the collapsed tower. Figure 5.14 shows the casting and some of the surviving attachments in store during 2016. The restored ball is now operated using a modern electrical and hydraulic arrangement in a reconstructed tower.

9 Longevity of Maudslay Time Balls

Bartky (1987) included a list headed "The last time balls (1987)", without identifying their suppliers. He noted four that were operational: Greenwich, Edinburgh, Deal and Lyttelton, NZ. He also noted two others on standby: Sydney, NSW and Göteborg in Sweden. The first five were those built originally by Maudslay, Sons & Field. All were listed by the Admiralty in 1880 and they outlived hundreds of others, leaving a remarkable tribute to their quality of design and construction.

References

Abell, L., & Kinns, R. (2010). 'Telegraph' Todd and the Semaphore time ball. *Journal of the Historical Society of South Australia, 38*, 42–57.

Bartky, I. R. (1987). The bygone era of time balls. *Sky and Telescope*, 32–35.

Bateman, D. (2013). The time ball at Greenwich and the evolving methods of control – Part I. *Antiquarian Horology, 34*(2), 198–218; Part II. *Antiquarian Horology, 34*(3), 332–346; Part III. *Antiquarian Horology, 34*(4), 471–488.

Christie, W. M. H. (1877). *Letter to W. Brown* dated, 30 June 1877. RGO 6 618, File 6, Leaf 199, Papers of George Airy, Cambridge University Library.

Foster Petree, J. (1934). Maudslay, Sons & Field as General Engineers, with written discussion. *Transactions of the Newcomen Society, 15*, 39–61. Paper read at the Science Museum, 12 December 1934.

Foster Petree, J. (1956). *Maudslay Brochure: Henry Maudslay, 1771–1831: And Maudslay, Sons & Field, Ltd: A commemorative brochure to mark the establishment of the Maudslay Scholarship, published by the Maudslay Society in 1949 and reprinted with amendments in 1956.* (Available on CD from the Kew Bridge Steam Museum.)

Gill, D. (1913). *A history and description of the Royal Observatory, Cape of Good Hope*. HM Stationery Office.

Gillett and Johnston. (1893). *Ledger entry showing an order for Table Bay Harbour* on 6 November 1893.

Kinns, R. (2011). The early history of the Edinburgh time ball and time gun. *International Journal for the history of Engineering & Technology, 81*(2), 264–290.

Kinns, R. (2014). Did the Edinburgh time ball really weigh 15 cwt? *International Journal for the History of Engineering & Technology, 84*(2), 160–174.

Kinns, R. (2009). Time keeping in the Antipodes: A critical comparison of the Sydney and Lyttelton Time Balls. *Journal of Astronomical History and Heritage, 12*(2), 97–107.

Kinns, R., & Abell, L. J. (2009). The contribution of Maudslay, Sons & Field to the development of time balls in Australia. *International Journal for the History of Engineering & Technology, 79*, 59–90.

Kinns, R., Fuller, P., & Bateman, D. (2021). Exploring the Portsmouth time balls. *Journal of Astronomical History and Heritage, 24*(3), 751–769.

Marshall, A. (2009). *Nelson Monument, Calton Hill, Edinburgh*, article by Gray Marshall and Associates, Edinburgh [responsible for the restoration].

Pickett, C., & Lomb, N. (2001). *Observer & observed; a pictorial history of Sydney Observatory and Observatory Hill*. Powerhouse Publishing. ISBN 1-86317-086-3.

Pole, W. (1888). *The life of Sir William Siemens*, reprinted and published by Siemens Ltd 1986.

Russell, H. C. (1899). *Letter to C. Todd*, 21 March 1899.

Scott, J. D. (1958). *Siemens Brothers, 1858–1958: An essay in the history of industry*. Weidenfeld and Nicholson.

Sells, C. (1842–1883). Notebooks: (1) 1842, also 1847; (2) April 1842 to 1849; (3) January 1849 to June 1856 (sic); (4) June 1853 to December 1855; (5) January 1856 to March 1859; (6) March 1859 to November 1861; (7) November 1861 to May 1865; (8) May 1865 to December 1867; (9) December 1867 to January 1871; (10) January 1871 to May 1875; (11) May 1875 to April 1878; (12) April 1878 to March 1883. The notebooks are now archived in the Science Museum Library at Swindon.

Siemens, C. W. (1871). *Letter to G.B. Airy*, dated 3 March 1871. RGO Archive 6 617 File 1, Leaf 37, Papers of George Airy, Cambridge University Library.

Smyth, C. P. (1853). Notice of the time ball at the Royal Observatory at Edinburgh. 'Given at the request of the Council on 12 December 1853. A working model and diagrams were exhibited'. *Later Published in Select Papers of the Society of Arts*, 4 (1856), 191–196.

Spencer Jones, J. (1993). Time and the sailor. *Waterfront Review*, 26–27.

The Lyttelton Time-Ball Station: An Account of its History, Operation and Restoration. (1979). New Zealand Historic Places Trust, Wellington.

Todd, C. (1899). *Letter to H.C. Russell*, 29 March 1899, SRSA GRG 31/1 0001, vols. 1–5, 345–346.

Wood, H. (1856). *Time Ball on Semaphore at H. M. Dock Yard Portsmouth: Answers to questions submitted by Professor Airy*, 4 September 1856. RGO 6.613, Section 7, Leaves 72–73.

Chapter 6
Admiralty Lists of Visual Time Signals After 1880

Roger Kinns

Abstract Admiralty lists, first published in 1880, included time signals that ranged from time balls, discs, lights and flags to time guns, whose flash was the precise signal. Preparatory signals and procedures to be followed in case of an inaccurate signal were described. After 1911, the list was widened to include other navigation signals and issued in different parts covering specific geographical areas, each including a chapter on visual time signals.

At first, signal times in Admiralty lists were specified as local (astronomical) mean time based on zero hours at noon and GMT. The latitude and longitude of the signal location were also specified. Standard times, governed by time zones, were introduced later and civil time, based on zero hours at midnight was used from 1925 onwards. Entries for St Helena, Vladivostok and China are used to demonstrate the detail of published time signal information and the chosen simplification in other chapters.

1 Introduction

The first edition of the Admiralty list of time signals was published in 1880. It combined details of signals worldwide into a single document. Before then, details of signals had been specified in notices, almanacs and pilot guides published by different authorities in different countries. Many early notices about time signals appeared in *The Nautical Magazine*, which was first published in 1834. The Admiralty lists included a wide range of devices, notably time balls or other objects that were dropped or collapsed at a specified time or times. They included flags that were hauled down at stated times and time guns whose flash was the signal. The lists also included signals that could only be seen by going ashore to an office or public building. In that case, an officer would usually use an intermediate watch, rather than

R. Kinns (✉)
The Maudslay Society, Pembroke College, Cambridge, England

R. Kinns (ed.), *The World of Visual Time Signals for Mariners*, Historical & Cultural Astronomy, https://doi.org/10.1007/978-3-031-57334-7_6

move a chronometer from its onboard location. The signals considered in this book are limited to those visible or audible to ships in harbour.

Decisions had to be made by the Admiralty about whether to include secondary signals that were not at major ports or were maintained and operated by private enterprise or local government. The 1880 list included a footnote about signals around the British Isles that were not in the main list, without giving details. Several were time guns and time balls, including small time balls on shop premises, and others were controlled clocks. Footnotes of this type were not included in lists for 1898 onwards. Intermediate lists published in 1888, 1892 and 1895 have not been seen by the author. Comments about the accuracy and reliability of time signals were often derived from reports by visiting ships. They were usually qualitative and served as a guide to subsequent visitors. The style of entries changed over the years but the level of information remained essentially the same.

Similar, often identical, information can be found in pilot guides issued by the United States and many other countries. Mistakes did occur and signals might be omitted accidentally or included after they had been withdrawn, but the Admiralty lists form the most comprehensive and reliable source available to the author.

2 Structure of Admiralty Lists

The Admiralty *List of time signals established in various parts of the World* was published as a single document for the whole World at intervals from 1880 to 1911. Two wireless time signals in North America were included in the 1908 and 1911 lists but wireless signals were later listed separately and are excluded from the present study. Later lists, covering lighthouses and other navigation lights as well as time signals, were issued in parts covering different geographical areas. The title used in 1912–1920 was *Admiralty list of lights and time signals*, which changed to *Admiralty list of lights, time signals, wireless direction finding stations and wireless meteorological signals* in 1921 and then to *Admiralty list of lights and visual time signals* in 1922–1928. These were issued in nine parts, at first annually. Later, each part was reissued every 3 years, with supplements in intermediate years to record changes, allowing a list of time signals to be constructed for any given year. The dates of signal introduction or modification were often noted in editions after 1911. The title was changed again in 1929–1971 to *Admiralty list of lights, fog signals and visual time signals*, issued in twelve parts. In a small number of cases, where a signal was on the boundary between two geographical areas, the same entry appears in two parts. The 1947 edition was effectively a fresh start, recording signals that still existed or had been introduced after the devastation of World War II.

At first, signal times in Admiralty lists were specified as local (astronomical) mean time, governed by the longitude of the signal location, and Greenwich mean time (GMT). Most, but not all, time balls were dropped at local noon or 1 pm. Some were dropped more than once per day. Local times were later changed to standard times, determined by time zone allocations. In a few cases, a local Standard time was specified as well as a national Standard time. One example is the use of Rangoon Standard time as well as Burma Standard time (see later).

It had been standard practice to use time based on 0 hours at noon for astronomical purposes, but this was changed to civil time in 1925, based on 0 hours at midnight (Howse, 1997).

3 Typical Admiralty List Entries

Admiralty lists were published in booklet form for specified geographical areas which changed over the years. Sample list entries are shown for 1880, 1922 and 1930. In these cases, the detailed references are specified. The 1880 list entry for St Helena is shown in Fig. 6.1 (Admiralty List, 1880, British Possessions, 8–9). Two time balls were dropped at 1 pm GMT, 22 minutes, 50 seconds ahead of 1 pm, mean time at the (discontinued) Ladder Hill Observatory.

Figure 6.2 shows the complete entry for China in 1922 when time for astronomical purposes was based on 0 hours at noon (Admiralty List, 1922, Part VI, China, 451–452).

Figure 6.3 shows the Admiralty list entries for Vladivostok in 1922 and 1930, before and after the change to Civil time in 1925 (Admiralty List, 1922, Part VI, Russian Tartary, 455–456; Admiralty List, 1930, Part VI, Manchuria, 537–538). The two entries also show the change of spelling from Vladivostock to Vladivostok, the change in location description from Russian Tartary to Manchuria, and the change in Vladivostok Standard time by 12 minutes, 25.3 seconds. The time ball was dropped and the gun was fired at noon Standard time.

In order to simplify references to Admiralty list entries, the year of the entry is stated for a given time signal in tables of results, without detail concerning page and part numbers. In a few cases, the entry is derived from a combination of the main list and corrections via supplements up to the stated year.

ST. HELENA - - - - - - - ST. HELENA.

15° 55′ 0″ S. - 5° 42′ 30″ W.	James Town -	Ball - -	Time Office west side of the lines, James Town Valley. 74 feet above high water. 40 feet above ground. (Drop 13 feet.)	h. m. s. 1 0 0	h. m. s. 0 37 10	Ball hoisted half way up as preparatory at 5 minutes before signal. Ball hoisted close up at 2 minutes before signal. Ball dropped at 1^h 0^m p.m. Greenwich mean time, which corresponds to 0^h 37^m 10^s p.m. St. Helena (Ladder Hill Observatory) mean time. [*Note.*—Signal not made on Sundays. Ships desirous of rating or comparing chronometers may send them to the Time Office, or, on application, may have the signal repeated at any hour.]
—	—	Ball - -	Yard arm of flag staff on Ladder Hill. 653 feet above high water. 45 feet above ground. (Drop 40 feet.)	1 0 0	0 37 10	Ball hoisted half way up as preparatory at 5 minutes before signal. Ball hoisted close up at 2 minutes before signal. Ball dropped at 1^h 0^m 0^s p.m. Greenwich mean time, which corresponds to 0^h 37^m 10^s p.m. St. Helena (Ladder Hill Observatory) mean time. [*Note.*—Signal is in sight of the whole anchorage, and can easily be observed by ships passing within a few miles of the harbour. Signal not made on Sundays.]

Fig. 6.1 Admiralty list entry for St Helena in 1880

451 TIME SIGNALS.

CHINA.

No.	Name.	Year established or last altered.	Position. Latitude N. Longitude E.	Signal adopted.	Time of signal being made. Greenwich Mean Time.	Standard Time.
			° ′ ″		h. m. s.	h. m. s.
5041	Hong Kong	1908 1920	Mast north-westward of the signal station, Blackhead hill 22 17 52 114 10 29	Red ball	14 00 00 20 00 00	22 00 00 4 00 00
		1920	Observatory wireless mast	Three white lights (vertical)	0 56 00 to 1 00 00	8 56 00 to 9 00 00
5042	Swatau	...	Flagstaff on the harbour master's office 23 21 27 116 40 13	Ball	16 00 00	0 00 00
5043		...	... ...	Gun*	16 00 00	0 00 00
5044	Amoy	...	Battery on north-eastern slope of Wellington Nose, north point of Kulangsu	Gun*	16 00 00	0 00 00
5045		...	Mast at the semaphore station, south-east point of island 24 26 40 118 04 04	Flag T, International code	16 00 00	0 00 00
5046	Shanghai	...	Semaphore at the French concession 31 14 07 121 29 08·7	Ball, diameter 6 feet	15 55 00 16 00 00	23 55 00 0 00 00
5047		...	Same position	Four white electric lights	0 55 00 to 1 00 00	8 55 00 to 9 00 00
5048	Chin Kiang	...	32 15 00 119 29 10	... ...	...	...
5049						
5050	Tsingtau (Kyau Chau)	...	36 04 — 120 19 —	Gun*	16 00 00	0 00 00
5051						
5052	Chifu	...	Mast eastward of Tower hill 37 32 56 121 23 30	Black ball	16 00 00	0 00 00
5053		...	... ...	Gun*	16 00 00	0 00 00
5054						

* See Note, page 441. † Local mean time.

TIME SIGNALS. 452

CHINA.

No.	ADDITIONAL DETAILS.
5041	166 feet above high water, 62 feet above ground. Ball hoisted half way up at 5 minutes before the signal, close up at 3 minutes before the signal, and dropped electrically from the observatory, Kowloon, at $22^h 00^m 00^s$ and $4^h 00^m 00^s$, Standard time. The $4^h 00^m 00^s$ signal will not be made on Saturdays, Sundays or holidays. On Saturdays the signal will be made at $1^h 00^m 00^s$, Standard time. Should the time ball be out of order, the above routine will be carried out with flag "Z" on the storm signal mast. On application to the harbour master by captains of ships leaving port, who have had no opportunity of obtaining error of chronometer by above time signal, the time ball will be dropped at other times of the day. The lights are extinguished momentarily at even seconds, excepting the 2nd, 28th, 50th, 52nd and 54th seconds of each minute, between $8^h 56^m 00^s$ and $9^h 00^m 00^s$, Standard time.
5042	121 feet above high water, 111 feet above ground. Ball dropped at Noon, Standard time. Signal made on Saturdays only, and is not to be depended on.
5043	Fired when the ball is dropped.
5044	Fired at Noon, Standard time. Reported unreliable.
5045	Flag hoisted at 5 minutes before the signal, and hauled down when the gun is fired.
5046	118 feet above high water, 102 feet above ground, drop 21 feet. Ball hoisted half way up at 10 minutes before the first signal, close up at 5 minutes before the first signal, and dropped at $23^h 55^m 00^s$, Standard time. It is immediately re-hoisted, and the signal repeated at Noon, Standard time. Should the signal fail, flag "U", International code, will be hoisted and kept up for about one minute.
5047	The lights are in the form of a diamond. The lights are extinguished simultaneously at each exact minute from $8^h 55^m 00^s$ to $9^h 00^m 00^s$, Standard time. (A short eclipse of 1 second is given 10 seconds before the signals to attract attention).
5048	A telegraphic time signal will be sent on Mondays from Zi-Ka-Wei observatory, if asked for.
5049	
5050	Fired at Noon, Standard time. Reported inaccurate 1915.
5051	
5052	213 feet above high water, 75 feet above ground. Dropped at Noon, Standard time. Stated not to be sufficiently accurate for comparing chronometers.
5053	Fired when the ball is dropped.
5054	

Part 6. 2 H

Fig. 6.2 Admiralty list entries for China in 1922

455 TIME SIGNALS.

RUSSIAN

No.	Name.	Year established or last altered.	Position. Latitude Longitude E.	Signal adopted.	Time of signal being made. Greenwich Mean Time.	Standard Time.
			° ′ ″		h. m. s.	h. m. s.
5065	Vladivostok	1912 1915	White mast at Observatory 43 06 48 N 131 53 39	Red ball diameter 3¼ feet	15 12 25·32	0 00 00†
5066		1912 1915	Platform at foot of above mast	Gun*	15 12 25·49	0 00 00†

TIME SIGNALS. 456

TARTARY.

No.	ADDITIONAL DETAILS.
5065	93 feet above the water, 46 feet above foot of mast. Ball hoisted at 5 minutes before the signal, and dropped automatically at Noon, Local mean time. Should the signal fail the ball will be immediately hoisted at half-mast, remain hoisted for one minute, and then slowly lowered. A more exact comparison of the chronometer may be obtained from the Astronomer, application being first made to the Naval Staff.
5066	Fired automatically when the ball is dropped.

537 TIME SIGNALS.

MANCHURIA.

No.	NAME. Year established or last altered.	Position— Latitude Longitude E.	Signal adopted.	Time of signal being made. Greenwich Mean Time (Civil).	Standard Time (Civil)
		° ′ ″		h. m. s.	h. m. s.
5077	Vladivostock 1912, 1915	White mast at Observatory 43 06 48·4N 131 53 40·2	Red ball, diameter 3½ feet	3 00 00	12 00 00
5078	1912, 1915	Platform at foot of above mast	Gun*	3 00 00	12 00 00

TIME SIGNALS. 538

MANCHURIA.

No.	ADDITIONAL DETAILS.
5077	93 feet above the water, 46 feet above foot of mast. Ball hoisted at 5 minutes before the signal, and dropped automatically at Noon, Standard time. Signal not made on Sundays and holidays. Should the signal fail, the ball will be immediately hoisted at half-mast, remain hoisted for one minute, and then slowly lowered. A more exact comparison of the chronometer may be obtained from the Astronomer, application being first made to the Naval Staff.
5078	Fired automatically when the ball is dropped. Not fired on Sundays and holidays.

Fig. 6.3 Admiralty list entries for Vladivostok in 1922 (top) and 1930 (bottom)

4 Specification of Latitude and Longitude

The latitude and longitude of a time signal were specified as accurately as possible. Longitude was relative to the prime meridian which passed through Airy's transit circle at Greenwich (see Howse, 1997). Modern GPS co-ordinates are not subject to this constraint and the prime meridian used for Google maps is positioned about 5 arc-seconds east of the Observatory (see Kinns et al., 2021). The difference is too small to influence locations on most maps in this book. There may be small location errors in larger-scale maps showing multiple signals in selected ports.

5 Warning About Use of Sound Signals

A cautionary note about time guns was always included in Admiralty lists of time signals, with minor variations of wording between editions. The following version is from the 1908 list.

> When the flash of a gun cannot be seen, the sound of the report may be made use of as an approximate indication of the time, by allowing for the necessary interval for the sound to travel the intervening space. Sound travels about 1,090 feet in a second of time, at the temperature of 32° Fahrenheit, and the speed increases at the rate of 1.15 per second for each degree of temperature above the freezing point. In fogs, however, the use of sound is not to be relied upon, as the path of the wave may not be direct on account of the interposition of impervious clouds.

The slow speed of sound led to the production of maps for individual time guns that showed the average time delay to different locations. However, the precise delay varied with weather conditions and wind direction. The accuracy of chronometer rating was thereby reduced relative to that attainable using a time ball or the flash of the gun, but a sound signal was entirely adequate for public use when errors of a few seconds were inconsequential. This qualification tended to be ignored or forgotten by supporters of time guns.

References: Admiralty Lists in Date Order

Lists from 1880 to 1911

Admiralty list, 1880. List of Time Signals, Established in Various Parts of the World 1880: Compiled for the Use of Seamen, as an Aid for Ascertaining the Errors and Rates of Chronometers. 1st Edition printed for the Hydrographic Department, Admiralty, London, 1880.

Admiralty lists were published in 1888 (2nd Edition), 1892 (3rd Edition) and 1895 (4th Edition). [Not seen by the Author.]

Admiralty list, 1898. List of Time Signals, Established in Various Parts of the World 1898: Compiled for the Use of Seamen, as an Aid for Ascertaining the Errors and Rates of Chronometers. 5th Edition printed for the Hydrographic Department, Admiralty, London, 1898.

Admiralty list, 1904. List of Time Signals, Established in Various Parts of the World 1904: Compiled for the Use of Seamen, as an Aid for Ascertaining the Errors and Rates of Chronometers. Printed for the Hydrographic Department, Admiralty, London, 1904.
Admiralty list, 1908. List of Time Signals, Established in Various Parts of the World 1908: Compiled for the Use of Seamen, as an Aid for Ascertaining the Errors and Rates of Chronometers. Printed for the Hydrographic Department, Admiralty, London, 1908.
Admiralty list, 1911. List of Time Signals, Established in Various Parts of the World 1911: Compiled for the Use of Seamen, as an Aid for Ascertaining the Errors and Rates of Chronometers. Printed for the Hydrographic Department, Admiralty, London, 1911.

Lists from 1912 to 1928

Admiralty list of lights and time signals. Issued in nine parts. Printed for the Hydrographic Office, Admiralty, by Wyman and Sons Ltd., London, 1912–1920.
Admiralty list of lights, time signals, wireless direction finding stations and wireless meteorological signals. Issued in nine parts (as in 1912–1920). Printed for the Hydrographic Office, Admiralty, by Wyman and Sons Ltd., London, 1921.
Admiralty list of lights and visual time signals. Issued in nine parts (as in 1921): Published for the Hydrographic Office, Admiralty, by HM Stationery Office, London, 1922–1928.

The following list of parts was specified in 1922, including sub-titles. There were variations between titles in different years.

I. British Islands.
II. North and Arctic Seas, except the British Islands (Eastern Shores of the North Sea, Norway, the White Sea, Iceland and the Færoes).
III. Baltic Sea, with Kattegat, Belts and Sound.
IV. Eastern sides of North and South Atlantic Oceans, South of Dunkerque (Western Coasts of Europe and Africa, including Açôres, Madeira, Canary and Cape Verde Islands, etc.).
V. Mediterranean, Black and Red Seas.
VI. Indian and West Pacific Oceans (South and East Africa, India, East Indies, China, Japan, Australia, New Zealand and West Pacific Islands).
VII. Western Side of South Atlantic Ocean and East Pacific Ocean (Eastern Coast of South America South of Cape Orange; West Coasts of South and North America, and East Pacific Islands).
VIII. Western Side of North Atlantic Ocean (Canada, Newfoundland and Labrador).
IX. Western Side of North Atlantic Ocean (United States of America, Gulf of Mexico, West Indies, and North Coast of South America to Cape Orange).

Lists from 1929 to 1971

Admiralty list of lights, fog signals and visual time signals. Issued in twelve volumes. Published for the Hydrographic Office, Admiralty, by HM Stationery Office, London, 1929–1971.

The following list of volume titles was specified in 1947, including sub-titles. There were variations between titles in different years.

1. British Isles.
2. North Sea except the British Isles (Southern and Eastern Shores of the North Sea, France, Belgium, the Netherlands, West Coast of Denmark, and South and West Coasts of Norway).

3. Baltic Sea, with Kattegat, Belts and Sound.
4. Eastern sides of North and South Atlantic Oceans, South of Dunkerque (Western Coasts of Europe and Africa, including Açôres, Madeira, Canary and Cape Verde Islands, etc.).
5. Mediterranean, Black, and Red Seas.
6. Arabian Sea, Bay of Bengal and North Pacific Ocean (East Africa, India, East Indies, China, Japan, Russia and North Pacific Islands).
7. Western Side of South Atlantic Ocean and East Pacific Ocean (Eastern Coast of South America South of Cape Orange, West Coasts of South and North America and Hawaiian Islands).
8. Western Side of North Atlantic Ocean (Greenland, Labrador, Newfoundland, Canada and Canadian Arctic).
9. Western Side of North Atlantic Ocean (United States of America, Gulf of Mexico, West Indies, and the Northern Coast of South America to Cape Orange).
10. Indian and South Pacific Oceans (South and East Africa, East Indies, Australia, New Zealand and South Pacific Islands).
11. North and Arctic Seas except the British Isles (West and North coasts of Norway, Spitzbergen, the Faroes and Iceland).
12. Arctic Ocean (North coast of Finland, White Sea, and Arctic Russia to Bering Strait).

Howse, D. (1997). *Greenwich time and the longitude*. Philip Wilson Publishers Ltd.

Chapter 7
British and Irish Time Signals for General Use

Paul Fuller

Abstract The aim of this chapter is to give a broad perspective of time signals in the British Isles, including signals that were not listed by the Admiralty and were intended for public use. It includes a table showing British and Irish time signals ordered by date of introduction. It is not exhaustive but it illustrates the range of signals that were available in practice. Almost all were either time balls or guns. Selected signals in different regions are described in detail, including reasons for their introduction, sources of funding and operational difficulties. Many were regulated using telegraph signals from Greenwich, but there was not usually a requirement to publish any errors. The total number far exceeded the number listed by the Admiralty for chronometer rating. Time signals that appeared as principal signals in Admiralty lists between 1880 and 1947 are described in more detail in Chap. 10.

1 Introduction

The aim of this chapter is to give a broad perspective of time signals in the British Isles, including signals that were not listed by the Admiralty.

2 Time Signals in Date Order of Introduction

Table 7.1 shows British and Irish time signals for use by mariners and the general public, ordered by date of introduction. It is not exhaustive but it illustrates the range of signals that were used in practice. It is a summary of results from a large number of sources which is subject to ongoing review and extension (Fuller, 2023). Only approximate dates are available in some cases. The date of withdrawal is indicated for each signal when known. Almost all in the list are either time balls or guns.

P. Fuller (✉)
Independent Researcher, Southampton, UK

R. Kinns (ed.), *The World of Visual Time Signals for Mariners*, Historical & Cultural Astronomy, https://doi.org/10.1007/978-3-031-57334-7_7

Table 7.1 Time signals in Great Britain and Ireland

Start Date	End date	Place	Signal type	Details	Reference
22 October 1829	1830?	Portsmouth	Ball	Notice of trial. Time ball on "high tower", dropped at 1 pm.	Chapter 3
28 October 1833	Still in operation	Greenwich	Ball	Notice of time ball service at the Royal Observatory.	Chapters 3, 4
January 1845	1867	Liverpool	Ball	City Observatory, Waterloo Dock. Augmented by time ball on Victoria Tower.	Chapter 10
August 1852	February 1866+	London	Ball	Electric Telegraph Company (later E&I TC), 488 Strand.	Clark, 1852; Airy, 1861
September 1853	1867	Liverpool	Ball	Electric Telegraph Company (later E&I TC), Castle Street. Replaced by Morpeth Dock time gun in 1867.	Airy, 1861 Chapter 10
20 March 1854	Restored and operable	Edinburgh	Ball	Erected by Charles Piazzi Smyth on Nelson's Monument, Calton Hill.	Chapters 5, 10
1854	20 December 1913	Aldershot	Gun	Return to barracks. Gun Hill from 1854, Thorn Hill from 1877.	
1 January 1855	25 February 1927 *(Replica 2004)*	Deal	Ball	Semaphore Tower. Ball dropped by electric telegraph signal from Greenwich with return signal to confirm drop. Museum display of original apparatus. Working replica in operation since 2004.	Chapters 5, 10
March 1856	February 1866+	London	Ball	City Observatory, Cornhill. Location of many chronometer manufacturers.	Airy, 1861
1856	20 December 1913	Portsmouth	Ball	On Semaphore tower. Burned down in 1913. Later miniature time ball at Portsmouth Observatory not visible to ships.	Chapters 3, 10
September 1857	1864	Glasgow	Ball	On Sailors' home. Still in place long after end of service. Replaced by controlled clocks along the Clyde.	Kinns, 2010; Clarke and Kinns, 2011
November 1858	1867	Wolverhampton	Ball	Lord Wrottesley's observatory.	
1858	?	Jersey, C.I.	Ball	Fort Regent. Uncertain existence.	
July 1859	?	Dover	Ball	Proposed at Archcliff Fort. Not mentioned in Admiralty lists. May not have progressed beyond planning in 1859.	Chapter 10
July 1859	1 November 1912	Dover	Gun	Dover Castle battery. Mentioned in footnote to 1880 Admiralty list. Not listed after 1898. Fired at noon and (later) at 9.30 pm.	Chapter 10
17 August 1860	April 1890	Great Yarmouth	Ball	Sailors' Home signal station. Time ball mentioned in footnote to 1880 Admiralty list.	Chapter 10
1860	December 1887	Plymouth	Gun	Military signal. Not used for chronometer rating. Drake's Island.	
1860	1867	Liverpool	Ball	On Victoria Tower. Replaced by Morpeth Dock time gun in 1867.	Chapter 10
June 1861	Still in operation	Edinburgh	Gun	Edinburgh Castle. Established by Charles Piazzi Smyth. Fired by special gun clock, regulated by telegraph from Nelson's Monument.	Chapters 3,10
June 1861	1886	Devonport	Cone	Mount Wise. Collapsible cone, replaced by time ball in 1886.	Chapters 3, 10.
1862	1893	Cardiff	Ball	Electric Telegraph Company. Ball changed location during period of operation.	Linnard, 2018
July 1863	?	Kingston	Ball	Initially, ball in Jacobs' jeweller's shop, 7 Whitefriar's Gate, Kingston upon Hull. Later at William Ushaw's shop at 50 Lowgate, likely to be the Hull time ball mentioned in the footnote to the 1880 Admiralty list.	Jessop, 2017
16 August 1863	Early 1874	Newcastle upon Tyne	Gun	Castle Keep.	
18 August 1863	31 August 1905	North Shields	Gun	Ballast Hill. Still listed by the Admiralty in 1908, apparently in error.	Chapter 10
26 August 1863	2 September 1866	Sunderland	Gun	Initially a small gun, fired for one week only. Later gun trials abandoned in 1866.	

(continued)

Table 7.1 (continued)

Start Date	End date	Place	Signal type	Details	Reference
1863	February 1864	Glasgow	3 Guns	Sauchiehall Street/Renfrew Street; St Vincent Place; Broomielaw. All fired from Edinburgh.	Kinns, 2010; Clarke and Kinns, 2011
November 1863	February 1864	Greenock	Gun	Fired from Edinburgh.	Kinns, 2010; Clarke and Kinns, 2011
1865	1916	Dublin	Ball	Dublin time ball mentioned in footnote to 1880 Admiralty list, later in main list at Port and Docks Board Building. Relocated in 1916.	Chapter 10
1865	1865	Belfast	Gun	Harbour Office. There were experiments on 3 March 1865 and in Sept 1867 but no permanent signal.	
21 September 1867	18 July 1969	Liverpool	Gun	Morpeth Dock, current from Bidston Observatory.	Chapter 10
January 1869	December 1897	Leigh, Lancashire	Rocket	Rockets fired at 10 pm GMT from Avenue House, The Avenue.	Eckersley, 2017
24 February 1871	November 1967	Carlisle	Ball	Wheatley's shop. Signal for mills and factories.	
21 March 1871	22 July 1872	Carlisle	Gun	Wheatley's shop. Signal for mills and factories.	
1871	1877?	Hereford	Gun	Broomy Hill, then Reservoir. Signal for mills and factories.	
3 September 1872	1916	Dundee	Gun	Initially Barrack Square, changed later	Chapter 10
1872	?	Dunecht, near Aberdeen	Gun	Observatory at the home of the Earl of Crawford. Gun introduced by David Gill.	
May 1873	1940	Kendal	Gun	Serpentine Walks. Primarily signal for mills and factories. Not mentioned in 1880 Admiralty list.	Airy, 1878
.1873	1913+	Guernsey, C.I.	Gun	Castle Cornet, Saint Peter Port.	
1873	1875	Middlesbrough	Gun	Dock Hill.	
1873	1909+?	Inverness	Ball	Mentioned in footnote to 1880 Admiralty list. P.G. Wilson's shop (short lived?) and Ferguson Brothers' shop on the corner of Union Street from June 1873 to 1879, resurrected January 1881 to c 1907. Possibly 15 April 1909+.	Airy,1878
October 1874	Early 1909	Swansea	Gun	South entrance to the Half Tide Basin of the North Pier, moved to East Pier in June/July 1886.	Chapter 10
1874	Late 1924	Hartlepool/West Hartlepool	Gun	Various locations.	Airy, 1878
1874	1896?	Whitehaven	Ball	Ball noted in footnote to 1880 Admiralty list. Probably a miniature ball in jeweller's premises at 21 and 22 Lowther Street.	Sutton, 1896
1874	?	Jersey, C.I.	Ball	T. Hollingshead premises in Brook Street.	
1875	?	Jersey, C.I.	Ball	Genge and Co's shop in Halkett Place.	
1875	1913	Woolwich	Gun	The Common. Also controlled clock. Return to Barracks.	
January 1876	1915	Queenstown	Gun	Quayside.	Chapter 10
February 1876	1939+	Cork	Gun	Victoria Quayside.	Chapter 10
1876	1883	Newport	Ball	Believed to have been operated by Robert Whitehall at his premises.	Linnard, 2018
11 September 1876	October 1886	Nottingham	Ball	Exchange Building. Signal for mills and factories.	
September 1877	?	Leeds	Ball	Briggate, Dyson & Son. Second ball added 1910.	
1877	1882	Bradford	Ball	Exchange Building.	
November 1877	1882? Isbell died 1885	Hereford	Gun	Described as "Isbell's private gun". Signal for mills and factories.	
1877?	1911?	Stockton	Ball	Above 99 High Street, dropped at 10 am and 1 pm. Shop owned by W.E. Harrison, who offered a chronometer rating service. Business closed in 1911.	Stockton Examiner, 1878

(continued)

Table 7.1 (continued)

Start Date	End date	Place	Signal type	Details	Reference
2 May 1878	1931	Blackburn	Gun	The Park.	
1878	1903	Blackburn	Ball	Market Place.	
Early October 1879	?	Romsey	Ball	Jeweller's shop, Market Place.	
November 1879	1882+	Sunderland	Ball	Time ball mentioned in footnote to 1880 Admiralty list. At premises of Gowland Brothers, 178 High Street West.	Chapter 10
1870s?	?	Jersey, C.I.	Ball	Time ball mentioned in footnote to 1880 Admiralty list. Possibly Genge and Co's shop in Halkett Street from May 1875.	
1880	1913	Chatham	Gun	Probably Fort Amherst. Not included in Admiralty lists.	Chapter 10
1884	?	Huddersfield	Ball	New Street, Pearce and Sons. Signal for mills and factories. Dropped at noon?	
1886	1928+	Devonport	Ball	Mount Wise. Time ball, operated using "the Devonport principle", replaced collapsible cone.	Chapter 10
1887	1924	Devonport	Gun	Mount Wise.	Chapter 10
Late 1887	11 October 1909	Wolverhampton	Ball	Queen Square.	
1887	1889?	Penmaen	Gun	Private initiative.	
30 January 1888	Late March 1904	Southampton	Ball	God's House Tower, South Castle, overlooking the Platform. Moved to South Western House on Canute Road.	Chapter 10
1888	1890s	Brighton	Ball	Clock Tower at Seven Dials. Queen Victoria Golden Jubilee memorial.	
24 May 1889	Restored 2014	Margate	Ball	Seafront. Queen Victoria Golden Jubilee memorial.	
September 1889	1907+	Dewsbury	Ball	Town Hall, northwest pavilion.	Dewsbury, 1889; 1907
25 March 1892	Early April 1893	Blackpool	Gun	By the destructor in the Corporation Yard on Princess Street.	
October 1894	1933	Sheerness	Ball	Garrison Point Fort near the Sheerness Dockyard.	Chapter 10
27 August 1897	1922+	Falmouth	Ball	Flagstaff on Pendennis Castle.	Chapter 10
July 1899	1939	Leigh, Lancashire	Rocket	Technical School in Railway Road. Rockets fired at 9 pm/10 pm GMT?	Eckersley, 2017
1900	1938?	Norwich	Ball/Gun	Castle. Signal for mills and factories.	
1901	1913	Colchester	Gun	Abbey Field. Return to barracks.	
31 March 1904	1933	Southampton	Ball	South Western House, Canute Road.	Chapter 10
1906	1928+	Chatham	Ball	Drill shed of the Royal Naval Barracks.	Chapter 10
1907	31 August 1933	Portland	Ball	Signal station over eastern end of Naval storehouse, south side of harbour.	Chapter 10
Erected by 1916	Not usable after 1922. Restored 2023	Kingston upon Hull	Ball	Humber time ball. Unreliable and inoperable by 1922. May never have provided a regular service. Not listed by the Admiralty.	Jessop, 2017
1916	1919+	Immingham Dock	Ball	Humber time ball listed by the Admiralty.	Chapter 10
1916	1926	Rosyth	Ball	Dockyard.	Chapter 10
1916	1940s+	Dublin	Ball	Eastern end of Sir John Rogerson's Quay. Change of location in 1916. Still included in 1947 Admiralty list.	Chapter 10
23 June 1919	1 August 1923	Dover	Gun	Castle Battery. Re-established time gun fired at noon and 9.30 pm for return to barracks. Not listed by the Admiralty.	Chapter 10

Controlled clocks, which were used widely by railway companies in particular, are not included in this table. Their use in Glasgow as signals for mariners is discussed later. Many other signals, including innovative devices like falling hats, were used for business promotion. Some of these have been described by Howse (1997).

Most signals described in this chapter were multi-purpose. Their accuracy and reliability are often unknown, although many used telegraph signals from Greenwich. There was not usually a requirement to publish any timing errors if signals were not listed by the Admiralty. Such errors might arise through failure of a ball to be released correctly or delayed gun detonation.

Time signals that appeared as principal signals in Admiralty lists between 1880 and 1947 are highlighted with a blue background in Table 7.1. These are described in Chap. 10, together with their history before 1880. Their accuracy and reliability were subject to official scrutiny. Some other signals were mentioned in a footnote to the 1880 Admiralty list. They are described in this chapter if they were not listed from 1898 onwards as principal signals. Signals which had little relevance to mariners or chronometer rating are highlighted in Table 7.1 with a green background.

Signals in England, subdivided into regions, are described first, followed by those in Scotland, Wales and Ireland. Some signals that were promoted actively never came to fruition. Examples are described towards the end of this chapter.

3 England

3.1 London

Two time balls in London were noted by Airy (1861), in addition to the Greenwich time ball which has been described in Chap. 4. By 1852, Airy had developed an automated time service at Greenwich which could send instantaneous time signals to several locations in central London. He did so with the assistance of the Electric Telegraph Company, one of the most important and innovative of the early telegraph companies.

3.1.1 The Strand

In 1852 the Electric Telegraph Company constructed a large ball on its premises at 448 The Strand in London. The ball was to be connected by telegraph with the Royal Observatory at Greenwich but it took more than 2 months before it became fully operational on 28 August 1852. The best-known image was published on 11 September 1852 (Electric time ball, 1852). It was included in Howse (1997, 100). As with many other early time balls, a large expectant crowd watched the ball drop in its early days. Unfortunately, it was never a great success because of technical difficulties and it appears to have been discontinued in the 1860s (Howse, 1997).

There had been various preliminary announcements about the Strand time ball, which were inconsistent so far as the dimensions and colour of the ball are concerned. Confusion abounded. An article in *The Illustrated London News* on 26 June (Regulation of time, 1852) stated that:

> The ball is nearly six feet high, and 16 feet (*sic*) in circumference. It is formed of zinc, painted black, with a broad white belt around it....

The accompanying engraving indicated that the ball was spherical. The circumference would have been 19 feet if the diameter was 6 feet.

Howse (1997, 100), quoting from another source, reported that the time ball had a diameter of 5 feet and was painted black with a broad white band around it. An article in *The Times* on 19 August (The electric time-ball, 1852) indicated a change of intended colour from black to red, but gave the same ball dimensions as *The Illustrated London News*:

> It is about six feet high and 16 feet (*sic*) in circumference, made of zinc, and painted of a bright red colour, so that it may be the more clearly discerned at a distance.... It has a broad white belt around it, thus having the appearance of a "great globe"....

A letter to the Editor of *The Times* on 8 September, after the ball started service, was signed by the engineer responsible for its design (Clark, 1852). It should be authoritative. He did not mention the colour but gave extensive details about the mechanical design and telegraph arrangements. The following extract indicated clearly that the ball diameter was 6 feet:

> The ball, six feet in diameter, is made of zinc and with its attached piston weighs about 2½ cwt. When fully raised it is 129 feet above the level of the Thames, and it falls through a space of 10 feet. The rod which supports it passes down the centre of the column and carries at its base a piston, which, in its descent plunges into a cast iron cylinder, 10 inches in diameter; the escape of the air being regulated so as to check at pleasure the momentum of the ball, and prevent any concussion from its weight....

The design had much in common with the arrangement at Greenwich (see Chap. 4), but the ball was larger and heavier, requiring the diameter of the air cylinder to be increased.

3.1.2 Cornhill

Another early time ball was constructed at Santiago French's City Observatory at Cornhill in the centre of London in early 1856. French was a chronometer maker who had previously practised his skills at the Royal Exchange. He had won an Admiralty prize in 1822 for making a chronometer which was used to calculate the precise longitudes of Falmouth and Madeira. The Cornhill ball received its current from Greenwich and at its inauguration "... the rather narrow thoroughfares of Bishopsgate Street and Cornhill are choked with spectators, all agape, with eagerness and expectations, until the critical moment, when they depart, apparently highly pleased with the result". It lasted until at least 1866.

3.2 Northeast England

3.2.1 Newcastle and North Shields

Great advances were made in 1863 when time guns were established at Newcastle upon Tyne and North Shields. The leading player in the call for time signals on the Tyne was James Mather, who thought nothing of venturing into dangerous mines to rescue miners after explosions or leading thousands of seamen to London to petition for their rights. In 1848, Mather produced an important work "Accidents in Coal Mines, their dangers and means of safety" which was widely debated in Parliament and led to improved safety measures. He collaborated with Thomas Sopwith, a well-known mining engineer and railway surveyor who worked with leading railway engineers such as George and Robert Stephenson, and knew Charles Wheatstone, the co-inventor of the electric telegraph, in the 1850s. Another member of this select group was the chronometer-maker David Reid from Grey Street who later claimed to have taken a leading role in the adoption and dissemination of Greenwich time on Tyneside as early as 1840.

Mather had collaborated with Airy during the latter's pendulum experiments at Harton Pit in 1854. After several attempts, he persuaded the Tyne Commissioners that the captains of hundreds of ships moored on the Tyne would benefit from having accurate Greenwich time, provided by Airy. He naturally assumed that the Electric Telegraph Company, chosen to provide the telegraph links, would be firing time guns from Greenwich. His enthusiasm for this solution led him into direct conflict with Charles Piazzi Smyth, who was equally keen to fire them from his observatory in Edinburgh. Piazzi Smyth employed a controversial character called Nathaniel J. Holmes to achieve this technological breakthrough and the theatre for this conflict was played out during the annual meeting of the British Association in August 1863.

Holmes began his working life with the Electric Telegraph Company but later worked for the Universal Private Telegraph Company. He was a genius when it came to promoting his company's technology. Alongside David Reid and Frederick Ritchie, who had worked on the Edinburgh time ball, Holmes spent weeks perfecting the telegraphy and gun-firing mechanism using the Electric Telegraph Company's precious telegraph lines. During the British Association's meeting Piazzi Smyth announced to his expectant audience that the Newcastle time gun was about to fire and unlike the embarrassing Edinburgh failures in 1861, the gun fired on time to great acclaim.

Firing three guns and dropping one ball from the same current was a great achievement but almost immediately a dispute broke out over the source of the time current and its funding. Holmes purchased the current from Edinburgh but James Mather believed that the current from Greenwich could be delivered more cost effectively. He undercut the Edinburgh charge and proposed to drop balls at different locations. Holmes retaliated by giving a lecture on his time gun system to the Newcastle Literary and Philosophical Society in which he explained that Captain Maury, the former Head of the US Naval Observatory who joined the Confederacy in 1861, intended to use time gun technology to sink Union ships. Holmes then claimed that he was the first person to think of firing a time gun from a distant observatory, a claim which clearly ignored the facts, and he added fuel to the fire by criticising the existing technology used by the Electric Telegraph Company (Fuller, 2023).

The accuracy of these time guns was subjected to considerable scrutiny but Holmes' Edinburgh system was clearly the more reliable as the Greenwich signal struggled over the longer distance. Piazzi Smyth intervened and ensured that his system was adopted. Months later, Airy had recaptured the Tyne guns via the Electric Telegraph Company. These guns were funded by five public agencies so

Airy's decision caused each authority to reconsider their funding. In 1874 the Newcastle gun ceased firing following a complaint but the North Shields gun continued firing from the Ballast Hill until 1905.

3.2.2 Sunderland

A small time-gun in Sunderland was located on a model training ship near the Orphanage Asylum at Moor Terrace and fired for the first time on 26 August 1863. It was introduced by Nathaniel J. Holmes as part of the Tyneside time signal system but it was only fired for a week. The gun was one of ten brass 6-pounders provided by the War Office to thank the port for its support during the earliest stage of the formation of the Naval Reserve (Fuller, 2023).

The success of the time guns at Newcastle and North Shields encouraged local solicitor William Scurfield Grey to call for the introduction of a public time gun at a meeting of the River Wear Commission in November 1863 but his proposal met with almost universal disapproval. He tried again in December but was again rebuffed. A gun arrived from Woolwich Arsenal in January 1864 but it was left in a brewery yard for over 2 years. In July 1866, a series of experiments were conducted in Mowbray Park but the cost of introducing the signal was a major obstacle and prominent members of the town council did not believe that the gun could be relied upon. They were made aware that there had been problems with the time guns at Glasgow and Devonport. They believed that the gun was dangerous and that there was no safe place for it to be located (Fuller, 2023).

The proposal for a time gun at Sunderland was abandoned in 1866, but a time ball was installed at a Sunderland business in November 1879. G.H. & C. Gowland were jewellers and chronometer makers who positioned a time ball on top of "our observatory" at 178 High Street West. It was regulated using the current from Greenwich and the ball was dropped daily, except Sundays, at 1 pm. This was the time ball included in a footnote to the 1880 Admiralty list.

3.2.3 Hartlepool/West Hartlepool and Middlesbrough

Hartlepool and Middlesbrough in the northeast of England both featured time guns from 1873. The twin towns of Hartlepool and West Hartlepool debated the need for a time gun to rival the ones at Newcastle, North Shields and Sunderland to meet the needs of their local shipping communities. The impetus at adjacent Middlesbrough was the local chamber of commerce but discussions in 1870 came to nothing due to the cost. In March 1873, Middlesbrough town council decided by 11 votes to 7 to press ahead with experiments and in June a time gun with only four pounds of gunpowder was tested at the Dock Hill. Not surprisingly, the experiment was a failure so the gun was moved to a new location.

It was these discussions which attracted the attention of nearby Hartlepool where George Pyman, a ship owner and magistrate, first raised the idea at a meeting of the

Improvement Commission in January 1873. He received a lukewarm reception even though he proposed that the cost of the gun be shared amongst a wider group of public agencies. The same day the Hartlepool Port and Harbour Commissioners discussed the time gun and were also less than enthusiastic. Their support was only carried after the intervention of Captain Young, a leading public figure who also held a commission in the 4th Durham Artillery Volunteers Regiment.

Gradually other agencies decided to contribute and the Improvement Commissioners took the lead, setting up a time gun committee which ordered a 12-pounder gun which was tested on 23 June. Despite this progress another commissioner brought a motion against the gun and the gun only survived when it became clear that many ratepayers had already contributed towards the cost of the gun and that arrangements were already well advanced to receive the current from Greenwich. The location of the gun and its perceived ineffectiveness were to haunt supporters of the gun as it was moved around on several occasions and its multi-agency funding repeatedly came under threat. These agencies are shown in Table 7.2.

In the Hartlepools, problems of cost and ineffectiveness caused the Corporation of Hartlepool to debate renewing its subscription of £10 a year. It took the intervention of another local ship owner, Jens Christian Nielsen, and a deputation from West Hartlepool to persuade the corporation to continue with its subscription providing the gun was moved to a more suitable location. In 1880 the original gun was replaced with a six-pounder bronze gun but this gun was only half the size of the original gun and it could not be heard across such a wide area. Critics of the gun trespassed into the time gun enclosure and filled it with stones in June 1882. In 1893 the second time-gun was replaced with another gun and both Hartlepool and West Hartlepool councils increased their funding. The gun was moved to another new location, this time near the Volunteer Artillery Battery at Newburn Bridge. By 1914 the gun had been moved to the refuse destructor yard on Burn Road. In 1924 the advent of wireless signals heralded the end of the Hartlepool time guns but there were several failed attempts to have the gun signal reintroduced.

Table 7.2 Agencies and individuals involved in funding and promoting the time guns of Hartlepool and West Hartlepool

Agency	Town
West Hartlepool Improvement Commission	West Hartlepool
The Hartlepool Port and Harbour Commission	Hartlepool
The West Hartlepool Town Council	West Hartlepool
The Corporation of Hartlepool	Hartlepool
Local Board of Middleton-in-Stranton	Hartlepool
North-Eastern Railway Company	Both towns
Seaton Carew Iron Company	West Hartlepool
Numerous local inhabitants	Both towns

A time gun at Middlesbrough was introduced in 1873. Problems with inaccuracy caused one councillor to describe the cost of the gun as "useless expenditure". He stated that "The gun did not keep correct time, and did not serve the purpose for which it was set—to allow the captains to regulate their chronometers by it". Another councillor described the gun as "a nuisance" which was "no use at all". Despite these criticisms, the council voted to continue firing the gun but its cost and unreliability meant that its days were numbered and at a council meeting on 9 February 1875 a motion of no confidence in the time gun was brought by councillor Thomas Hugh Bell. Several other councillors were scathing about the gun, which was often 8–10 seconds in error. Alderman Edward Williams said that the gun was "an utter waste of public money" whilst another said that the £100 a year would be better spent than being "puffed away in a cloud of smoke". The vote was taken and the gun lost by 13 votes to 9.

3.2.4 Kingston upon Hull

The history of time signals at Kingston upon Hull has been described by Jessop (2017). For many decades, time was signalled using time balls in shops. The following summary is taken from Jessop's paper.

> The first to do so was Bethel Jacobs at 7 Whitefriargate, a jeweller, gold and silversmith; he advertised in December 1865: 'Chronometers rated and tested by the Time Ball, which is in direct communication with Greenwich daily'. Jacobs may have started his time-ball service a few years before this, in July 1863. Jacobs did not keep his monopoly for long, being joined in October 1871 by the Hull Observatory at 50, Lowgate, which was not an actual observatory but a shop selling and repairing chronometers, watches, clocks, and sextants and offering chronometer rating. They advertised thus: 'Public time ball signals, direct from the Royal Observatory, Greenwich, every morning, in the Lowgate Window, at Ten o'Clock.'

The footnote in the 1880 Admiralty list is likely to be a reference to the Lowgate time ball.

The history of the unsuccessful time ball at the Guildhall has been described in detail by Jessop. The decision to install a time ball was made in 1914 and its erection was largely complete by the end of 1915, but it may not have entered regular service until much later, if at all. The costs of operation and the value of the signal became causes for concern and parts of the operating arrangement were removed during 1922, preventing further use. The Admiralty-listed time ball at Immingham Dock fulfilled the requirement for a Humber time ball from 1916 until after the end of the First World War (see Chap. 10).

The Guildhall is an important building, well-regarded for its architectural design, and the decision was made to bring the time ball back into operation as a tourist attraction. The new gilded time ball was dropped for the first time on 30 June 2023 (Hull's time ball, 2023).

3.2.5 Leeds and Bradford

Although Leeds never had a public time signal, the watch maker John Dyson erected a large time ball on his jeweller's shop in Briggate just to the north of the River Ayre in 1876. It received the current from Greenwich and was claimed to be the largest time ball in the north of England. A second time ball was added in 1910.

Bradford became a major wool and textile manufacturing town in the middle of the nineteenth century. The main proposer of a time signal was Henry William Ripley who was the owner of a large dyeing company and who was also a founding member of the chamber of commerce. He was frustrated at the series of miniature time balls in the town so in 1877 he paid for a public time ball which was erected on the Exchange Buildings in the town centre. After he died in 1882 the time ball fell into disuse as it was known to be inaccurate. In 1886 the chamber of commerce decided to resurrect the idea of a public time ball but the idea was dropped when it became clear there was no enthusiasm for its funding.

3.2.6 Dewsbury

The town of Dewsbury on the River Calder in West Yorkshire included a time ball on its new town hall, which operated for the first time in September 1889. It was appreciated by local inhabitants: "The falling of the Town Hall time-ball (given by Mr. Chaley Fox) has been watched by spectators each day at noon, some of whom always set their chronometers by it with an air of having successfully performed a great feat." (Dewsbury, 1889). Major Chaley Fox (1844–1909) was a local councillor and wool merchant, who served as mayor of Dewsbury from 1904 to 1906. The time ball apparatus was repaired in January 1907 (Dewsbury, 1907), but its date of final operation has not been identified.

3.3 Northwest England

3.3.1 Carlisle, Cumbria

Carlisle was an unusual location for a time ball. It had a small port but it was goldsmith and jeweller James Atkinson Wheatley who decided to establish a time ball at his premises in English Street in the centre of the town. He was concerned about the inaccuracy of the town's public clocks so he persuaded the Postmaster General to lower his annual charge for the current from Greenwich but the cost of the ball was still considerable for a private individual to bear. He approached the town council and at first, they were non-committal, but later Wheatley suggested that a cheaper alternative was to have an illuminated clock outside his window which could be controlled by a miniature time ball inside his shop. The council agreed to pay for the cost of the gas if Wheatley paid for the cost of the clock and time ball. After another debate the council changed their mind about paying for the cost of the gas but they agreed a nominal cost for the telegraph line to the Post Office. Wheatley paid for the ball and it was inaugurated by the mayor on 24 February 1871. The event was a great success and whilst an immense crowd celebrated the arrival of accurate Greenwich time Wheatley held a special *déjeuner* for his closest friends. He later added a miniature time gun on top of his building and this gun was fired for the first time as the ball dropped on 21 March.

Whilst many people were pleased with Wheatley's efforts a few objected. They were led by an elderly magistrate, Joseph Fergusson, who suffered from "delicate" health. Fergusson tried to bring an action against Wheatley for the alleged nuisance caused by the firing of his time gun and he collected a petition from seventeen people living in English Street, Lowther Street and Devonshire Street. His first attempt to bring proceedings failed but a court case was held on 22 July when Wheatley unwisely conducted his own defence. The case was heard before the mayor, who had supported the introduction of the time ball only 16 months before, but Wheatley lost and he was fined two shillings and sixpence. The decision caused an outcry in the town and many questioned the dubious application of the local bye-laws, which were drawn up years before time guns had ever been thought of. An amusing debate took place in the Council on 27 August but there was no reprieve for the time gun although the ball continued dropping until November 1967.

3.3.2 Whitehaven, Cumbria

Whitehaven is a small port at the mouth of the Solway Firth. It was noted in a footnote to the 1880 Admiralty list as having a time ball that was operated by private enterprise and dropped at 10 am. A public time ball at Whitehaven had been suggested (Time Ball, 1862), but no evidence has been found that it was ever erected. A local jeweller and silversmith, Robert Sutton, is believed to have "made arrangements for obtaining Greenwich time daily" in 1874 (Obituary, 1903). His premises were in Lowther Street and he advertised "Greenwich time daily at 10 a.m." in his business notices over a long period (Sutton, 1896). It appears likely from the 1880 Admiralty list footnote that he used a miniature time ball inside his shop.

3.3.3 Kendal, Cumbria

Kendal is a fine example of a series of time signals at an inland town. It was included by Airy (1878) in the list of signals he prepared for the German Government, but it was not mentioned in Admiralty lists. The signals arose following a dispute between a local watchmaker and the son of the local postmaster. Thomas Rhodes had gone to the Post Office to obtain the Greenwich signal but no one was there to supervise him. He walked behind the counter and was reprimanded for being in the wrong place. Rhodes had only gone to the Post Office because of the unreliability of the town's clocks and was understandably outraged. He raised the matter with his local

MP John Whitwell, who happened to be a Lieutenant-Colonel in the 1st Battalion Westmorland Rifles. Not surprisingly Whitwell favoured a time gun over a ball, and another supporter of the signal just happened to own a local factory which manufactured gun powder.

Kendal town council and local businesses paid for the time gun which was situated in Serpentine Walks overlooking the town in May 1873. The gun was maintained by Thomas Rhodes and his brother Edmund and on good days it could be heard 14 miles away. By April 1874, the gun was being funded by eleven subscribers who represented the town's mill owners and merchants. The original gun was replaced in 1882 and again in 1888. Following complaints about the retort the gun was rotated 180° in 1905 but this decision was quickly reversed when the gun could not be heard in the town. In 1908 a smaller gun was introduced but by 1914 this gun had become unserviceable. During the First World War a fourth time-gun was introduced and this carried on firing until 1940.

3.3.4 Blackpool

In 1892, Blackpool introduced a time gun as part of the seaside town's modernisation project. The gun was located at the destructor in Princess Street but in April 1893 it blew itself apart after its firer Thomas Hornby misguidedly packed the gun with additional cladding to increase its retort.

3.3.5 Blackburn

Blackburn town council were the prime movers in the establishment of a time ball on the Market House Tower in 1877. The ball was made by Reid Brothers, the first telegraphic engineering company in the world, who also made the Nottingham time ball. In 1878 the tower supporting the ball gave way and the ball fell onto a market stall narrowly missing its proprietors and sending their oranges flying. A time gun was added in May 1878 but by 1882 it was thought that the gun was causing injury to the town hall and there had been many complaints about the sound of the gun so it was moved to the corporation yard. In September 1892, there was a panic when the gun accidentally fired during the night. "Hundreds of people left their beds and ran half-dressed into the streets, while the police station was besieged by crowds clamouring for information". The time ball mechanism failed in 1903 and was not repaired until 1922. It was not dropped again until 1924. The time gun continued firing until 1931 but afterwards it only fired on Armistice Day and to welcome in the New Year. In 1964 the clock tower and time ball were demolished to make way for a new shopping centre (Thomas & Thomas, 2015). Figure 7.1 shows the Blackburn time ball.

Fig. 7.1 Blackburn time ball. (Courtesy Klaus Hülse collection)

3.3.6 Leigh, Lancashire

A fine example of Victorian time signal innovation took place at Leigh in Lancashire by Dr. George Henry Evans in 1869. Dr. Evans served during the Crimean War. He had his own private observatory with a transit telescope at Avenue House and had visited the Royal Observatory at Greenwich to ensure that his time system was accurate. He fired a time rocket at 10 pm every evening providing he knew it would be accurate to within 2 seconds of true time. He also encouraged local watchmen to fire their pistols to make the system more audible. After Dr. Evans died in 1897 a member of the Local Board William Edward Marsh took over the project. He had visited Bidston observatory and hoped to recreate Dr. Evans' level of accuracy, but he faced almost 2 years of frustration before the new time rocket could be fired. This unique signal continued until 1939. The use of accurate time at this inland town was to benefit local mill owners and the general population.

3.4 The Midlands

3.4.1 Wolverhampton

One of the earliest privately-owned time balls was constructed by John Wrottesley, the second Baron Wrottesley, who was a founding member of the Royal Astronomical Society. Between 1829 and 1831 he erected a private observatory at Blackheath close to the Royal Greenwich Observatory. His early astronomical work was assisted by a young John Hartnup, who would later become a key player in the development of the Liverpool Time Service. In 1854 he was elected President of the Royal Society and the following year he would become President of the Royal Astronomical Society.

Following the death of his father in 1841, Wrottesley subsequently transferred his Blackheath observatory to his ancestral residence 4 miles to the north-west of Wolverhampton in Staffordshire. In November 1858, Wrottesley constructed a large time ball which was hoisted 60 feet above the ground. Unusually, the ball fell at 3 pm on Tuesdays and Saturdays. The time ball ceased falling after Wrottesley's death in 1867.

A second time-ball was erected on top of the Birmingham, Dudley and District Bank in Queen Square, Wolverhampton in the English Midlands in 1887. It was donated to the town by Charles Tertius Mander, an industrialist who had been educated at Rugby School and Corpus Christi College, Cambridge. He was a partner in the family's paint and varnish works who would go on to be mayor of the town in 1892–96. Figure 7.2 shows the Wolverhampton time ball in the early twentieth century.

Fig. 7.2 Wolverhampton time ball. (Courtesy Klaus Hülse collection)

3.4.2 Nottingham

Nottingham introduced a time ball to the top of the Exchange Building in the Market Place in August 1876. This was another inland location where the signal was not used to set marine chronometers. It was made by Reid Brothers of City Road, London and it lasted until 1886 but was never considered reliable and was the butt of many jokes. A plan to mount a time gun on Nottingham Castle in 1876 was never implemented.

3.4.3 Hereford

A time gun was introduced in 1871 by the town council following pressure from the Reverend Henry Cooper Key and his brother Astley, who had been one of the members of the Royal Commission on the Defence of the United Kingdom in 1860. Astley was Director of Naval Ordnance from 1866 to 1869 so knew all about time signals and gunnery whilst Henry was a grand amateur who had begun astronomical observations whilst still in his teens. He had visited Aberdeen in 1870 and had struck up a friendship with David Gill, who had failed to persuade his own town to adopt a time gun but would later establish one at the Earl of Crawford's observatory at Dun Echt in 1873. Hereford town council obtained a gun from the dockyard at Devonport and the time gun was located at Broomy Hill to the west of the town centre. It was moved to an enclosure by the town reservoir but was only used for a few years before being replaced by a privately-owned time gun which was fired by the surgeon Edwin James Isbell, a sun dial enthusiast whose son had studied at Edinburgh when the city sported both a time gun and a time ball. Hereford is inland and its signal was used by mill owners and factory workers.

3.5 Southern England

3.5.1 Romsey

The little Hampshire town of Romsey featured a time ball in 1879. It was introduced by jeweller John Tuck who later became Mayor of Romsey. Tuck was probably encouraged to install his time ball by the Reverend Edward Lyon Berthon who was a Fellow of the Royal Astronomical Society and the vicar of Romsey between 1860 and 1891. Berthon designed a small working observatory which was widely used by amateur astronomers.

3.5.2 Brighton

Brighton and Margate both introduced time balls to commemorate Queen Victoria's Golden Jubilee of 1887. The Brighton time ball was positioned on top of the Jubilee Clock Tower in the North Street Quadrant. The tower was paid for by the

Fig. 7.3 Margate time ball. (Courtesy Klaus Hülse collection)

advertising entrepreneur James Willing and the tower and time ball were inaugurated in front of a large crowd on 28 June 1888. The ball fell every hour during daylight hours and was controlled by the clock. It fell into disuse in the 1890s following complaints about the noisy mechanism.

3.5.3 Margate

The Margate Jubilee Clock Tower and Time Ball were inaugurated on 24 May 1889 (Jubilee clock-tower, 1889) and this time ball fell at 1 pm. Figure 7.3 shows the arrangement. The time ball had been withdrawn by 1920. It has now been restored and can be operated again (Margate, 2014).

3.6 Eastern England

3.6.1 Norwich

Norwich seems an unlikely site for a time signal but in 1899 the town council decided to erect a signal on Norwich Castle following the demise of an earlier time ball outside Thomas Dixon's jeweller's shop in London Street. A series of experiments were conducted by the city engineer and a combined time ball and time gun was established in 1900. At first the signal was unreliable and in 1903 it had to be repaired. It is uncertain how long the signal was in use but the gun signal could be heard over the whole city.

3.6.2 Great Yarmouth

A time ball was erected on the newly built Sailor's Home at Great Yarmouth in Norfolk in 1860. The home was used by destitute sailors and its origins lay with the Shipwrecked Mariners' Society and senior members of the British Association. It was removed in 1890 and in 1904 a call for a time gun did not attract sufficient support.

3.7 Army Time Guns

One of the earliest British Army time signals of the 1860s was located on Drake's Island in Plymouth Harbour. It was situated on a Napoleonic Fort and its introduction was a direct result of the invasion scare of the previous 2 years. It was introduced as part of the defence improvements for the port which were recommended by the Royal Commission on the Defence of the United Kingdom.

A series of time guns was established by the army to summon soldiers to their barracks. The first was at Aldershot in Hampshire in 1854 but others followed at Colchester, Dover, East Tilbury, Pembroke Dock, Shorncliffe (Folkestone), Weymouth and Woolwich. There was another at Stirling Castle in Scotland. These time guns were axed by the Army Council during economy drives between 1909 and 1915, although the Dover time gun was temporarily re-introduced between 1919 and 1923.

The Dover time gun, which had wider application, was included in the Admiralty list for 1898. Its history is considered in more detail in Chap. 10.

4 Scotland

The signals at Edinburgh and Dundee were listed by the Admiralty over an extended period. Those for Glasgow and Inverness were mentioned in a footnote to the 1880 list (see Chap. 10), but not as principal signals from 1898 onwards. They are described below.

4.1 West Coast

4.1.1 Glasgow

The history of visual time signals in Glasgow reflects the tension between technical requirements and political sensitivities that often featured in time signal provision. The relative prestige of the Glasgow and Edinburgh observatories was an important issue at a time when the industrial city of Glasgow was often described as the second city of the Empire (Kinns, 2010; Clarke & Kinns, 2011). There was no telegraphic link between Glasgow Observatory and the City until the end of 1863, but it had been demonstrated as early as September 1855 that a time ball could be dropped by telegraph from Edinburgh (see Kinns, 2010). The complete history of Glasgow astronomy has been described by Clarke (2013).

Sir Thomas Brisbane (1773–1860), former Governor of New South Wales and amateur astronomer, had retired to Edinburgh and observed initial operation of the Edinburgh time ball in 1853 (Morrison-Low, 2004). He recommended that similar time balls should be erected in Glasgow and Greenock. A time ball on the Sailors' Home on the Broomielaw on the River Clyde near the centre of Glasgow was erected in 1857 but had been discontinued by 1864. It was noted by Airy in his list of time balls (Airy, 1861). The design of the time ball machinery had much in common with the Edinburgh apparatus, using electrical connections to a mean time clock in the Home. This clock required adjustment by hand each day to compensate for its losing rate. The accuracy of the transit telescope at nearby commercial premises was also criticised. Such manual intervention and lack of independent verification of accuracy undermined the authority of the signal. A growing problem was that the time ball could not be seen by large ships west of the city centre when general visibility was being reduced by industrial smoke pollution and construction of intervening buildings. Low water depth also forced large ships to be berthed further out. Nevertheless, the ball remained in place for decades after the service had been discontinued (see Kinns, 2010). Figure 7.4 shows the time ball in 1876.

An unsolicited Edinburgh initiative in September 1863 was instigated by Nathaniel J. Holmes of the Universal Private Telegraph Company. He located time guns at three locations in the City, with another at Greenock, and fired them by telegraph from Edinburgh. They caused offence, as well as some damage to property, in Glasgow and the trials were terminated in February 1864. Holmes had also fired guns in the northeast of England from Edinburgh in 1863. Professor Grant (1814–1892), Director of Glasgow Observatory, argued successfully that a system of slave clocks controlled from Glasgow Observatory would be far superior to either a time ball or time guns which only provided a signal once per day. He won the debate in March 1864 and his use of slave clocks was well-regarded internationally (Ellery, 1868). It appears that controlled clocks were not seen by the Admiralty to offer the authority of a time ball which had a single purpose, although they had been included in a list of signals prepared for the German Government (Airy, 1878).

Fig. 7.4 Glasgow time ball in 1876. (Courtesy Graham Lappin collection)

4.2 *Northeast Coast*

4.2.1 Inverness

A time ball at Inverness was noted in a footnote to the 1880 Admiralty list (see Chap. 10) as being dropped at 10 am, except on Sundays and general holidays. It was described by Airy (1878) as being "a small time ball in a private shop" operated using "current in the post office wires" from Greenwich.

The time ball service was in existence by June 1873 (Time signals for Inverness, 1873), when it was announced that the jewellery businesses of P.G. Wilson and Ferguson Brothers would receive the 10 am signal from Greenwich and that it would be used in both cases to operate a time ball and to strike a bell. Ferguson brothers positioned the time ball above their shop door and dropped it at both 10 am and 1 pm. It appears to have become the dominant signal by August 1873 (Greenwich Time, 1873):

> We notice that Messrs Ferguson Brothers, jewellers, have made a useful and somewhat novel addition to the time ball erected over their business premises in Union Street. The ball is now connected with a large bell, and as it falls punctually at ten and one o'clock each day, there is a simultaneous stroke, loud enough to be heard at a considerable distance around....

The company was renamed in 1881 and continued to provide the time ball service (Ferguson and Macbean, 1881). It moved to larger premises later that year. The service continued until 1907, discontinued following the deaths of the partners, and was then re-established on 15 April 1909 (Greenwich Time, 1909):

> The enterprise of Messrs Fraser, Ferguson, and Macbean, jewellers, Union Street, in having introduced a time gong, worked directly from Greenwich, will be greatly appreciated by the public. It is over two years since there was a similar connection. At ten a.m. and one p.m. the large ball above the door of the premises will fall and a gong rings. Yesterday was the first day of its introduction.

The time ball continued to operate until at least 1913.

5 Wales

The time gun at Swansea was listed between 1898 and 1911 as a principal signal and is described in Chap. 10. A time ball at Cardiff was included in a footnote to the 1880 Admiralty list as maintained by private enterprise, but not in lists from 1898 onwards. A time ball at Newport was not mentioned at all. Both were on the south coast of Wales in the Bristol Channel.

5.1 Bristol Channel

5.1.1 Cardiff

The Cardiff time balls were a civic undertaking. A time ball had been erected on the Town Hall in 1862 by the Electric Telegraph Company (Linnard, 2018). The apparatus had to be repaired frequently and was never seen as a reliable signal. It was replaced in 1887 by another time ball but that also acquired a reputation for unreliability and was finally withdrawn from service in 1893. It had been a chastening experience for Cardiff. Figure 7.5 shows the Cardiff time ball at an uncertain date, probably after 1887.

5.1.2 Newport

A time ball at Newport appears to have been the personal initiative of Robert Whitehall, a local entrepreneur who traded as a jeweller and clockmaker (Linnard, 2018). He erected the ball on his premises and it is believed to have been in operation from 1876 to 1883, but its visibility to ships is uncertain. Whitehall ceased trading in 1884 and there is no further record of a Newport time ball.

Unusually, the balls at Cardiff and Newport were both dropped at 10 am GMT.

Fig. 7.5 Cardiff time ball, uncertain date. (Courtesy William Linnard)

6 The Channel Islands

The history of time signals in the Channel Islands is presently unclear and requires further research. They are known to have existed on the islands of Jersey and Guernsey.

6.1 Jersey

It is believed that Newell and Company paid for the cost of installing a time ball on Fort Regent overlooking St Helier in 1858. The company had laid the first successful telegraph from England to the Channel Islands but this wire was fraught with technical problems and had to be repaired repeatedly. Other time balls are believed to have been installed by T. Hollinshead in Brook Street in 1874 and on the roof of Genge and Co's shop in Halkett Place in 1875.

6.2 Guernsey

In 1873 morning and evening guns were fired from Castle Cornet in Saint Peter Port but these were not connected to an accurate clock. They were still firing in 1893. By 1871, J. Le Lacheur was receiving Greenwich time at his shop in the High Street.

7 Failed Attempts to Introduce Time Signals

The failed attempts by Astronomers Royal George Airy and William Christie to establish time signals at the Lizard and Start Point after 1855 are discussed briefly in Chap. 10. Other signals throughout the British Isles that are known to have been under consideration but were never implemented are considered in this chapter (Fuller, 2023).

7.1 England

7.1.1 Bath

Following criticisms of the accuracy of the Abbey Clock the Rector of Bath, the Reverend Charles Kimble, proposed placing a time ball on the Abbey tower in 1862. He was supported by a former mayor, Dr. Randle Wilbraham Falconer, who was author of two books promoting the benefits of the mineral waters of the town. During a council meeting on 4 February it was announced that a local watchmaker Alexander Mowatt had recommended constructing a globe 9 feet in diameter which would cost £200 to make and £20 per year to drop. The Reverend Kimble offered to pay the £200 if the council would pay the annual charge, but this proposal was lost by a single vote because the council had other competing priorities and it was suggested that a more accurate clock would better meet the town's needs. Accurate Greenwich time was already being provided by several local watchmakers.

7.1.2 Burnley

Alderman William Miller Coultate proposed that the town introduce a time gun at a council meeting on 9 November 1863. Coultate was a doctor who received his medical training at Dublin. He was also an early promoter of the local mechanics' institution and had been surgeon to the 5th Royal Lancashire Regiment in 1853. He suggested that the time gun be situated on Healey Heights about a mile to the south of the town centre. The gun would be triggered by the current from Bidston and this would cost about £80 a year.

His proposal met with a mixed response during the council meeting despite the support of Alderman Thomas Turner Wilkinson who was a mathematical tutor at Burnley Grammar School and a Fellow of the Royal Astronomical Society. Wilkinson tried to obtain the support of Joseph Baxendell, the Astronomer to the Corporation of Manchester, but Baxendell was surprisingly negative about time guns, stating that they were "much over-rated by its advocates". Wilkinson also read a supportive letter from John Hartnup at the Bidston Observatory, but Wilkinson concluded that Burnley should base their decision on the Edinburgh time gun, but this had only been in use for 2 years so perhaps it was better to wait and see if it was a success. The vote for a time gun was lost by 15 votes to only 2.

7.1.3 Bristol

It is surprising that a port of Bristol's importance never had a public time signal despite its prominent role in the trans-Atlantic passenger service and extensive trading around the globe. In 1862 a local opticians, Hill and Price, were receiving the 10 o'clock current from Greenwich, at their shop by the quay. In 1873 the Reverend Richard Glover and The Western Daily Press called for a time gun to be established as the town's clocks reported both local time and Greenwich time. Bristol's location meant that local time was 10 minutes behind Greenwich time but this caused endless confusion as exemplified by the clock on the Exchange which had 2 minute hands, one set to Greenwich time and one set to local time. As a substitute for accurate time mariners depended on three local chronometer makers who were receiving the Greenwich current so no public signal was ever adopted.

7.1.4 Barrow

In 1876 Barrow had a major ship building industry and the largest steel works in the world but the proposal by the timber merchant David Robert Charles in 1880 to establish a time gun ultimately failed because of cost and hesitation by the town council. Councillor Charles had business interests in Liverpool so would have been aware of the advantages of having a time signal for Barrow but he ran into financial problems and moved away from the town in 1881.

7.1.5 Wrexham

In 1878 problems with the accuracy of the town hall clock and confusion over whether to use Greenwich time or local time led to the suggestion by the General Post Office that they supply accurate time to the town hall and a time gun on the barracks via their 10 o'clock current. James Edgar, the local postmaster, had also tried but failed to get the corporation's clock to report accurate Greenwich time but it was not until 1881 that a local timber merchant Edward Meredith Jones added to the call for a time gun when he reported on his investigations into the Kendal time gun. Both towns were of a similar size and the current for the gun would have cost just £43 5s per year. Despite the perceived advantages to the town very few people contributed to the fund but a local watchmaker, D.D. Pierce, was paid by the town council to supply Greenwich Time to the church and the town hall via the Greenwich current.

7.1.6 Shirehampton

Shirehampton was another example of a time signal in the making which had all the key ingredients but which never happened. The main impetus came from the grand amateur Edward Joseph Lowe who was one of the founding members of what would become the Royal Meteorological Society and who was also a member of the Geological, Linnean and Royal Astronomical Societies. Lowe spent most of his life in Nottingham where he founded a dual-purpose observatory collecting detailed meteorological readings for decades.

In 1882 Lowe retired to Shirehampton near Chepstow in Monmouthshire but he found it difficult to adjust. The location was only 6 km from the sea, had fabulous views of the Bristol Channel and Lowe thought that with his social connections and scientific background he would easily be able to attract funding for a new observatory. He thought that towns like Bristol, Swansea and Cardiff would be eager to contribute towards a system which provided storm warnings as well as a reliable time signal for the mariners. His observatory would have been an impressive edifice to Victorian science and would have been comparable to Bidston. Lowe even offered to pay the estimated capital costs of £ 8000. Despite enthusiastic support from leading scientific personalities, he was never able to attract the additional funding which would have kept the observatory running. His Bristol Channel Observatory was never built.

7.1.7 Exeter

Exeter, the scene of two titanic "uniformity of time" debates in 1851 and 1852, never had a time signal, despite the arrival of the telegraph in 1852 and the presence of scientific men who understood the importance of adopting standard time. The low number of vessels using the Exeter ship canal was probably another reason why a time signal was never established or even proposed. The canal was too small and as ships became larger the canal was never modernised so captains found other ports to import and export Devon's goods.

7.1.8 Gravesend

Gravesend considered establishing a time ball in 1860. The associated correspondence between Stephen Leach of the Thames Conservancy and George Airy, who suggested the alternative of a semaphore signal, was reproduced in Chap. 3. It was not pursued. Gravesend wanted to celebrate Queen Victoria's Golden Jubilee with a time ball in 1887 but the proximity of the Greenwich time ball and cost were deciding factors. Ultimately, the town built an impressive clock tower minus a time signal.

7.1.9 Ipswich

At Ipswich, the main promoter Charles Foote Gower died before his proposal to establish a time ball on the town hall in 1863 was achieved and the annual cost was expected to be as high as £150 a year.

It is possible that smaller towns like Ipswich and Gravesend simply could not afford to introduce and maintain time signals and both were too close to Greenwich to compete.

7.2 *Scotland*

7.2.1 Aberdeen

At Aberdeen, Nathaniel Holmes of the Universal Private Telegraph Company and then David Gill, later Her Majesty's Astronomer at the Cape, proposed the establishment of a time gun to the Harbour Board in 1864. Holmes claimed he was already in consultation with Aberdeen during his lecture to the *Newcastle Literary and Philosophical Society* on December 11, 1863. There was also a plan to mount a time ball on top of the New Exchange buildings but these proposals came to nothing.

Aberdeen could have obtained accurate time from Edinburgh or from Dun Echt, the nearby country home of the Earl of Crawford which housed a fine observatory. Gill introduced a time gun there in 1872. There were further calls to introduce public time signals at Aberdeen in the 1870s, 1880s and 1890s but these were thwarted by the perceived costs and alternative sources of accurate time.

7.3 Wales

7.3.1 Milford Haven

Milford Haven is a strange case where all the catalysts for a time signal were present—a deep-water harbour, the presence of the Royal Navy and its Channel Fleet, a nearly-complete astronomical observatory with the telescope already constructed and the package service to Ireland. John Lort Stokes, an admiralty surveyor who had sailed with Darwin on the Beagle, called for a time signal at Milford in the light of the Royal Charter storm of 1859 but this never happened. It may be because Milford was considered too remote to attract considerable commercial investment and the railway and telegraph did not arrive until 1863 (Fuller, 2023).

7.4 Ireland

7.4.1 Belfast

Belfast is a good example of a port which should have had a permanent public time signal but never did. The port benefited from Harland and Wolff's innovative and technologically advanced ship building from 1861. In March 1865 Robert Symington of the Universal Private Telegraph Company fired a time gun at the harbour office in a failed attempt to persuade the authorities to install a time signal. He had been invited to do so by representatives of the chamber of commerce who had attended the British Association's meeting at Newcastle in 1863 and it is a mystery why Belfast did not introduce a public time signal at this date or in the 1890s when further attempts were made to persuade public bodies who should have been eager to fund a ball or a gun. It is likely that mariners used time signals in jewellers' shops to rate their chronometers.

8 Benefits and Costs

Benefits and costs can be measured in several different ways but not all time signals can be considered successes. British and Irish time signals did not include the numerous variations of time lights and discs which were introduced abroad.

8.1 Saving Lives

How many lives and ships were saved by British and Irish time signals? It is almost impossible to say, but in the 1850s and 1860s large ships with hundreds of people aboard would regularly disappear in the Atlantic in the winter storms and no one would know where they had foundered. In the early decades of British and Irish time signals marine compasses and admiralty charts were not always reliable and many ship's captains were known to make rough guesses of their positions rather than take proper astronomical readings. In 1856 one of the Royal Mail Steam Packet Company's captains simply pointed his paddle steamer in the direction of the West Indies and counted the number of revolutions of the ship's wheels, with the inevitable result that he eventually lost his ship and its valuable cargo.

In Liverpool in the 1850s, underwriters often reduced the rate of insurance for ships with two or three good chronometers. Courts of enquiry into losses of ships frequently found that failure to question chronometer readings was a critical error by a ship's master, leading to loss of his certificate. That alone helped to raise standards. Accurate rating of chronometers may have saved many thousands of lives world-wide during the decades that time balls and time guns were the principal means of calibration. It is a difficult statistical challenge to estimate the number of lives and ships that were indeed saved by providing accurate visual time signals. This would benefit from further research.

8.2 Costs of Purchase and Maintenance

Time balls varied considerably in cost, ranging from the expensive apparatus supplied by Maudslay, Sons & Field, described in Chaps. 4 and 5, to much cheaper arrangements that tended to be less reliable. They all needed trained operators and careful maintenance to ensure accuracy. Time guns needed checking and replacing on a regular basis. They could also cause damage to property and injury to operators if they were not managed carefully. The cost of either system was prohibitive for smaller towns and ports, such as Durham (1868), Taunton, Somerset (1872), Arbroath (1891) and Llanelly (1899). Local jewellers and watchmakers often had time signals in their shops, using telegraph signals from Greenwich, which offered an alternative source of accurate time. Many of those are included in Table 7.1.

8.3 Deaths and Injuries

Despite the benefits, time signal histories often ignore the deaths and injuries caused by signalling systems. Some of these are listed in Table 7.3 for the British Isles.

Table 7.3 Deaths and injuries due to time signals

Name	Location	Date	Signal	Outcome
John Knox	Barrack Hill North Shields	18/06/1871	Gun	Badly burnt on neck and face.
Edward Huddy	Edinburgh Castle	25/09/1871	Gun	Struck in face by gun tube, forefinger of right hand severely damaged. Later sectioned due to criminal behaviour. Eventually committed to Broadmoor Asylum.
Francis Baildon Oerton	Edinburgh Castle	20/07/1872	Gun	Death attributed to standing too close to the time gun at the moment of firing a few weeks previously.
Alfred Scott	Platform Saluting Battery Portsmouth	15/06/1874	Gun	Accidentally killed. The gun was not a classic time signal but was still used to indicate rough time.
Ephriam Maskell	Cattewater Plymouth	26/06/1880	Gun	Accidentally killed by a privately-owned time gun.
George Copus	North Pier Swansea	01/11/1887	Gun	Badly burnt on face. Eyesight permanently damaged.
John Hartnup jnr	Bidston Observatory	21/04/1892	Gun	Accidentally killed. John had an attack of vertigo whilst attending to the observatory's roof top meteorological instruments. He fell off the roof and was killed instantly.
Michael Cronin	South Jetty Cork	06/04/1892	Gun	Accidentally killed. Fell off ladder whilst attending to the time gun. Death probably caused by a stroke.
Sydney Osborne + his sister	In a tram on a Lochee Road, Dundee	08/01/1896	Gun	Showered with glass from broken window caused by the retort of the time gun.
Thomas Lewis	Royal Observatory Greenwich	??/08/1913	Ball	Accidentally broke his arm whilst attending to the time ball.
Unnamed	On a tram on Lochee Road, Dundee	06/02/1913	Gun	Passengers on tram alarmed by the shattering of glass caused by the retort of the time gun.
2 unnamed ladies	On a tram on Lochee Road, Dundee	08/11/1913	Gun	Passengers greatly startled by the smashing of two panes of glass on the lower deck when the gun was fired. Possibly the same event as 06/02/1913.

9 Popular Culture

The introduction of time signals often resulted in public fascination and they were causes of celebration as if having a time signal put a town on the map of forward thinking towns. In some places, there were public *déjeuners* to which a select few were invited, and usually the inauguration of a new signal was attended by large crowds. At Liverpool, the press was taken on a tour of the new City Observatory and shown how the time ball was operated whilst at Brighton, a time ball perfume was marketed. Charles Dickens lived near to and knew about the time ball at the Royal Greenwich Observatory. He wrote about it and its chronometer rating function in "Household Words" in 1850. He probably based Wemmick's time gun (on his "castle") in his fictional story *Great Expectations* (1861) on the time gun at Edinburgh castle. Amazingly, there was even a coursing dog called Time Gun which was owned by G.A. Thompson and which raced in the 1860s.

Not everyone was a supporter of Britain's time signals. At Liverpool, cotton merchant John Taylor launched a vituperative campaign against the Waterloo Dock time ball, saying how it could "... only serve as a signal to the very few ships that ever lie in the river; nor can anyone venture, in earnest, to trust it for the regulation of a chronometer, although it may serve to give the hour for dinner" (Schmidt & Dearden, 2019). His unkind comments could not have been more misplaced. At Glasgow and Carlisle, court cases were fought over the perceived nuisance of noisy time guns, with differing outcomes. At Dundee, the time gun was inconveniently located close to the local infirmary and in 1876 the doctors complained bitterly about the effect of the gun on their nervous patients. The medical students poked fun at the gun during their Hogmanay celebrations to welcome in 1897 whilst at Glasgow and Hartlepool the time guns were "spiked" by irate residents. At North Tyneside, persons unknown cut the telegraph wire to the time gun and the council offered a reward of 2 guineas to anyone who could identify the perpetrators if there was a successful conviction.

Some of the time gun operators were interesting characters deserving of an epitaph. Dundee's Sergeant-Major Edward Ponton was described as "two and a half yards and a bit of integrity, honesty, temperance, and love of Queen and country". Less successful was Edward Huddy who was injured firing the Edinburgh Castle time gun in 1871. Because of his accident, he was disabled and disfigured. He became unfit for regular service and was discharged from the army before descending into episodes of serious mental illness. He was sectioned following his firing of a gun outside the Mansion House in London, then later sent to Broadmoor Asylum for attempting to murder a police officer and for accusing various members of the Royal Family of stealing his property.

Sceptics of time signals were usually people who had nothing to do with their home port. At Southampton, the dentist Councillor Payne thought that the time ball was "a scandalous waste of money" whilst the herring expert J.W. de Caux dismissed the Norwich time ball as "a useless toy".

10 Concluding Remarks

Many British and Irish time signals have a rich economic and social history which can be established from surviving records. Some signals were created because of rivalry between ports or because proponents had seen signals operating successfully at Greenwich, Edinburgh or Tyneside. Leading proponents often had to battle ignorance and prejudice within their local shipping communities to gain support for a signal and it is not always clear why some ports had a time signal whilst others never did.

The funding available for signals varied considerably and it was not always the harbour board or the local chamber of commerce which took the lead. Many public agencies contributed towards the cost of establishing a time signal and ordinary people often paid for accurate time. In some towns, local businesses and mills benefited from having a time signal and in others a signal was perceived as putting a town on the map.

Time signals were perceived by some as symbols of colonialism and oppression. Some promoters of time signals, such as the Shand and Tobin families of Liverpool (and later at Cork) gained much of their wealth from their families' involvement in slavery prior to its abolition in Britain in 1807.

In 1899 a French anarchist accidentally blew himself up whilst attempting to bomb the Royal Greenwich Observatory. Then in 1913 the Royal Observatory of Edinburgh was also attacked, this time by suffragettes who symbolically managed to destroy a clock (Rooney, 2017).

Acknowledgements I am glad to acknowledge that quotations from newspapers in this chapter have been accessed using the British Newspaper Archive. I am also grateful to Mark Simmons, Curator (Museum Development) at Hartlepool Borough Council for providing additional information.

References

Airy, G. B. (1861). *List of time balls in actual operation*, May 1861. RGO 6 615, Item 2, Papers of George Airy. Cambridge University Library.

Airy, G. B. (1878). *Exhibition of time signals on or near the British coasts.*, table signed by Airy on 22 January 1878. RGO 6 618, File 4, Papers of George Airy. Cambridge University Library.

Clark, E. (1852, September 8). The electric telegraph company, letter to the Editor. *The Times*.

Clarke, D. (2013). *Reflections on the astronomy of Glasgow: A story of some five hundred years*. Edinburgh University Press.

Clarke, D., & Kinns, R. (2011). Some new insights into the history of the Glasgow time ball and time guns. *Journal of Astronomical History and Heritage, 15*(1), 59–67.

Dewsbury. (1907, January 4). Article published in *The News*, 7.

Dewsbury Town Hall Opening. (1889, September 21). Odds and ends. *The Dewsbury Reporter*, 5.

Eckersley, Y. (2017). Keeping Leigh time, Leigh's two time signals, PastForward. *Wigan Museums and Archives*, *74*, December 2016–March 2017, 30–31.

Electric time ball and clock, West Strand. (1852, September 11). *Illustrated London News*, 205.

Ellery, R. J. L. (1868). On a plan for maintaining true time throughout the city, and on the railways. *Transactions and Proceedings of the Royal Society of Victoria*, 8, Art. VIII, 66–71.

Ferguson and Macbean notice. (1881, January 12). *The Northern Chronicle*, 8.

Fuller, P. (2023). *Unpublished study concerning the origins of time signals in the British Isles.*

Gowland, G. H., & C. (1882, December 30). Advertisement. *Sunderland Daily Echo and Shipping Gazette*, 2.

Greenwich time. (1873, August 15). *The Inverness Advertiser*, 2.

Greenwich time. (1909, April 16). *The Inverness Courier*, 4.

Howse, D. (1997). *Greenwich time and the longitude*. Philip Wilson Publishers Ltd.

Hull's time ball. (2023). https://www.hullccnews.co.uk/hcc_videos/hulls-time-ball-drops-for-the-first-time-in-100-years/
Jessop, L. A. (2017). The time ball, the Guildhall, Kingston upon Hull: History, context and significance. *Historic England research report 16–2017.*
Jubilee Clock-tower, Margate. (1889, May 25). *The Illustrated London News*, 663.
Kinns, R. (2010). Time balls, time guns and Glasgow's quest for a reliable local time service. *Journal of Astronomical History and Heritage, 13*(3), 194–206.
Linnard, W. (2018). Welsh time balls and time guns. *Antiquarian Horology, 39*(1), 98–102.
Margate jubilee clocktower timeball restored. (2014). https://www.kentonline.co.uk/thanet/news/historic-margate-clocktower-timeball-restored-17768/
Morrison-Low, A. (2004). The soldier-astronomer in Scotland: Thomas Makdougall Brisbane's scientific work in the northern hemisphere. *Historical Records of Australian Science, 15*, 151–176.
Obituary. (1903, March 14). Obituary notice for Robert Sutton. *The Maryport Advertiser*, 5.
Regulation of time by the electric telegraph. (1852, June 26). Article in *The Illustrated London News*, 516.
Rooney, D. (2017, January 19). *The politics of precision: Standard time and violent protest, 1884–1914*. Antiquarian Horological Society London Lecture.
Schmidt, R. E., & Dearden, P. (2019). The Liverpool Observatory at Waterloo Dock Part 1: Origins and controversy. *The Antiquarian Astronomer, 13*, 9–10.
Stockton Examiner. (1878, April 6), 4.
Sutton, R. (1896, November 12). Notice in *The Cumberland Pacquet*, 4. [This is the latest date seen for the notice].
The electric time-ball in the Strand. (1852, August 19). Article in *The Times*.
Thomas, S., & Thomas, D. (2015). The long and expensive pursuit of an accurate timekeeper in Blackburn, Lancashire. *Antiquarian Horology, 36*(2), 225–238.
Time Ball. (1862, February 13). Letter to the editor of *The Whitehaven News*, signed using the pseudonym Telegraphic Progress, 4.

Part II
Visual Time Signals for Mariners Around the World

Chapter 8
Introduction to Part II

Roger Kinns

Abstract The limitations of Part II are described in this chapter. The principal aims of subsequent chapters are to explore the broad evolution of visual time signals for mariners and to identify the locations of important signals in different countries which were erected from the 1830s onwards. There is much greater clarity after 1880, when lists of signals were first published by the British Admiralty and other authorities worldwide. The development of time signals in different countries or groups of countries is described in succeeding chapters. Results are presented using maps of signal locations and tables showing changes in the availability and types of signals.

Mauritius, New Zealand and South Africa used some extraordinary time signals that have merited close examination. In other countries, information may have been destroyed during wartime or has yet to be studied by historians. Short chapters are limited by information available, not because time signals were less significant.

1 Aims of Part II

The principal aims of Part II are to explore the broad evolution of visual time signals for mariners and to identify the locations of important signals worldwide. Signals were erected from the 1830s onwards. Some have been described in detail, but original records of many signals have not been found and may have been destroyed. Their dates of introduction and removal are often uncertain.

There is much greater clarity after 1880, when the Admiralty lists of signals were first published. Large numbers of signals were not included in Admiralty lists because they were either inland for domestic use or of reduced significance for marine navigation. In many cases, their accuracy was uncertain. A complete listing of visual time signals for all purposes would be at least as large again. Other lists of

R. Kinns (✉)
The Maudslay Society, Pembroke College, Cambridge, England

R. Kinns (ed.), *The World of Visual Time Signals for Mariners*, Historical & Cultural Astronomy, https://doi.org/10.1007/978-3-031-57334-7_8

time signals have been published, but they are derived from many different sources and include signals that were not visible to ships in port (Hite, 2014; Hülse, 2023). The present study also includes electric light signals.

Wireless signals became widely available after 1911 and were then listed separately from visual signals. A few time balls included in this study were operated using wireless signals.

2 Division into Countries and Geographical Regions

Time signals have been explored in considerable depth in some countries where it has been possible to access national archives and to obtain images of time signals, often taken from postcards. In other countries, information may have been destroyed during wartime or has yet to be studied by historians. Information then has to be restricted to details found in notices to mariners or Admiralty lists. Time and language limitations have made it impossible for the author to access archives in China and Japan, for example, which might be a rich source of design information. Future collaboration with historians overseas will be essential to complete a balanced worldwide picture.

Table 8.1 shows the organisation of chapters in Part II. Many countries have changed boundaries since the era of visual time signals. Names appearing in Table 8.1 represent modern boundaries and descriptions. For example, Ceylon is

Table 8.1 Part II division into countries and regions

Chapter	Continental region	Countries or regions
10	British Isles	England, Scotland, Wales, Ireland
11	Western Europe	France, Spain, Portugal, Gibraltar
12	Northern Europe	Germany, Poland
13	Northern Europe	Netherlands, Belgium, Denmark, Norway, Sweden
14	Eastern Europe, Asia	Russia, Finland, Latvia, Ukraine
15	Mediterranean	Italy, *Sicily*, Croatia, Albania, Greece, Malta, Syria and Egypt
16	Africa	South Africa
17	Africa, Asia	West and East Africa, and Yemen
18	Africa	Mauritius
19	Indian sub-continent	India, Pakistan, Sri Lanka
20	Southeast Asia	Myanmar, Indonesia, Vietnam, Singapore, Philippines
21	Asia	China, Japan, Korea
22	Australasia	Australia, *Tasmania*
23	Australasia	New Zealand
24	North America	Canada, *Newfoundland*, Bermuda
25	North America, Pacific	United States of America, *Hawaii*
26	Central and South America, West Indies	Mexico, Panama, Guyana, Suriname, Argentina, Brazil, Uruguay, Cuba, WI Islands
27	Atlantic Islands	St Helena, Ascension, Cape Verde Islands, Azores

now Sri Lanka, Burma is now Myanmar, Poland now includes Danzig, while Italy and Croatia include areas that were once part of Austria. In other cases, countries that were listed separately in the era of visual time signals are now federated. These are shown in italics after the larger country that now includes them. These include Tasmania, Newfoundland and Hawaii.

Chapter lengths vary considerably. Although Mauritius is a small island, it used some extraordinary time signals that have merited close examination. The history of time signals in New Zealand has also been explored in depth. Short chapters are limited by information available, not because time signals were less significant.

3 Tables Showing Signal Locations

In order to show where and when signals of different types were recognised by the Admiralty between 1880 and 1947, lists are presented in tabular form using colour coding. These are for selected years: 1880, 1898, 1904, 1911, 1915, 1919, 1922, 1928, 1934, 1939, and 1947. Rows indicate the location of the signal in the stated geographical region. Some signals were in use before or after they were listed by the Admiralty. In the early years, their inclusion or deletion would have to await publication of a new list, usually prepared using information available up to the end of the previous year. Other reasons for exclusion are likely to have varied from doubts about signal reliability to accidental omission. In a few cases, signals were still listed after they are known to have been withdrawn.

Various country boundaries, particularly in Europe, changed after World War I and World War II. For example, Trieste and other locations in the Adriatic were listed first as being in Austria and later in Italy or Croatia. Beirut was in Syria and is now in an independent Lebanon, while Helsingfors (now Helsinki), was in Russia and is now in an independent Finland. Changes like these are noted against the signal location.

Table 8.2 shows the colour coding used in the tables of signals. Time balls are highlighted in the tables using a blue background, with text showing the colour of the ball, if stated in Admiralty lists. Time guns and other flashes from explosions are highlighted in green. Lights are highlighted in yellow, with coloured text to indicate signals that were not white lights. Discs, semaphores, cones and cylinders are highlighted in grey; flags in red; clocks, whistles and sirens in brown.

Table 8.2 Colour coding of signals in tables showing signal evolution

Type of signal	Background colour
Time ball	
Gun or powder explosion	
Disc, semaphore, drum, cone, cylinder	
Flag	
Clock, whistle, siren	
Time light	

The aim has been to give an immediate visual impression of how signals evolved with time in different geographical areas. Each table has a summary at the bottom showing the number of time balls, time guns, time lights and other signals in each of the selected years. The locations in each list are primarily in alphabetical order within each country. Spellings of place names varied and were usually anglicised, but modern place names are indicated in the individual tables. The latitudes and longitudes of signals have not been specified in this simplified study. They can be found in Admiralty lists together with details of the signal and measures taken in case of signal failure. Details of this type have been given in journal papers about time signals in individual countries but have been avoided here to allow inclusion of over 200 signals in tabular form.

Some signals have been subject to detailed investigation by the present and other authors. In a few cases, the signals appear to have either been operational for longer than previously thought, or been included in error. One example is the time ball at East London in South Africa, which was still listed in 1939 when all other time balls in South Africa had been discontinued by the end of 1934. The time ball at Greenwich remained in the list during the 1940s although it was not operational for a long period up to 1951 (see Chap. 4). A few others were short-lived, such as a Lisbon time ball which was introduced in 1914 but had been superseded by a time light in 1916. Time balls at some cities in the United States, such as San Francisco, were modified or moved from one building to another (Bartky, 1983). This is also apparent from details such as signal latitude and longitude stated in Admiralty lists. Unless two balls were operated together, these are shown in one row.

4 Maps Showing Signal Locations

At least one map is included in each chapter. These show the locations of visual signals in Admiralty lists between 1880 and 1947 for the geographical area covered by the chapter. Some locations used multiple signals that changed over time. Not all existed together. The colour coding in Table 8.3 is used to describe the type of signal that was either in use in 1947 or at the latest date when a signal was listed at the specified location. In a few cases, earlier signals have been included using different colours.

Table 8.3 Colour coding of signal locations on maps

Type of signal	Font colour	Pin colour
Dropping object		
Gun or powder explosion		
Dropping object, with gun or powder explosion		
Rotating disc(s)		
Flag or siren		
Flag and gun		
Time light(s), with or without other signals		

The same colour code is used for all types of signal that were made using a dropping object. Most were time balls. The colour coding exception is Greenwich, which is highlighted in red to distinguish it from nearby time balls in England. Some locations, such as Edinburgh, used a combination of a falling object and a time gun or explosion. Signals at many locations changed to time lights in the twentieth century, sometimes as a complement or alternative to a time ball or other signal. Only Shanghai is recorded as reverting to a time ball in 1947, having used a time light as well as a time ball between at least 1911 and 1939. The evolution of signals at a specified location is shown in the tables.

Signals were more concentrated in Europe than other geographical areas. Map scales and text fonts have been adjusted according to the density of signals.

References

Bartky, I. R. (1983). Time dissemination, sky with ocean joined. In S. J. Dick & L. E. Doggett (Eds.), *Proceedings of the Sesquicentennial Symposia of the U.S. Naval Observatory*, December 5 and 8, 1980. US Naval Observatory.

Hite, L. (2014). *How time balls worked*. Featuring the Cincinnati Observatory, Birthplace of American Astronomy. http://www.voamuseum.org/media/1355/how_time_balls_worked

Hülse, K. (2023). *Themen-Seite Zeitsignale: Zeitbälle oder Zeitkanonen*. http://www.leuchtturm-welt.net/HTML/TIMEBPK/TIMEBALL.HTM

Chapter 9
Worldwide Evolution of Visual Time Signals After 1880

Roger Kinns

Abstract This chapter gives an overview of the numbers and types of signal that were described in Admiralty lists between 1880 and 1947. They show when time signals were introduced or changed over a long period. The introduction of wireless time signals eventually made visual signals redundant but it was a gradual change.

The number of visual signals peaked in the period after the First World War and declined markedly after the Second World War. Electric time lights first appeared as principal signals in about 1909 and replaced or complemented many time balls. Time guns were operated widely, many in parallel with time balls. In 1880, the Admiralty recognised 82 visual signals: 53 time balls, 10 time guns and 19 other signals. In 1922, there were 220 listed signals: 129 time balls, 45 time guns, 21 time lights and 25 other signals. The distributions of signal types within and outside Europe are presented.

1 Introduction

The Admiralty lists from 1880 onwards show how time signals were introduced or changed over a long period. The introduction of wireless time signals eventually made visual signals redundant but it was a gradual change, giving a service for mariners that could be used by ships without radio equipment or where wireless signals were unavailable for any reason. Time balls and time guns remaining now are essentially working memorials to a bygone age. Some types of signal, such as discs that rotated about a horizontal axis to simulate a disappearing ball and time lights that were extinguished as the time signal, no longer exist but played an important role in several countries.

The development of time signals in different countries or groups of countries is described in succeeding chapters that cover the whole world. The aim of this chapter is to give an overview of the numbers and types of signal that were described in

R. Kinns (✉)
The Maudslay Society, Pembroke College, Cambridge, England

R. Kinns (ed.), *The World of Visual Time Signals for Mariners*, Historical & Cultural Astronomy, https://doi.org/10.1007/978-3-031-57334-7_9

Admiralty lists. This has developed from earlier versions (Kinns, 2022, 2023b) that exposed the worldwide dominance of time balls over a long period.

2 Evolution of Listed Time Signals from 1880 to 1947

In order to study the evolution of time signals for mariners, the following rules and constraints have been applied:

1. Signals must appear in the main Admiralty list(s).
2. Signals must be visible or audible to ships in harbour (signals requiring a visit ashore are not included).
3. There is occasional ambiguity between discs and balls, notably at Port Elizabeth in South Africa.
4. Time signals listed for France as "balloons" are included as time balls (see later).
5. Duplicate signals (e.g. time balls at different locations in a port or a combination of signals such as a time gun and a time ball) are included. An array of time lights was sometimes used, but those are counted as one signal only.
6. Alternative signals (e.g. either time discs or time lights on a given day) are both included.
7. Although not visual signals, sirens and whistles in Admiralty lists have been included as "other" signals.

Many time balls or other devices for chronometer calibration had short lives, so the total number that had been listed by the Admiralty at one time or another exceeds the maximum number identified in any 1 year.

Figure 9.1 shows how the numbers of different signals in Admiralty lists changed between 1880 and 1947. The number of visual signals peaked in the period after the First World War and declined markedly after the Second World War. Electric time lights first appeared in about 1909 and replaced or complemented many time balls. Time guns were operated widely, many in parallel with time balls. In 1880, the starting year for the present study, the Admiralty recognised 82 visual signals: 53 time balls, 10 time guns and 19 other signals. Of these, about 55% were outside the British Empire. In 1922, there were 220 listed signals: 129 time balls, 45 time guns, 21 time lights and 25 other signals. The proportion outside the British Empire had risen to over 65%. Many were destroyed or discontinued during World War II. The number in operation during the war years is uncertain although lists continued to be published. A review in 1947 showed that there were then 64 listed visual signals: 26 time balls, 19 time guns, 14 time lights and 5 other signals. The signals listed in 1947 were those believed to be in commission. Some of these, including the time ball at Greenwich, were not in use in 1947 but could still be operated.

Time lights first appeared as principal signals in the 1911 list and reached a maximum of 41 in 1934. A light in Hamburg was mentioned in the 1908 list, with a second mentioned in 1911, but time balls were then the principal signals in Germany.

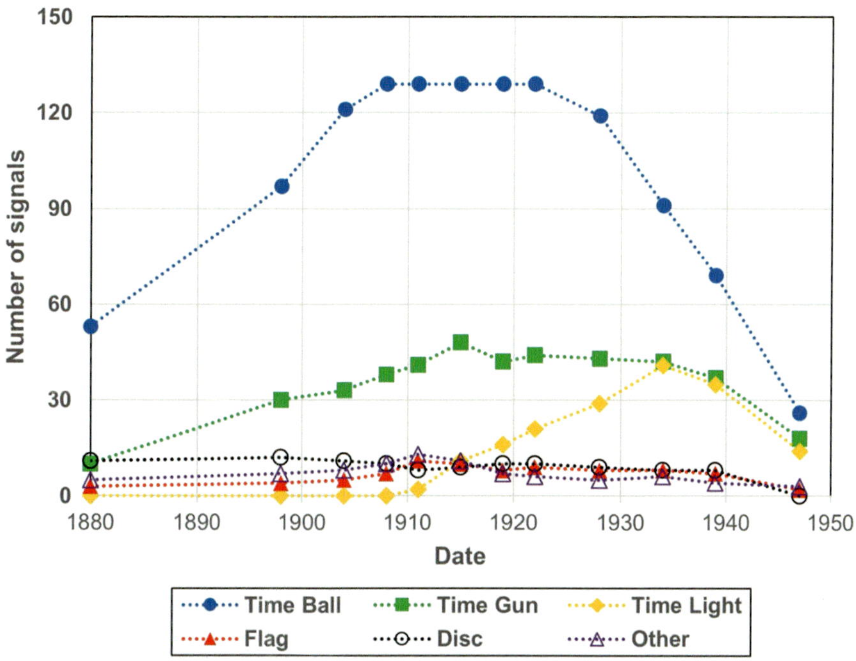

Fig. 9.1 Types of signals worldwide in Admiralty lists

Many were simple white lights switched on for a few minutes and then extinguished as the signal, but arrays of flashing lights were also used. The unique arrangement of coloured time lights in New Zealand, which replaced time balls at Wellington and Auckland, is described in Chap. 23. Time lights worldwide are described in more detail in recent papers (Kinns, 2022; 2023a, b) and in later chapters of this book.

The number of time guns in the Admiralty lists was always smaller than the number of time balls and reached a maximum of 48 in 1915. They were usually regarded as secondary signals, for the reasons noted in Chap. 6.

3 Evolution in Europe and the Rest of the World

Signals evolved differently within and outside Europe. Signals in Europe were more concentrated than elsewhere. Figure 9.2 shows that the total number of signals in the two regions were similar. Figures 9.3 and 9.4 show although time balls were dominant signals in both regions, the numbers of other signals differed significantly. Notably the number of flag signals outside Europe was significant, but there was only one listed flag in Europe and that only for a short period in the 1930s.

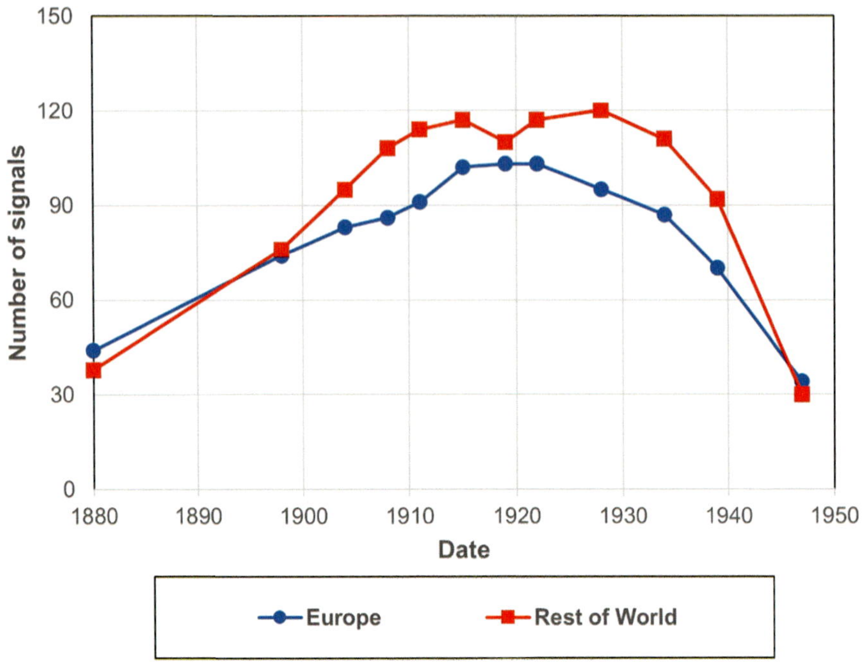

Fig. 9.2 Total numbers of listed signals in Europe and the Rest of the World

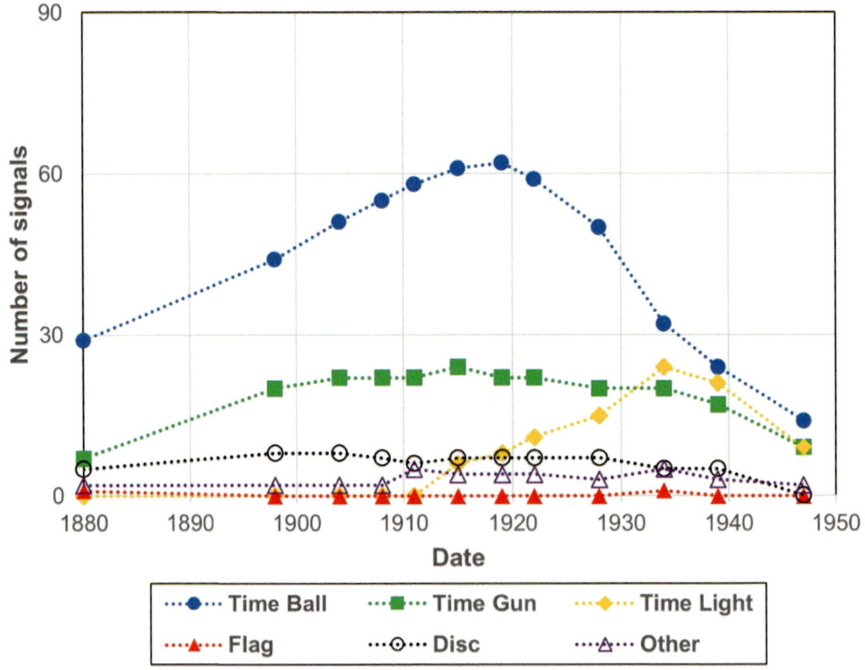

Fig. 9.3 Types of signals in Europe in Admiralty lists

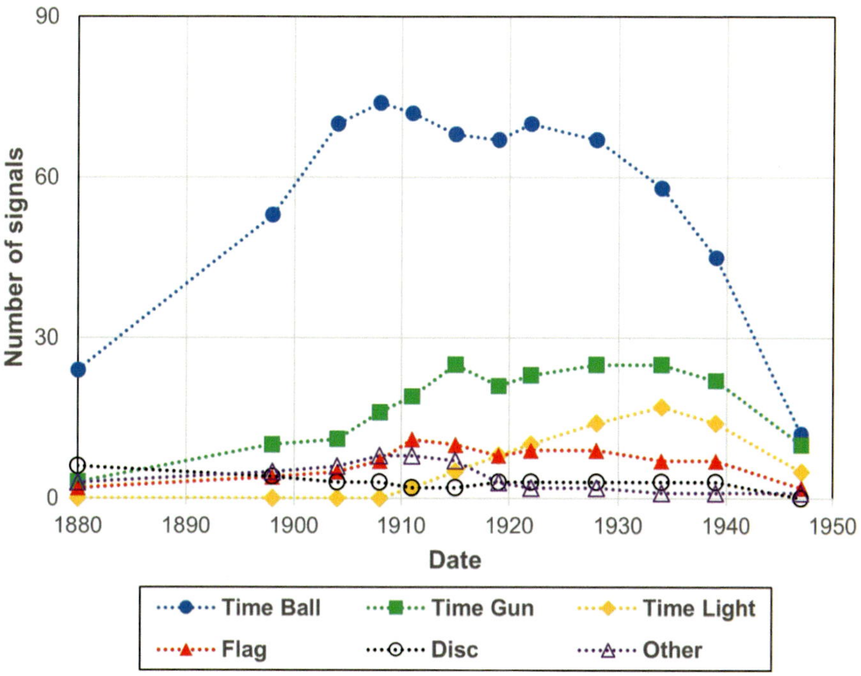

Fig. 9.4 Types of signals outside Europe in Admiralty lists

References

Kinns, R. (2022). Visual time signals for mariners between their introduction and 1947: A new perspective. *Journal of Astronomical History and Heritage, 25*(4), 601–713.

Kinns, R. (2023a). The emergence of time light signals for rating of marine chronometers. *International Journal for the History of Engineering & Technology, 93*(1), 16–46.

Kinns, R. (2023b). Evolutions in the history of visual time signals for mariners. In *Essays on astronomical history and heritage: A tribute to Wayne Orchiston on his 80th birthday*. Springer.

Chapter 10
British Isles

Paul Fuller and Roger Kinns

Abstract The emphasis in this chapter is on signals in England, Wales, Scotland and Ireland that were listed by the Admiralty from 1880 onwards and on signals that preceded them at listed locations.

Time balls at Greenwich, Edinburgh, Deal and Portsmouth were introduced before 1860. A collapsible cone signal at Devonport had been introduced in June 1861. It was replaced by a time ball in 1886. The number had been increased to nine in 1898 and reached a maximum of twelve in 1919. These were located at Chatham, Deal, Devonport, Falmouth, Greenwich, Immingham Dock, Portland, Portsmouth, Sheerness and Southampton in England; Edinburgh and Rosyth in Scotland; Dublin in Ireland.

Precision time guns at Edinburgh and Liverpool were followed by several others around the British Isles, so that nine were listed in 1898. Most had been withdrawn before or during the First World War, but the Liverpool gun was only withdrawn in 1969 and the Edinburgh gun remains as a tourist attraction.

1 Introduction

The emphasis in this chapter is on signals that were listed by the Admiralty from 1880 onwards and on signals that preceded them at listed locations. These signals were in England, Wales, Scotland and Ireland. Figure 10.1 shows the locations of signals that were included in Admiralty lists between 1880 and 1947.

The symbols used for time signals have been described in Chap. 8. An exception has been made for Greenwich, which is shown in red to distinguish it from signals at Sheerness and Deal on a small-scale map.

P. Fuller
Independent Researcher, Southampton, UK

R. Kinns (✉)
The Maudslay Society, Pembroke College, Cambridge, England

R. Kinns (ed.), *The World of Visual Time Signals for Mariners*, Historical & Cultural Astronomy, https://doi.org/10.1007/978-3-031-57334-7_10

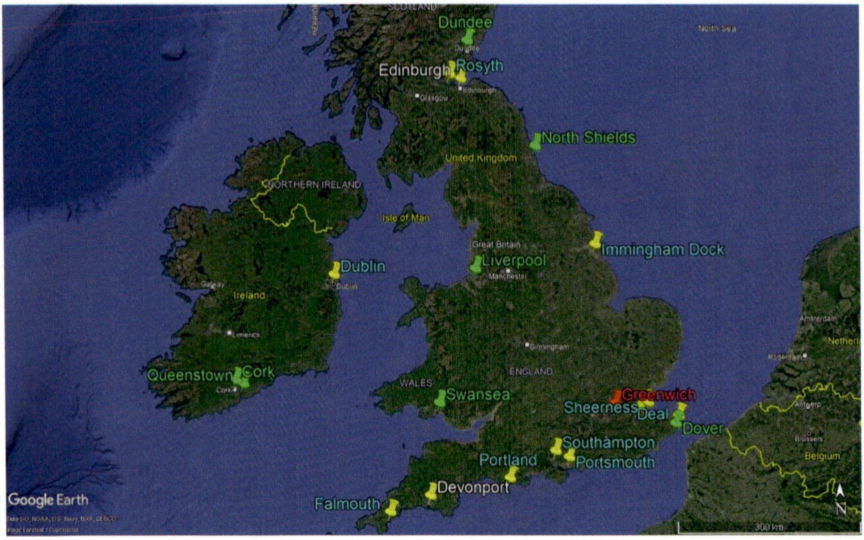

Fig. 10.1 Signals in the British Isles (Google Maps)

Table 10.1 shows the signals listed in specific years. The locations within each country are in alphabetical order. Admiralty lists did not always mention time ball colours. The colour is indicated where stated in lists.

1.1 Signals Mentioned in Footnotes and by the Board of Trade

The 1880 Admiralty list included footnotes concerning various secondary signals for mariners, without giving their detailed locations or procedures to be followed in case of an incorrect signal. They are transcribed in Table 10.2. Footnotes of this type were not included in lists from 1898 onwards. Some, but not all, of these signals were included in lists for 1898 onwards with procedural details.

A supplementary list was also published by the *Board of Trade* (Time signals, 1881):

> The Board of Trade are given to understand that a Time Signal from the Greenwich Observatory is at present regularly transmitted by the General Post-office to the undermentioned bodies and persons, who afford facilities to Shipmasters for ascertaining Greenwich time. The Board cannot, however, vouch for the accuracy of the information afforded by the bodies and persons in question.

The list of ports (with the number of recipients in brackets if more than one) was: Barrow-in-Furness, Bristol (2), Cardiff, Cork, Dublin, Dumbarton, Dundee, Glasgow (2), Gloucester, Greenock, Guernsey, Hartlepool West, Hull, Inverness (2), Jersey (2), Limerick, London (8), Newcastle-on-Tyne (4), Queenstown, North

Table 10.1 Signals in the British Isles

Country	Location	1880	1898	1904	1911	1915	1919	1922	1928	1934	1939	1947
England	Chatham				TB	TB	TB	TB	TB			
	Deal	TB	TB	TB (black)	TB (black)	TB (black)	TB (black)	TB (black)				
	Devonport	Cone										
			TB	TB (black)	TB (black)	TB (black)	TB (black)	TB (black)	TB (black)			
			Gun	Gun	Gun	Gun	Gun	Gun				
	Dover		Gun									
	Falmouth		TB	TB (black)	TB (black)	TB (black)	TB (black)	TB (black)				
	Greenwich	TB	TB	TB (black)	TB (black)	TB (black)	TB (black)	TB (black)	TB (black)	TB (black)	TB (black)	TB (red)
	Immingham: Dock						TB (black)					
	Liverpool	Gun	Gun	Gun	Gun	Gun	Gun	Gun	Gun	Gun	Gun	Gun
	North Shields		Gun	Gun								
	Portland				TB	TB	TB	TB	TB			
	Portsmouth	TB	TB	TB (black)	TB							
	Sheerness		TB	TB (black)	TB (black)	TB (black)	TB (black)	TB (black)	TB (black)			
	Southampton		TB	TB	TB	TB	TB	TB	TB			
Wales	Swansea		Gun	Gun	Gun							
Scotland	Dundee		Gun	Gun	Gun	Gun						
	Edinburgh: Monument	TB	TB	TB	TB	TB	TB	TB	TB	TB	TB	TB
	Edinburgh: Castle	Gun	Gun	Gun	Gun	Gun	Gun	Gun	Gun	Gun	Gun	Gun
	Rosyth						TB (black)	TB (black)				
Ireland	Cork		Gun	Gun	Gun	Gun	Gun	Gun	Gun	Gun	Gun	
	Dublin		TB	TB	TB	TB	TB	TB	TB	TB	TB	TB
	Queenstown		Gun	Gun	Gun	Gun						
Signal type totals		1880	1898	1904	1911	1915	1919	1922	1928	1934	1939	1947
Time ball, including collapsible		4	9	9	11	10	12	11	8	3	3	3
Gun, including powder flash		2	9	8	7	6	4	4	3	3	3	2
Lights		0	0	0	0	0	0	0	0	0	0	0
Other: disc, drum, flag, etc.		1	0	0	0	0	0	0	0	0	0	0

Shields, Southampton, Stockton-on-Tees, Sunderland, Swansea, Waterford and Whitehaven.

The list concluded with the statement:

> The particulars given in this notice are additional to those contained in the List of Time Signals recently published by the Admiralty, which gives detailed particulars of some most reliable signals in use at certain ports in the United Kingdom.

Table 10.2 Signals mentioned in 1880 footnotes

Locations and management	Type of signal
At Cork, Dover (about gun flash as signal), Dundee, Hartlepool West, North Shields, Queenstown, Swansea [as maintained by public bodies]	Gun fired daily (except Sundays and general holidays) at 1h 0m p.m. Greenwich mean time.
At Dublin [as maintained by public body.] At Sunderland [as maintained by private enterprise]	Ball dropped (except Sundays and general holidays) at 1h 0m p.m. Greenwich mean time.
At Cardiff, Hull, Inverness, Jersey, Southampton, Stockton-on-Tees, Whitehaven, Yarmouth (Great), as maintained by private enterprise	Ball dropped (except Sundays and general holidays) at 10h 0m a.m. Greenwich mean time.
At Dumbarton, Glasgow, Leith and Waterford	Clocks are maintained by public bodies, regulated by electric current received from London 10h 0m a.m.

The types of signals provided by recipients of the Greenwich signal were not stated, in contrast to Table 10.2.

1.2 Signals Described in Chap. 7

An overview of listed and unlisted signals for mariners, as well as inland signals for general public use, has been included in Chap. 7. A time ball and time gun trials in Glasgow that had been discontinued in 1864 have been described, as well as the controlled clocks along the Clyde that replaced them. Chapter 7 also includes discussion of time signals that were unsuccessful or proposed but never implemented.

2 Types of Signals in Admiralty Lists for the British Isles

The Admiralty lists for the British Isles included only one exception to time balls and time guns in principal entries. It was a collapsible cone.

2.1 Time Balls

Time balls at Greenwich, Edinburgh, Deal and Portsmouth were in the main list for 1880. The number had been increased to nine in 1898 and reached a maximum of 12 in 1919. These were located at Chatham, Deal, Devonport, Falmouth, Greenwich, Immingham Dock, Portland, Portsmouth, Sheerness and Southampton in England; Edinburgh and Rosyth in Scotland; Dublin in Ireland.

2.2 *Time Guns*

Precision time guns at Edinburgh and Liverpool were followed by several others around the British Isles, so that nine were listed in 1898, as shown in Table 10.1. These were at: Devonport, Dover, Swansea, Liverpool and North Shields in England; Dundee and Edinburgh in Scotland; Swansea in Wales; Cork and Queenstown in Ireland. Most had been withdrawn before or during the First World War, but the time gun at Liverpool continued until 1969 and that on Edinburgh Castle still continues as a tourist attraction.

2.3 *Collapsible Cone*

A collapsible cone signal at Devonport had been introduced in June 1861 (see Chap. 3). It appeared in the 1880 list but was replaced by a time ball in 1886. It is the only list entry in Table 10.1 that was not a time ball or gun. Time lights were never listed for the British Isles.

3 Reports by the Astronomers Royal

Every year from 1836 to 1964, the Astronomer Royal prepared an annual report to the *Board of Visitors* on the occasion of the Board's June visit to Greenwich (Dolan, 2023). Those up to 1880 were prepared by George Airy, those between 1881 and 1911 by William Christie, those between 1912 and 1933 by Frank Dyson and those from 1934 to 1940 by Harold Spencer Jones. They usually included entries about time signals and the management of chronometers at Greenwich. Only those time signals for which Greenwich had direct responsibility were described. These were the time ball at Greenwich, the time ball at Deal which was operated by direct telegraph from Greenwich, and the regulated clocks at "Admiralty Signal Stations".

3.1 *Admiralty Signal Stations*

The Admiralty signal stations were at Devonport (1885–1936), Portsmouth (1892–1936), Portland (1907–1933) and Rosyth (1920–1927). Time balls at these signal stations used Dent clocks fitted with auxiliary pendulums that were corrected by daily telegraph from Greenwich at 10 am and used to operate time balls. One of these clocks is on display at Portsmouth (Kinns et al., 2021). Its operation will be described in the following section concerning Devonport (from Christie, 1885).

4 Signals in England

4.1 Greenwich

The design of the original time ball apparatus at Greenwich and its evolution as a precise signal over almost two centuries to the present day, have been described in Chap. 4. The Greenwich ball is known to have been painted red in 1924 and 1925/6. There is some doubt that the black colour indicated for years between 1922 and 1939 in Table 10.1 is actually correct. Changes from black to red and then again to black in 1928 and 1934 seem unlikely. It was listed as red by the Admiralty from 1940 onwards.

4.2 Liverpool

Funding for an observatory at Liverpool was finally agreed in 1842 after many years of agitation for an observatory that could provide a service for rating chronometers (Fuller, 2022). The first operational time ball in the British Isles after Greenwich was established at Waterloo Dock in Liverpool in 1845. It is shown in Fig. 10.2 and was one of three Liverpool time balls that were operational in 1866.

Fig. 10.2 The first Liverpool time ball at the Observatory (Mersey Gateway)

John Chapman Hartnup (1806–1885) was appointed as Director of Liverpool Observatory in 1843 and the observatory was fully operational in 1846. He had been elected a Fellow of the Royal Astronomical Society in 1845. His work was highly regarded internationally, with many contributions to *Monthly Notices of the Royal Astronomical Society*. Hartnup developed a multi-purpose observatory which combined its time service with the collection of important astronomical and meteorological data. He offered exceptional services for in-house chronometer calibration as well as accurate external time signals. His work in establishing the effect of temperature on chronometer rates has been described in Chap. 3.

A second Liverpool time ball was erected in 1853 on the Electric Telegraph Company's building in Castle Street, primarily for domestic use. It was painted red with a white band, like the same company's time ball in the Strand, London, and is shown in Fig. 10.3. It was operated by the Electric Telegraph Company (ETC), later named the Electric and International Telegraph Company (E&ITC), using signals from Liverpool Observatory (Hartnup, 1853). It was listed by Airy (1861).

Hartnup had foreseen the need for another time ball, owing to reduced visibility of the Observatory time ball from ships as the harbour developed. A time ball with a diameter of 1.6 m was installed on top of the new Victoria Tower in 1860 and was controlled by electrical signals from Liverpool Observatory. It is shown in Fig. 10.4 and was also listed by Airy (1861). The Waterloo Dock time ball was omitted from Airy's list, but it was still operating many years later in parallel with the ball on the Victoria Tower (Time signal balls, 1866).

Fig. 10.3 Liverpool time ball on the Electric Telegraph Company office building (Mersey Gateway)

Fig. 10.4 Liverpool time ball on the Victoria Tower (Mersey Gateway)

In 1864 the Mersey Docks and Harbour Board debated whether to move their time signal as the area needed redevelopment to cater for larger steamships. The decision to move the observatory to Bidston on The Wirral followed long public debate. Hartnup suggested that a time gun could be controlled from the new observatory by telegraph and in 1867 a time gun based on the Edinburgh system was established at Morpeth Dock on the Mersey estuary. By that time, the Waterloo Dock and Castle Street time balls were deemed to be "practically useless" to mariners (The new Mersey Docks time gun, 1867). All three time balls at Liverpool were discontinued, so never appeared in Admiralty lists from 1880 onwards. This time gun was to become world famous for its extraordinary accuracy. Excluding wartime years, the time gun fired every day from 21 September 1867 until 18 July 1969.

4.3 Deal

The original intent had been to install a time ball on the South Foreland Lighthouse, but it was decided that it should instead be installed on the Semaphore Tower at the Naval Yard in Deal (Dolan, 2023). The mechanical design was almost identical to that used at Edinburgh in most respects (see Chap. 5 for a comparison of the arrangements). A major innovation for Deal was its control using an electric telegraph connection from Greenwich to London Bridge Station and from there to Deal using the South Eastern Railway telegraph. The total distance was over 100 km. The arrangement also provided a return signal to confirm that the ball had dropped.

The following extract concerning Deal is from the Astronomer Royal's report for 1854 (Airy, 1854).

> I alluded in my last Report, to the erection of a Time-Signal Ball at Deal, to be dropped every day by a galvanic current from the Royal Observatory. The Ball has now been erected by Messrs, Maudslays and Field, and is an admirable specimen of the workmanship of those celebrated engineers. The galvanic connexion with the Royal Observatory (through the telegraph wires of the South-Eastern Railway) is perfect. The automatic changes of wire-communications are so arranged that, when the Ball at Deal has dropped to its lowest point, it sends a signal to Greenwich to acquaint me, not with the time of the beginning of the fall (which cannot be in error), but the fact that it has already fallen. The Ball has several times been dropped experimentally with perfect success; and some small official and subsidiary arrangements alone are wanting for bringing it into constant use....

Airy commented on the enthusistic support he had received from Charles Walker and Latimer Clark, principal Telegraph Officers of the South-Eastern Railway Company and the Electric Telegraph Company. The Deal time ball started official operation on 1 January 1855. Airy's next report illustrated some of the practical difficulties encountered in achieving reliable operation in all conditions (Airy, 1855). His report was presented on 2 June 1855, so only described issues encountered up to that date.

> The Time-Signal Ball at Deal was brought into regular use at the beginning of the present year. In a short time, however, its action was interrupted, partly by derangement of the apparatus, and partly by the severity of the weather, which froze the sulphuric acid to the state of jelly. I sent an Assistant and workman to put it in order, and since that time it has generally acted very well. Since March 2 there have been three failures; one of these arose from the Ball hanging on the clips, which were nor properly oiled, and one from the turning off of the current on the Railway line; the cause of the third has not been traced out with certainty....

Each report thereafter included information about the number of telegraph and other failures that had occurred. Occasionally, high winds prevented dropping of the ball. The mast had to be repaired in 1871 (Airy, 1871). A clock was available to drop the ball should there be no signal from Greenwich (Airy, 1872).

Later reports included statistics concerning receipt of the return signal at Greenwich and weakness in the drop signal which required the ball to be dropped by hand. A required change to the return signal arrangement was heralded in 1883 and implemented in 1884 (Christie, 1884):

> As regards the Deal time-ball, after various delays the arrangement, referred to in the last report, for sending a current to Deal and receiving a return-signal through the Chronopher of the Post Office telegraphs, was brought into operation on February 29, and has worked well since....

A reduction in the number of signal transmission failures was reported. The detailed changes have been described by Beresford and Combridge (1990).

There was a major overhaul of the electrical arrangements in 1913 (Dyson, 1913):

> The electrical apparatus used for dropping the Time Ball at Deal has been completely overhauled. The wiring has been renewed, the old instruments in use since 1853 replaced, and the arragements in general brought more into line with modern methods.

The Deal Time Ball became obsolete with the advent of reliable radio time signals and the time ball service ended on 25 February 1927 (Dyson, 1927). The original apparatus is on display in the Time Ball Tower, now operated as a museum, and a modern arrangement is used to operate a replica ball. Figure 10.5 shows the tower in 2004.

4.4 *Portsmouth*

George Airy had tried in 1855 to persuade the Admiralty to erect time balls at Portsmouth and other dockyards that would be similar to the arrangement at Deal, with operation by telegraph from Greenwich. He had sought an estimate for supply by Maudslay, Sons and Field for the mechanical arrangement and consulted Portsmouth dockyard about its possible location, without any reference to the 1829 installation (Kinns et al., 2021). In the event, it was decided to use an obsolescent approach, where the Observatory signalled time to the main ball operator using a

Fig. 10.5 Deal time ball (https://commons.wikimedia.org/wiki/File:Deal_Timeball_Tower.jpg#/media/File:Deal_Timeball_Tower.jpg)

miniature pilot time ball. The time ball location was on a semaphore tower that had been built in 1833. It was introduced in 1856, over 26 years after the first time ball trial there. Airy noted an operational Portsmouth ball in his 1861 time ball list (Airy, 1861).

In his 1859 report, Airy stated that "I have called the attention of the Government to the advantage of daily time-signals at Portsmouth and Devonport...." In 1860 he included the comment:

> No further step has been taken by the Admiralty for the exhibition of Daily Time-Signals at Portsmouth and Devonport, or of Hourly Time-Signals at the Start Point. All these schemes are, in my opinion, important; and I propose, when opportunity presents itself, again to urge them.

It appears that a manually operated time ball was in operation at Portsmouth after 1856 until the late 1870s, but it used local observatory facilities of doubtful accuracy. These were much criticised by Airy (Kinns et al., 2021).

Airy reported in 1874 that:

> A proposal has been made to me to drop a ball at Portsmouth by direct current from the Royal Observatory, but no further action appears to have been taken in the matter by the Admiralty.

Christie reported that a clock corrected daily from Greenwich with an auxiliary seconds pendulum had been established at Portsmouth in November 1892 for dropping the time ball (Christie, 1893). This was similar to the arrangement that had been used at Devonport and gave a return signal after the time ball drop to confirm the clock accuracy. The reported error in return signals from Portsmouth was ±0.17 seconds in 1912 (Dyson, 1912).

The Portsmouth time ball for mariners burned down in 1913 and was deleted from Admiralty lists of visual time signals for mariners (see Table 10.1). Subsequent references to a time ball at Portsmouth must be to a small ball at Portsmouth Observatory that was used as a dockyard reference signal. Its accuracy was included in reports up to 1937.

4.5 Devonport

A time ball dropped using electromagnets at Devonport near Plymouth in Devon had been under consideration as early as 1847 (Dockyard, 1847):

> ... Estimates are being prepared for the building of an Observatory in this yard, to be connected with a time ball on the top of the column – by electro-magnetism; the object of which is to give ... the most correct means of determining the rates of the chronometers and Greenwich mean time....

The early plan to install a time ball appears to have been abandoned. Instead, Devonport installed a collapsible cone in 1861 (Washington, 1861). The same device was included in the 1880 Admiralty list and was noted in 1885 as a possible type of signal for use at other locations (Christie, 1885). This and later signals were on Mount Wise.

A new type of time ball apparatus was introduced at Devonport in 1886, which was simpler and cheaper than earlier designs and became widely used at other locations worldwide (Lewis, 1910). The time ball had been discontinued by 1934. The apparatus is often described as using the Devonport Principle. The black ball with a diameter of 1.1 m was raised using halyards and held in position by triggers that were released electrically. There was no need for a special building and it was dropped from a yard arm at Devonport. An iron frame was used at other locations. The drop height varied considerably between installations. After initial free fall, the descent of the ball could be arrested using an arrangement of pulleys, weights, springs and dampers or a friction brake. There are no surviving examples in commission.

Airy had failed to convince the Admiralty in 1856 that other time balls should replicate the Deal arrangement. He continued to promote the idea for many years afterwards, but never secured the required Admiralty funding. Christie resurrected the idea of hourly signals at the Start Point or the Lizard in 1885, working with Captain (later Rear-Admiral) Wharton, the Hydrographer of the Navy based at Devonport. The suggested arrangement for the signal was to use a controlled clock with an auxiliary pendulum (Christie, 1885).

> In connexion with the establishment of hourly time-signals at the Start or Lizard, which was long advocated by Sir G. B. Airy, I have received from the Committee of Lloyds', in answer to my inquiry, an assurance that that corporation would be willing to undertake the maintenance of hourly time-signals at any of their signal stations, provided the Government would supply the necessary apparatus. After consultation with Capt. Wharton, it has been thought better that, before taking further steps, some preliminary trials should be made of a collapsible cone, as an hourly signal, facilities for doing which exist at Devonport. As regards ball-drop or other time-signal, I would propose that it should be made automatically by a local clock, to be corrected daily by the help of a time-signal from Greenwich at 10 a.m., which should automatically start an auxiliary seconds pendulum, suspended freely just behind the clock pendulum. The attendant would then accelerate or retard his clock pendulum (by electro-magnetic action as in the Greenwich mean solar clock) so as to make it pass through the middle of its vibration at the same time and in the same direction as the auxiliary pendulum, and thus to indicate accurately Greenwich mean time. A return signal to Greenwich sent by the local clock at the next hour (11 a.m.) would show that the observation of coincidence of vibration of two pendulums can be made with great certainty, and Messrs. E. Dent and Co. are now arranging for the mounting of an auxiliary pendulum on one of the Transit of Venus clocks, and for adapting it to give hourly time-signals.

In his report for the following year, Christie confirmed that the Dent clock had been modified, but that "No further action has been taken as regards the establishment of hourly time-signals at the Lizard or the Start, as the arrangements for preliminary trial of a collapsible cone at Devonport are not yet completed." (Christie, 1886). Successful return signals from the clock were reported in the following year (Christie, 1887) but it was described as "the Devonport time-ball clock" in his 1888 report and the idea of hourly signals at the Start or Lizard was no longer mentioned.

It appears that a daily time ball had replaced the collapsible cone at Devonport (the year was noted as 1886, in Admiralty lists of time signals). It was the first to use the "Devonport Principle". Collapsible cones were never used elsewhere in the British Isles. Christie's report for 1889 included the observation "No further action has been taken as regards the proposal to establish hourly time-signals at the Lizard, as with the present pressure on the staff there would be great difficulty in making the necessary arrangements for securing the efficiency of such a time-service." It was not mentioned again in later reports.

The accuracy of return signals from Devonport was included in reports up to 1937. The usual accuracy was better than 0.2 seconds. Return signals were not always received, especially on Sundays. The same problem occurred at the other signal stations. Those that were received showed an accuracy that was usually better than 0.2 seconds. The average accuracy was stated to be ±0.13 seconds in 1912 (Dyson, 1912). The time ball was no longer listed in 1934, so it appears to have been withdrawn before the clock.

A time gun at Devonport on the slope of Mount Wise, near the time ball, was introduced in 1887 and continued operation until 1924.

4.6 Southampton

Southampton's time ball was a late arrival, given the importance of the port. The key promoter was Captain Tom Augustus Bevis, the superintendent of the Royal Mail Steam Packet Company. Bevis joined the Royal Mail Steam Packet Company in the 1840s and by the following decade captained a series of steam ships across the Atlantic to the West Indies. He became superintendent of the company in 1872. Over a 43-year period the company suffered heavy shipping losses which were often the result of poor or faulty navigation (Fuller, 2023).

A time ball at Southampton was included in the footnote to the 1880 Admiralty list (Table 10.2), when it was dropped at 10 am. It was then operated "by private enterprise", but the following history of port development suggests that an authoritative time ball was not in operation until 1888.

By 1884 Southampton's port had grown substantially but it was losing trade compared with rival ports. The P & O company had relocated to London in 1881 and Southampton periodically had to defend its status as a mail packet station against rival ports. The key player in this battle was Joseph Rankin Stebbing who provided accurate Greenwich time from his Nautical Observatory in Canute Road from 1851. Stebbing was the major driving force in the development of the port and he took a leading role to ensure that Southampton retained its mail packet status. He died in 1874 and his business was bought out by John Blount Thomas, a chronometer maker from Cardiff. Thomas was a town councillor and a member of the chamber of commerce, an institution which was founded largely due to Stebbing's foresight to defend the interests of the port. Thomas was elected mayor in 1878 and in 1884 following the International Prime Meridian Conference Captain Bevis was encouraged by the shipping companies to write to the town council to propose a time ball for the port. John Blount Thomas, chairman of the important Commerce of the Port committee, took a leading role in the negotiations with the harbour board. This was the start of nearly 4 years of painful discussions which culminated in the erection of a time ball on South Castle, next to the docks, in 1888 (Fuller, 2023). This was the location specified in the 1898 Admiralty list, more usually known as God's House Tower. It is shown in Fig. 10.6.

In 1904 the harbour board withdrew their financial contribution as they believed that the time ball was inaccurate. Its location had been changed to "Flagstaff on roof of South Western Hotel" in the 1904 Admiralty list, continuing until its withdrawal in 1933. The ball was hoisted 5 minutes before being dropped "by electricity from Greenwich" at 1 pm, at both locations.

4.7 *Sheerness*

A time ball at Sheerness was erected in 1894. The black time ball had a diameter of 3.5 feet (1.05 m) and a drop height of 10 feet (3 m). It was on the "Garrison Fort flagstaff" and was withdrawn in 1933. Like other listed time balls that had been introduced after 1880, it was "hoisted at 0h 55m 00s p.m., and dropped daily at 1 p.m. by telegraph from the Royal Observatory, Greenwich".

Fig. 10.6 Southampton time ball on God's House Tower, South Castle. (© Southampton Cultural Services)

4.8 Falmouth

A series of protracted negotiations took place at Falmouth after earlier calls for a time signal came to nothing in 1878 (Fuller, 2023). In 1895 the second call for a time ball was made by the jeweller and clock-maker Harry Louis Owen who provided accurate Greenwich time free of charge to the port. The harbour commissioners agreed to contribute to the funding of the time ball but most of the impetus for the ball came from the chamber of commerce. Admiral Wharton, Hydrographer of the Navy, took a leading role in negotiations with Falmouth and the Admiralty. The time ball arrangement used the design that he had developed for Devonport and was maintained by the local coastguard. Due to its exposed position, it needed regular maintenance and the harbour board was unhappy with the unexpected and unpredictable expenditure which came their way. In 1897 a local electrician, John James Blight, had to defend the ball's accuracy in the local press, blaming the unreliable current from Greenwich (Fuller, 2023). More complete details of the Falmouth arrangement and its problems have been described by Hawkins (2012). The ball fell into disuse in the 1920s.

Admiralty lists confirm that the Falmouth time ball was introduced in 1897. The black ball was located on the "Flagstaff on Pendennis Castle". It had a diameter of 1.2 m and a drop height of 4.2 m. and was "dropped by electric telegraph from the Royal Observatory at Greenwich". It is shown Fig. 10.7. It was still listed in 1922, but was discontinued before the time balls at Chatham, Devonport, Portland, Sheerness and Southampton.

Fig. 10.7 Falmouth time ball. (© Crown Copyright. Historic England Archive)

4.9 *Chatham*

In 1869 the Chatham Board of Health appealed to the War Office for a time gun for the benefit of the navy, army and civilian population of the town. The main proposer was Thomas Aveling, the mayor of nearby Rochester, who was also the co-inventor of the first self-propelling tractor. The Board's appeal was turned down, as was an attempt to erect a time ball in 1869, but by January 1880 the royal dockyard sported a time gun situated on top of Fort Amherst (Fuller, 2023).

The gun was replaced in 1906 with the time ball that was listed by the Admiralty. It was located on the "Drill Shed of the Royal Naval Barracks" and was dropped twice daily at 10 am and 1 pm throughout its life. Its date of withdrawal is uncertain but it was still listed in 1928.

4.10 *Portland*

The time ball at Portland was introduced in 1907. It was located at the "H.M Dockyard signal station over eastern end of Naval storehouse, south side of harbour" (1928 list). The controlled clock there was reported to the *Board of Visitors* (Christie, 1908). It was similar to that used at Devonport. Return signals from the clock were first received on 29 May 1907. As at Devonport and Portsmouth, the usual accuracy was better than 0.2 seconds, giving an average accuracy of ±0.17 seconds (Dyson, 1912). The Portland time ball was dismantled and return signals from the clock were discontinued on 31 August 1933 (Spencer Jones, 1934).

4.11 *Immingham Dock*

A black time ball with a large diameter of 1.8 m was introduced at Immingham Dock on the River Humber in 1916 (Notices, 1916). It was notified that the ball would be dropped electrically twice daily at 10 am and 1 pm GMT. It was also notified that the clock time would be 1 hour later up to 30 September, owing to the introduction of BST. It was still listed in 1919 but had been discontinued before 1922. A time ball at nearby Kingston upon Hull (noted as Hull in Table 10.2) was unreliable and was not included in Admiralty lists from 1898 onwards. It has been described in Chap. 7.

4.12 *Time Guns as Single Signals in England*

Time guns at Dover and North Shields were not associated with time balls in Admiralty lists, although there had been an early proposal to erect a time ball at Dover.

4.12.1 Dover

The time gun at Dover was included in 1880 footnotes, when it was fired at 1 pm (Table 10.2). It was described in 1898 as "Near the Drop Battery" and fired at noon GMT, except on Sundays or general holidays. It did not appear in Admiralty lists from 1904 onwards.

A Dover gun and time ball were proposed in July 1859 (Signal and alarum guns, 1859)

> Arrangements have been made for announcing Greenwich time at this garrison every day at noon. A ball upon the flag staff at Archcliff Fort will be connected by telegraph with the Royal Observatory, and each day at 12 o'clock it will descend, when a gun will be fired from the saluting battery at the Castle....

There are no known records of a Dover time ball in operation, so it appears that it was never brought into service. The gun at Dover Castle was fired regularly until 1 November 1912. It was initially fired at noon only but a second firing was introduced at 9.30 pm as a signal to troops to return to barracks (When Dover had a noon gun, 1938). The second firing was not mentioned in the 1898 Admiralty list.

A preliminary announcement in May 1919 indicated that the gun would be reintroduced and again fired twice daily: "Dover Castle time gun, which was fired at noon and 9.30 p.m. in peace time, is to resume firing." (Dover's time gun, 1919). The gun resumed service on 23 June 1919, but was abandoned again on 1 August 1923 (When Dover had a noon gun, 1938). The re-established gun was not listed by the Admiralty.

4.12.2 North Shields

Time guns for the Tyne had a history dating back to the 1860s, with firing by telegraph first from Edinburgh and ultimately from Greenwich. The extended history has been described in Chap. 7.

The time gun at North Shields was mentioned in a footnote in the 1880 Admiralty list and appeared in the main lists from 1898 to 1908. There is, however, clear evidence that the time gun service had been withdrawn in 1905 (The time gun, 1905), with subsequent correspondence regretting its demise as a public signal. It was described from 1898 onwards as "On a hill close to the river, and a little east of Albert Edward Dock" and was fired at 1 pm "by electricity through General Post Office". It is shown in Fig. 10.8.

Fig. 10.8 North Shields time gun on Ballast Hill. (Courtesy, Newcastle Libraries)

5 Signals in Scotland

The time ball and time gun at Edinburgh still exist in 2023, although operation of the time ball was discontinued during the Covid-19 crisis and has yet to be reinstated. The time ball at Rosyth and the time gun at Dundee were also listed by the Admiralty but no longer exist. The history of Glasgow time signals has been summarised in Chap. 7.

5.1 Edinburgh

Robert Wauchope had proposed a time ball for Edinburgh as his first recommendation after the success of the Portsmouth time ball trial (Wauchope, 1830), but no action was taken until many years after the Greenwich time ball had been introduced in 1833. There was renewed interest in an Edinburgh time ball from 1841 with increasing support from the City and Admiralty. It gained momentum after Charles Piazzi Smyth (1819–1900) became the second Astronomer Royal for Scotland from 1846 to 1888 (see Kinns, 2011).

5.1.1 The Time Ball

The plan to have a time ball on the Nelson Monument, rather than the nearby observatory, was agreed by the City and Admiralty in 1846. The apparatus was finally ordered from Maudslay, Sons & Field of London in 1852. Smyth commented

(Smyth, 1853b) that the apparatus was procured by the Officers of Woods and Forests who:

> knowing well how admirably the Greenwich time-ball had performed, applied to Mr Field to make a similar one for Edinburgh, with all the additional improvements that had been suggested by the Astronomer Royal, Mr Airy, who took much personal interest in the execution of the machine.

Details of the mechanical design are described in Chap. 5. Smyth emphasised the quality of the arrangement for arresting the descent of the ball using an air cylinder (Smyth, 1853a, b). The apparatus arrived in Leith during September 1853 (The time-ball, 1853a). An extract from the newspaper announcement is shown below.

> We are just informed of the arrival at Leith of this long-expected and useful piece of machinery. It has been constructed by the well-known engineering firm of Maudslay and Field in London, and is an admirable specimen of the substantial as well as accurate character of their workmanship, and its consequential practical efficiency during a long period of time. Doubtless it might have been as well made by some of our own engineers, and it was no slighting of their merits that a London house was employed; but the fact that that house had already made similar time-balls for the Royal Observatory of Greenwich, and other establishments in the south, and still preserved the various moulds and patterns prepared on the first occasion, enabled them to furnish on this, equally good workmanship, at a less cost than any other manufacturer could, entering upon it de novo.... Accordingly, we have now as good, perhaps a better, Time-Ball than that of the Greenwich Observatory....

It was erected under Smyth's leadership and extended trials were undertaken to prove its reliability in all weathers before it became an official signal on 20 March 1854. From the outset, the ball was released by an electric telegraph signal from a clock at the nearby Royal Observatory. Frederick James Ritchie, clockmaker of Edinburgh, made important contributions to the accuracy and reliability of the signal. Smyth kept meticulous records of time signal accuracy that were open to public inspection and can still be seen at the Observatory in Edinburgh.

Sir Thomas Brisbane, who had been Governor of New South Wales and was a noted amateur astronomer (Morrison-Low, 2004), made checks on time ball operation during December 1853, using his own chronometer. He published his findings at a meeting of the *Royal Society of Edinburgh* (The time-ball, 1853b). He was particularly interested in the correctness of the intervals between raising the time ball halfway, raising it to the top of the mast and then the drop. The first two stages required manual operation of a windlass, so timings were not expected to be precise, but Smyth had introduced preparatory electrical signals for the benefit of the operator. Brisbane's opening remarks confirmed the status of the system:

> On my arrival in the city on 5th December, I was much gratified to find that the time-ball which had been memorialised for so often had at length been erected, and though not yet completed in all its parts, is being experimentally tried nearly every day at one p.m.

He then described the process that was followed when there had been a failure to drop at 1 pm and presented his timing observations.

> ... This shows that the beginning of the ascent, if well observed, never deviating more than three seconds from the truth, is quite accurate enough for most purposes, and will give a very close idea of the true time on any occasion, when perhaps an unforeseen accident may prevent the descent from taking place.

5.1.2 The Time Gun

The time gun on Edinburgh Castle was introduced in 1861, with automatic firing by a Castle clock provided by Ritchie that was controlled by electric telegraph from the Observatory. It set new standards for accuracy and became the preferred public signal. The gun clock incorporated a special mechanism that allowed for the fraction of a second that it took for powder to explode after receipt of the electrical spark (Ritchie, 1861; see Kinns, 2011). A 1.3 km long wire suspended between the Monument and the Castle was used from 1861 until December 1873. It had been changed from a light to a heavier wire in 1865 after expression of concerns about safety, but this increased the load on the Monument. The wire was changed again in 1873 to one using support from intervening buildings. The gun's location on Edinburgh Castle high above the City meant that it was audible over a long distance without causing damage to local property. This was in marked contrast to Glasgow and Dundee.

Smyth promoted the time gun as his preferred signal, arranging for the publication of maps that showed the time delay in sound propagation to different locations in Edinburgh and Leith. In a letter to Airy, Smyth (1878) stated:

> With the Time-Ball the first instant of the fall is recommended to observers, but is in reality always about 0.15 sec. too late, on account of the time necessarily taken up in the action of the trigger. With the Time-Gun the fire is 0.05 sec. too soon, owing to the difference of instants at which the escapement of the clock concerned is liberated, and at which the electrically controlled pendulum arrives at the end of the arc at each second.

Although the Edinburgh time signal service under Smyth's leadership was impressive, the Observatory suffered from chronic under-funding for many years, leading to concerns about the state of instruments and buildings (see Kinns, 2011). The observatory was relocated from Calton Hill to Blackford Hill in 1895 and the time signals were afterwards controlled from there using extended electrical connections. The gun has been replaced several times but continues to give the one o'clock signal as a tourist attraction.

The original time ball apparatus was carefully restored in 2009 and is still operable manually (Marshall, 2009). The telegraph link to the Observatory no longer exists. The time ball had weighed about 90 kg when first installed, but its mass increased with repairs to its frame and zinc skin so that the restored ball weighs almost 200 kg (Kinns, 2014). The myth that the original ball weighed 15 cwt. has been discussed in Chap. 5. The original air spring arrangement to arrest the ball's descent is still in use but requires careful maintenance to avoid unacceptable impact loads on Nelson's Monument.

Figure 10.9 shows the restored time ball above Nelson's Monument. Figure 10.10 shows the 105 mm field gun that was in use at Edinburgh Castle in August 2019. Ritchie's gun clock can be seen just above the muzzle of the gun.

Fig. 10.9 Edinburgh time ball after restoration(https://upload.wikimedia.org/wikipedia/commons/a/a5/Edinburgh_Calton_Hill.jpg)

Fig. 10.10 Edinburgh time gun in August 2019. (Photograph by Paul Fuller)

5.2 *Dundee*

The Dundee time gun was mentioned in a footnote to the 1880 Admiralty list and last appeared in the list for 1916. It had a chequered history, in common with other time guns where it was difficult to achieve audibility without causing damage and complaints.

After negotiations with the Harbour Board, a time gun station was built in the Barrack Square. The gun was fired by telegraph for the first time from the Royal Observatory at Edinburgh on 3 September 1872 but there were early criticisms about its indistinct retort and inaccurate firing of the gun. Another problem was the effect of the gun's retort on the patients of the Dundee Infirmary, which was located only 200 m from the gun. In March 1894, these complaints finally caused the town council to move the gun to a new location at Dudhope Barracks but the complaints continued. The council experimented by moving the gun to Law Hill and Balgay Hill but results were still unsatisfactory and the gun returned to Dudhope Barracks. It ceased firing on 24 February 1916 following more complaints from the Royal Infirmary, which was then full of wounded soldiers. After the end of the First World War there were unsuccessful calls for the gun signal to be reintroduced.

5.3 *Rosyth*

A time ball was introduced at Rosyth Dockyard in 1916. It had a small diameter of 0.9 m, half that of the ball at Immingham Dock in England which was introduced in the same year.

The controlled clock giving a return signal at Rosyth was introduced in October 1920 (Dyson, 1921). The signal accuracy was reported between 1921 and 1926, but not in 1927 or later (Dyson, 1927). As at the other locations, the usual accuracy was better than 0.2 seconds. The time ball at Rosyth is presumed to have ceased operation in 1926. It was listed in 1922 but not in 1928.

6 Signals in Wales

A time ball at Cardiff was noted in an 1880 footnote (Table 10.2) but proved unreliable. The history of time signals in Wales has been described by Linnard (2018) and summarised in Chap. 7. Only the Swansea time gun appeared in Admiralty lists from 1898 onwards.

6.1 Swansea

A time gun was established at Swansea in South Wales in October 1874. It had been mentioned in a footnote to the 1880 Admiralty list as one of several guns maintained by public bodies. It was funded by the local council and was situated on the North Pier at the entrance to the harbour. It was criticised by local chronometer makers for its inaccuracy and was replaced in 1882 (Linnard, 2018). In 1886 the "nervous old gentlemen" who worked at the Guildhall complained about the gun which was disrupting court cases. As a result, the time gun was moved to the East Pier on the other side of the Rive Tawe (Fuller, 2023). In 1898 its location was given as "On Old Eastern Pier", when it was fired "by electricity from the General Post Office". The same description was still included in the 1908 and 1911 lists, although operational problems in 1904 had suggested that it might soon be discontinued (Linnard, 2018). The gun fired for the last time in early 1909, so it should have been removed from the 1911 list.

Another Swansea time gun had existed for a short period in the 1870s (Fuller, 2023). Robert Capper was appointed General Superintendent and Harbour Master of Swansea docks in March 1877 and he introduced his own private time gun at his country residence at Penmaen, 9 miles to the west of Swansea on the Gower peninsular. It was an isolated location and the gun was situated in front of his house. It probably fired for only 2 years.

7 Signals in Ireland

7.1 Dublin

Dublin established a time ball on top of the Port and Docks Board Building in 1865, but its location was some distance from the harbour. It was mentioned in a footnote to the 1880 list as maintained by a public body (see Table 10.2). The time ball was included in the 1898 list as being "dropped by electricity from Dunsink Observatory", with similar entries up to 1916 when the ball was moved to the eastern end of Sir John Rogerson's Quay. It was then "dropped electrically using a clock in the ballast office which is controlled by Dunsink Observatory" but continued to have limited visibility from harbour locations. By the 1920s the ball was being ridiculed in the local press, but it continued to be listed by the Admiralty throughout the 1920s and 1930s. It was still listed by the Admiralty in 1947, when it was dropped electrically using the wireless time signal from Rugby.

A plan for a second time ball to be installed on top of a new hotel on the seafront at Dún Laoghaire in 1866 was abandoned due to financial constraints. A further plan to position a time ball on Tuskar Rock off the south-east coast of County Wexford in 1866 also foundered for financial reasons.

7.2 Cork and Queenstown

The earliest call for a time gun at Cork was made in 1864 but the necessary finance was not then available. In 1871 Airy reported to the *Board of Visitors* at Greenwich that "A proposal has been made to drop a ball … at Queenstown, which can be done without difficulty." The proposal was to replicate the arrangement at Deal (Airy, 1871).

The Cork time gun proposal was resurrected and accepted in July 1874. After some debate about whether to install a ball or a gun the Harbour Board decided to establish time guns for Cork and neighbouring Queenstown (Fuller, 2023). They were advised by Frederick James Ritchie who had played a major role in implementing the Edinburgh time signals. The Queenstown gun fired for the first time in January 1876 (The time-gun, 1876). The Cork gun started slightly later and it was also decided that the Cork gun should not be fired on Sundays (Cork Harbour Board, 1876; Meeting, 1876).

The Cork and Queenstown guns were included in a footnote to the 1880 Admiralty list and appeared in the main lists from 1898 until they were discontinued. Inconsistent operation of the Cork time gun was criticised in Parliament in 1887 (Post Office Ireland, 1887) but it remained in Admiralty lists until 1946. This contradicts the observation that the Cork gun had been discarded after the 1922 Anglo-Irish Treaty in an otherwise informative article (McCarthy, 2016). The time gun at Queenstown was not listed after 1915.

Acknowledgements The authors are glad to acknowledge that quotations from newspapers in this chapter have been accessed using the British Newspaper Archive.

References

Airy, G. B. (1854). *Report to the Board of Visitors*.
Airy, G. B. (1855). *Report to the Board of Visitors*.
Airy, G. B. (1861). *List of Time Balls in Actual Operation*, May 1861. RGO 6 615, Item 2, Papers of George Airy. Cambridge University Library.
Airy, G. B. (1871). *Report to the Board of Visitors*.
Airy, G. B. (1872). *Report to the Board of Visitors*.
Beresford, C. F. C., & Combridge, J. H. (1990). The Deal time ball. *Antiquarian Horology, 19*(1), 33–43. [The paper includes a description by Beresford that was published in *The Military Telegraph Bulletin* in September 1884, with additional explanatory material by Combridge].
Christie, W. H. M. (1884). *Report to the Board of Visitors*.
Christie, W. H. M. (1885). *Report to the Board of Visitors*.
Christie, W. H. M. (1886). *Report to the Board of Visitors*.
Christie, W. H. M. (1887). *Report to the Board of Visitors*.
Cork Harbour Board. (1876). *The Cork Constitution*, 13 January 1876, 2.
Dockyard. (1847). Article in *West of England Conservative*, 12 May 1847.
Dolan, G. (2023). The Royal Observatory. *Greenwich*. http://www.royalobservatorygreenwich.org/articles.php?article=933
Dover's time gun. (1919). *Pall Mall Gazette*, 31 May 1919, 7.
Dyson, F. W. (1912). *Report to the Board of Visitors*.

Dyson, F. W. (1921). *Report to the Board of Visitors*.
Dyson, F. W. (1927). *Report to the Board of Visitors*.
Fuller, P. (2022). *Unpublished review of events leading to the Liverpool time ball in 1845*.
Fuller, P. (2023). *Unpublished study concerning the origins of time signals in the British Isles*.
Hartnup, J. (1853). The time-ball in Castle-Street. *The Liverpool Mercury*, 23 December 1853, 10.
Hawkins, R. S. (2012). The Falmouth Time Ball. *TROZE, The Online Journal of the National Maritime Museum of Cornwall, 3*(2), 1–24.
Kinns, R. (2011). The early history of the Edinburgh time ball and time gun. *International Journal for the History of Engineering & Technology, 81*(2), 264–290.
Kinns, R. (2014). Did the Edinburgh time ball really weigh 15 cwt? *International Journal for the History of Engineering & Technology, 84*(2), 160–174.
Kinns, R., Fuller, P., & Bateman, D. (2021). Exploring the Portsmouth Time Balls. *Journal of Astronomical History and Heritage, 24*(3), 751–769.
Lewis, T. (1910). *Notes for the Astronomer Royal dated 21–25*. April 1910. RGO 7 257, Papers of William Christie. Cambridge University Library.
Linnard, W. (2018). Welsh time balls and time guns. *Antiquarian Horology, 39*(1), 98–102.
Marshall, A. (2009). *Nelson Monument, Calton Hill, Edinburgh.*, article by Gray Marshall and Associates, Edinburgh [responsible for the restoration].
McCarthy, K. (2016). Kieran's our city, our town. *Cork Independent,* 2 June 2016.
Meeting of Cork Town Council. (1876, February 17). *The Cork Constitution*, 3.
Morrison-Low, A. (2004). The soldier-astronomer in Scotland: Thomas Makdougall Brisbane's scientific work in the Northern Hemisphere. *Historical Records of Australian Science, 15*, 151–176.
Notices. (1916). *East Coast, Humber River – Time signal established*. Notices Jul–Dec, No. 2431, 778.
Post Office (Ireland). (1887). *Entry in Hansard* (Vol. 320). Debate on 8 Sept 1887.
Ritchie, F. J. (1861). Description of the clock-trigger for the time-gun signal in Edinburgh Castle, dated 22 July 1861. In *Select papers of the Society of Arts* (Vol. VI, 152–159). Read before the Society, and illustrated experimentally.
Signal and alarum guns. (1859). *The Dover Express and East Kent Intelligencer*, 2 July 1859, 2.
Smyth, C. P. (1853a). Remarks on the Erection of the Time-Ball of the Royal Observatory, Edinburgh. *Monthly Notices of the Royal Astronomical Society, 14*(1), 23–25.
Smyth, C. P. (1853b). Notice of the Time Ball at the Royal Observatory at Edinburgh. 'Given at the request of the Council on 12 December 1853. A working model and diagrams were exhibited.' Later published (1856) in *Select Papers of the Society of Arts, 4*, 191–196.
Smyth, C. P. (1878). *Letter to G B Airy*, 11 January 1878. RGO 6 618, Item 2, Papers of George Airy. Cambridge University Library.
Spencer Jones, H. (1934). *Report to the Board of Visitors*.
The new Mersey Docks time gun. (1867). *The Liverpool Daily Post*, 5 September 1867, 5.
The time gun. (1905). *The Shields Daily News*, 21 September 1905, 4.
The time-ball at the Royal Observatory. (1853a). Article following arrival of the time ball apparatus at Leith. *The Scotsman*, 1 October 1853, 2.
The time-ball at the Royal Observatory. (1853b). Article describing a paper read by Sir T.M. Brisbane to the Royal Society of Edinburgh. *The Scotsman*, 24 December 1853, 3.
The time-gun at Queenstown. (1876). *The Cork Constitution*, 18 January 1876, 2.
Time signal balls. (1866). *The Liverpool Journal of Commerce*, 28 July 1866, 3.
Time Signals. (1881). *Shipping and Mercantile Gazette*, 16 August 1861, 6.
Washington, J. (1861). *Time Signal for Plymouth Sound, Notice to Mariners [No. 21]; England – South Coast*, dated 18 June 1861, Hydrographic Office, Admiralty, London. [Washington was the Hydrographer of the Navy].
Wauchope, R. (1830). Plan for ascertaining the rates of chronometers by signal. *Edinburgh New Philosophical Journal, 8*, 160–162 and 289–291. [The first part was submitted on 23 November 1829 and the second part, which included the diagram, was submitted on 1 February 1830].
When Dover had a noon gun. (1938). *The Dover Express and East Kent News*, 21 January 1938, 9.

Chapter 11
France, Spain, Portugal and Gibraltar

Roger Kinns

Abstract French time signals were located on the English Channel; Atlantic and Mediterranean coasts, These included a rotating disc at Cherbourg, a flag at Brest and time balls at Lorient, Fouras, Rochefort and Toulon. These are believed to have been introduced in the 1870s. There is known to have been a much earlier but unregulated time ball and gun at Nice. Some of the time balls in France, described as "balloons" in Admiralty lists, were collapsible.

Only two signals were listed for Spain. These were time balls at Cadiz and Vigo. Portuguese time signals were all at locations in Lisbon. They changed from time balls to time lights. A time ball in Gibraltar was erected in 1903. No time ball was listed before that date and earlier illustrations may have been of a ball signal that was used for a different purpose.

1 Introduction

Table 11.1 shows time signals in southern Europe between 1880 and 1947. Their locations are shown in Fig. 11.1.

2 France

French time signals were located on the English Channel, Atlantic and Mediterranean coasts, Signals in the 1880 Admiralty list were a rotating disc at Cherbourg, a flag at Brest and time balls at Lorient, Fouras, Rochefort and Toulon. These are believed to have been introduced in the 1870s. There is known to have been a much earlier but unregulated time ball and gun at Nice. Some of the time balls in France, described as "balloons" in Admiralty lists, were collapsible.

R. Kinns (✉)
The Maudslay Society, Pembroke College, Cambridge, England

R. Kinns (ed.), *The World of Visual Time Signals for Mariners*, Historical & Cultural Astronomy, https://doi.org/10.1007/978-3-031-57334-7_11

Table 11.1 Signals in France, Spain, Portugal and Gibraltar

Country	Location	1880	1898	1904	1911	1915	1919	1922	1928	1934	1939	1947
France	Brest	Flag										
			TB (black)	TB (black)	TB (black)	TB balloon	TB balloon	TB balloon	TB balloon			
	Cherbourg	Disc	Disc	Disc	Disc	Disc	Disc	Disc	Disc			
	Dunkerque			TB	TB	TB	TB	TB				
	Fouras	TB	TB balloon	TB balloon	TB balloon	TB	TB	TB				
	Ile d'Aix								TB (black)	TB	TB	
	Lorient	TB (black)	TB (black)	TB (black)	TB (black)	TB (black)	TB (black)	TB (black)	TB (black)	TB (black)	TB (black)	
	Rochefort	TB	TB balloon	TB balloon	TB balloon	TB	TB	TB				
	Toulon	TB	TB (black)	TB (black)	TB (black)	TB (black)	TB (black)	TB (black)	TB (black)	TB (black)	TB (black)	TB (b&w)
Spain	Cadiz	TB (black)	TB (black)	TB (black)	TB (black)	TB (black)	TB (black)	TB (black)	TB (black)	TB (black)	TB (black)	TB (black)
	Vigo				TB (black)	TB (black)	TB (black)	TB (black)				
Portugal	Lisbon: Observatory	TB										
	Lisbon: Naval School		TB (black)	TB (black)	TB (black)	TB (black)						
	Lisbon: Nat..Ropery						Light	Light	Light	Light	Light	
	Lisbon: wharf						Light	Light	Light	Light	Light	
Gibraltar				TB	TB	TB	TB	TB	TB	TB		
Signal type totals		1880	1898	1904	1911	1915	1919	1922	1928	1934	1939	1947
Time ball, including collapsible		6	7	9	10	10	9	9	6	5	4	2
Gun, including powder flash		0	0	0	0	0	0	0	0	0	0	0
Lights		0	0	0	0	0	2	2	2	2	2	0
Other: disc, drum, flag, etc.		2	1	1	1	1	1	1	1	0	0	0

2.1 *Early Signals at Nice*

Although not official signals, a time ball and a time gun were operated at Nice in the 1860s by Thomas Coventry, an English amateur astronomer. They were described in a letter by him to the *Horological Journal* with an attachment which suggested that the signals had been in use since 1861 (Coventry, 1865). He had "established a time-ball on the roof of his hotel, where he has a small observatory" and was supported by "a retired artillery-man" who fired a gun at a ruined castle when the ball was seen to drop at noon. The extract below indicates that the signals had been opposed initially by French officials and that Coventry had to pay for the signals from his own resources throughout their lives (Coventry, 1865). It also suggests that time guns were used in many seaside towns as approximate signals for domestic use.

> … The advantage of this standard of time to a population that has hitherto been regulated by rough sundials is immense, and is at length, after a period of four years, duly appreciated.

Fig. 11.1 Signals in France, Spain, Portugal and Gibraltar. (Google Maps)

> Much opposition was at first experienced at the introduction of mean time; and the discharge of a gun for non-military purposes was considered a great infringement of conservative principles. It was only on appeal to high authorities in Paris that this boon was conceded, and then only on condition that every expense connected with it should be defrayed by the owner of the time-ball, which has been cheerfully complied with. This is certainly the first introduction of anything of the kind in France, beyond a morning and evening gun in seaport towns, but it is only wonderful how so great a public benefit can be so generally neglected.

Coventry died in 1869 (Obituary, 1869). The Nice time service ceased when he no longer funded its operation, possibly some years before he died. The cannon at the "Fort de Mont Alban", high above the city, is believed to have operated from 1861 to 1866. Another cannon was operated by municipal authorities after 1875 (Découverte des horloges, 2022). There was no mention of the Nice time ball in that reference: those noted all appeared in Admiralty lists.

2.2 *French Time Signals in Admiralty Lists*

France used a wide range of time signal types, some of which were not used elsewhere. A conventional time ball at Brest was replaced by a balloon, a term used to describe a collapsible ball which could be expanded in preparation for the signal. Earlier time signals at Fouras and Rochefort, listed as balloons by the Admiralty, may have been operated as conventional time balls after expansion. A disc which could be rotated about a horizontal axis was used at Cherbourg.

Before 1911, times were defined by Observatories in France, usually Paris. The times changed to GMT in March 1911, specified in France as Paris time retarded by 9 minutes 21 seconds (Howse, 1997, 151). Most signals in France were repeated 2 minutes after the first signal at 10 am throughout their lives. Time guns and time lights were not listed for chronometer rating in France between 1880 and 1947.

2.3 *Brest*

In 1880, a flag was used for the time signal at Brest. It was at the "Observatory of the Nautical Schools" and was hauled down at noon, Brest mean time. This had been replaced by a black ball with a diameter of 1.8 m by 1898, dropped at 10 am and then again at 10.02 am, Paris mean time. It was still listed as being at the Observatory. The same details were listed in 1911. The signal changed again to a black "balloon" in that year, with a 1.8 m diameter, apparently at the same location as the earlier time ball. It was expanded 5 minutes before collapse as the signal at 10 am GMT (Standard time at Brest), then expanded again and collapsed at 10.02 am GMT. The unusual arrangement is shown in Fig. 11.2. It appears to have consisted of a set of interconnected rings, rather like a Chinese lantern. The likely procedure

Fig. 11.2 Brest time balloon. (Courtesy Klaus Hülse collection)

was that a halyard would expand the framework. Release of the halyard would collapse the ball.

2.4 Cherbourg

Admiralty list entries for Cherbourg specified a disc signal with a diameter of 0.9 m from 1880 to 1928 at the "Marine Observatory, Vigie des Onglet, Quay Napoleon". It was removed from the 1929 list. Figure 11.3 shows how it was located. The mast to the right of the steps has an inverted 'U' support for the time disc, whose horizontal axis is supported at the tips of the 'U'. The disc was rotated to the vertical position 5 minutes before the signal. Its return to the horizontal signified the exact time. When not in use, the disc was set at 45°. In 1880, the signal was at noon only, Paris mean time. In 1898 and later lists, the signal was at 10 am, repeated at 10.02 am.

2.5 Dunkerque

The Dunkerque time ball first appeared in the 1904 list and had been removed by 1927. It was on Lengenaer Tower and had a diameter of 0.9 m and drop height of 3 m. The signal was at 10 am, repeated at 10.02 am, Paris mean time, changed to

Fig. 11.3 Cherbourg time disc. (Courtesy Klaus Hülse collection)

GMT in 1911. The arrangement in Fig. 11.4 suggests that the "ball" may have been a double cone.

2.6 Fouras, Rochefort and Île d'Aix

The time balls at Fouras (Basque Roads) and on the St. Louis Tower at Rochefort were included in the 1880 list. They were dropped then at the same time, 2 minutes 50.4 seconds behind GMT noon (close to local noon at both places) repeated 2 minutes later. Both time balls had been listed as "balloons" in the 1898 list, dropped at 10 am and 10.02 am Paris mean time. The times later changed to GMT. The associated descriptions are confusing, because they refer to balls with drop heights of 13 m and 11 m. They were again listed as balls from 1915 onwards with the same drop heights. This suggests that collapsible and rigid balls were interchangeable at Fouras and Rochefort and that dropping of the ball was the signal. At Brest, the signal was ball collapse. The Fouras time ball was replaced by a time ball at Île d'Aix in 1924, which was still listed in 1939. This was dropped at 10 am and 10.02 am throughout its life. The nearby Rochefort time ball was not listed after 1927.

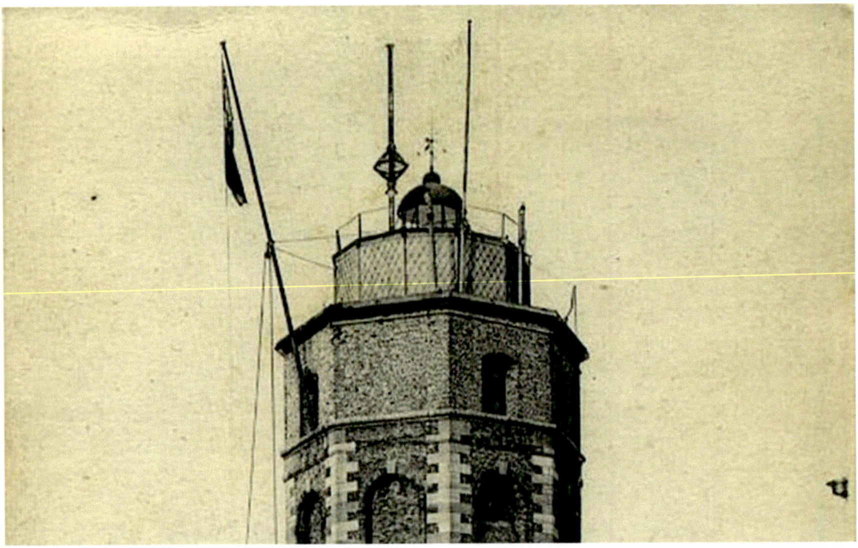

Fig. 11.4 Dunkerque time ball. (Courtesy Klaus Hülse collection)

2.7 Lorient

A time ball at Lorient (L'Orient) was listed from 1880 until 1939. Initially, it was dropped at noon, L'Orient mean time. Up to at least 1908 it was located on a "Mast at the harbour tower" and had a diameter of 0.9 m. It was then 16.5 m above ground with a drop height of 13.5 m and was dropped at 10 am and 10.02 am Paris mean time. From 1911 onwards, its location was given as "peak of harbour tower mast", 43 m above ground with a drop height of 6 m. Its elevation had been increased substantially, but its drop height was much smaller than before. The ball was dropped at 10 am and 10.02 am GMT. A replica working time ball is now in place on the tower at Lorient (Prévost, 2020).

2.8 Toulon

The time ball at the Mediterranean port of Toulon was located on "Signal Staff at the Naval Observatory". It had a small diameter and drop height of 0.6 m and 1.8 m and was dropped initially at 11 am and 11.02 am Toulon mean time. This had been changed to 10 am and 10.02 am Paris mean time in the 1898 list, with a further change to 10 am and 10 am GMT in 1911. It was still extant in 1947. The time ball is shown in Fig. 11.5.

3 Spain

Only two signals were listed for Spain between 1880 and 1947. These were time balls at Cadiz and Vigo.

3.1 Cadiz

The time ball at Cadiz had been erected before 1880 and was still in operation at the San Fernando Observatory in 1947. The black ball had a diameter and drop height of 1.5 m and 3 m respectively, like most of the German time balls and the time balls at Edinburgh and Deal. Its supplier has not yet been identified. Figure 11.6 shows the position of the time ball at the Observatory.

Fig. 11.5 Toulon time ball. (Courtesy Klaus Hülse collection)

Fig. 11.6 Cadiz time ball. (Courtesy Klaus Hülse collection)

3.2 Vigo

The black ball at Vigo was at the Meteorological Observatory and had a smaller diameter of 0.9 m but a larger drop height of 4.2 m. It was listed in 1911 but withdrawn before 1927.

4 Portugal

All the listed time signals in Portugal were at Lisbon.

4.1 Lisbon

A time ball on the flagstaff of the Marine Observatory was listed in 1880. It was replaced by a time ball at the Naval School which was erected during the 1880s and listed until 1916. This second ball was painted black and had a diameter and drop height of 1 m and 3.6 m. It was then withdrawn and replaced by time lights at two other locations in Lisbon. These were at the National Ropery and the Custom House Wharf. Each signal was "made from a lantern mounted on piles, and with three faces, 5 feet (1.5 m) high and 6½ feet (2 m) wide; in the middle of each face an horizontal line of light". The signal was made every 2 hours from 8 am to 8 pm GMT and the lamps were respectively 35 m and 30 m above ground level (1919–1939 lists).

5 Gibraltar

5.1 Gibraltar

The 1.8 m diameter time ball at Gibraltar was dropped from a yard arm at the Naval Signal Station high on "Windmill Hill flats", with a large drop height of 18 m. The listed time ball had been erected in 1903 and initially dropped at noon GMT (1904 list). It was dropped every hour during daylight from 1905 until its withdrawal in the 1930s. There were no entries for Gibraltar in the 1880 and 1898 lists. Figure 11.7 shows the arrangement on a postcard that was mailed in December 1903. The much earlier engraving in Fig. 11.8 shows a small ball that predates Admiralty list entries. It may have been used for a different signalling purpose.

Acknowledgements I am grateful to David Moss, then resident in France, for helping me to find references to French time signals.

Fig. 11.7 Gibraltar time ball. (Courtesy Klaus Hülse collection)

Fig. 11.8 Early Gibraltar signal. (Courtesy Klaus Hülse collection)

References

Coventry, T. (1865). Time-ball and time-gun at Nice, France, letter to the editor of *The Horological Journal*, dated 15 February 1865, published on 1 March 1865.

Découverte des horloges. (2022). http://www.patrimoine-horloge.fr/hab-france.html

Howse, D. (1997). *Greenwich time and the longitude*. Philip Wilson Publishers Ltd.

Obituary, Thomas Coventry. (1869). A useful man gone home: Obituary notice for Thomas Coventry in *The Commercial Journal and Family Herald*, 15 May 1869, 2.

Prévost, S. (2020). À Lorient, le time-ball de la tour de la Découverte attend son heure, *Le Télégramme*, 7 February 2020. [In French]. https://www.letelegramme.fr/morbihan/lorient/a--lorient-le-time-ball-de-la-tour-de-la-decouverte-attend-son-heure-07-02-2020-12498113.php

Chapter 12
Germany and Danzig

Roger Kinns

Abstract Time balls in Germany were first established in 1875; following correspondence with Greenwich about the nature of British signals. The first time balls were erected in 1875 at the Baltic ports of Neufahrwasser and Swinemunde, and at Cuxhaven and Bremerhaven on the North Sea Coast. Some used special towers while others were positioned on existing buildings. Additional time balls at other locations gave a total of eleven German time balls in 1915. Many were replaced or complemented by time light signals which first appeared as principal signals after 1911. A single time gun was located first at Kiel and then on a training ship at Mürwick.

Neufahrwasser; later known as Gdansk, was first in Germany, then in Danzig, then Germany again and now in Poland. The 1876 Neufahrwasser time ball was replaced by another at a different location in 1894, replaced in turn by a time light in 1927. Another more powerful light was introduced in 1930 at nearby Gdynia.

1 Introduction

Table 12.1 shows how time signals in Admiralty lists for Germany and Danzig were distributed between 1880 and 1947. Figure 12.1 shows their locations. The Kiel time balls were close to the Observatory. The number of time balls reached a maximum of 11, reducing to a single time ball at Hamburg in 1939. Many were replaced or complemented by time light signals which first appeared as principal signals after 1911. Two at Hamburg had been noted previously as supplementary signals. A single time gun was located first at Kiel and then on a training ship at Mürwick.

R. Kinns (✉)
The Maudslay Society, Pembroke College, Cambridge, England

R. Kinns (ed.), *The World of Visual Time Signals for Mariners*, Historical & Cultural Astronomy, https://doi.org/10.1007/978-3-031-57334-7_12

Table 12.1 Signals in Germany and Danzig

Country	Location	1880	1898	1904	1911	1915	1919	1922	1928	1934	1939	1947
Germany (North Sea)	Bremen		TB (black)	TB (black)	TB (black)	TB (black)	TB (black)	TB (black)	TB (black)			
										Light		
	Bremerhaven	TB (black)	TB (black)	TB (black)	TB (black)	TB (black)	TB (black)	TB (black)	TB (black)			
										Light	Light	
	Cuxhaven	TB (black)	TB (black)	TB (black)	TB (black)	TB (black)	TB (black)	TB (black)	TB (black)			
										Light)		
	Emden					TB	TB	TB	TB			
						Light	Light	Light	Light	Light	Light	
	Hamburg	TB (black)	TB (black)	TB (black)	TB (black)	TB (black)	TB (black)	TB (black)	TB (black)	TB (black)	TB (black)	
	Hamburg: Kaiser Wil'n					Light	Light	Light	Light	Light	Light	
	Hamburg: St Pauli					Light	Light	Light	Light	Light	Light	
										Clock	Clock	Clock
	Hamburg: Clock-house									Clock	Clock	Clock
	Wilhelmshaven	TB (black)	TB (black)	TB (black)	TB (black)	TB (black)	TB (black)	TB (black)	TB (black)	TB (black)		
Germany (Baltic)	Kiel: Observatory	TB (black)										
	Kiel, Fjord		TB (black)	TB (black)	TB (black)	TB (black)	TB (black)					
	Kiel, Wik				TB (black)	TB (black)	TB (black)	TB (black)	TB (black)			
	Kiel		Gun	Gun	Gun							
	Mürwik: Training Ship					Gun	Gun	Gun				
	Stettin					TB (black)	TB (black)	TB (black)	TB (black)			
										Light	Light	
	Swinemunde	TB (black)	TB (black)	TB (black)	TB (black)	TB (black)	TB (black)	TB (black)	TB (black)			
										Light		
Gdansk (was Germany then Danzig, now Poland)	Neufahrwasser	TB (black)	TB (black)	TB (black)	TB (black)	TB (black)	TB (black)	TB (black)				
									Light	Light	Light	
	Gdynia									Light	Light	Light
Signal type totals		1880	1898	1904	1911	1915	1919	1922	1928	1934	1939	1947
Time ball, including collapsible		7	8	8	9	11	11	10	9	2	1	0
Gun, including powder flash		0	1	1	1	1	1	1	0	0	0	0
Lights		0	0	0	0	3	3	3	4	10	7	1
Other: disc, drum, flag, etc.		0	0	0	0	0	0	0	0	2	2	2

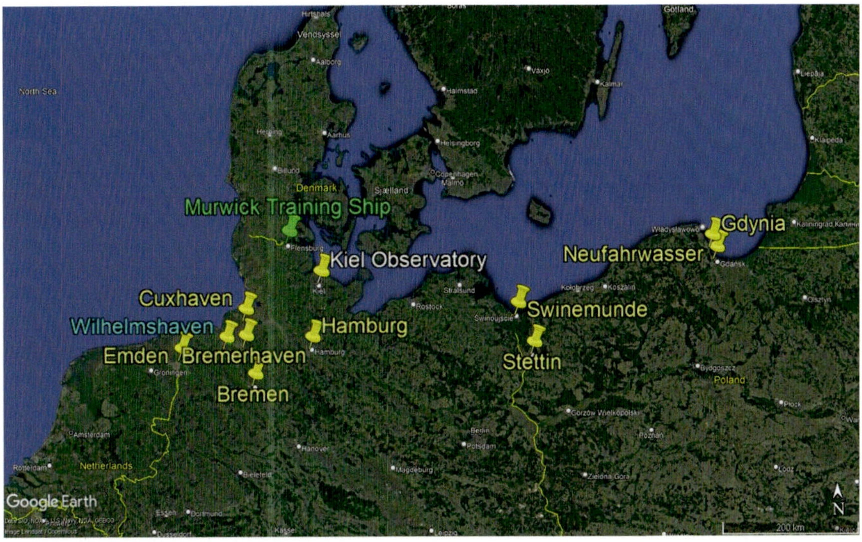

Fig. 12.1 Time signals in Germany and Danzig. (Google Maps)

1.1 *Initial Development*

The development of German time signals followed a request for information about British time balls from the Consul General of the German Empire in London which was forwarded to Astronomer Royal George Airy at Greenwich (Wilkes, 1873).

> I have received a communication from the Foreign Department of Berlin notifying that it is the intention of the Imperial Government to erect Time-balls at certain places hereafter to be decided upon, similar to those in use in this country at Greenwich, Liverpool, and Portsmouth, and I am directed to ask that the Board of Trade will be pleased to furnish me for the use of the Imperial Government with copies of plans at the places before mentioned.

This was followed by exchanges concerning the costs of copying plans, which led to a decision that only the plans for Deal need be sent. Details about the capital cost and running costs of the Deal installation were also provided, apparently to the satisfaction of the German government (Malcolm, 1873):

> With reference to former correspondence on the subject of procuring plans of the different time-balls for the use of the German Government, I am directed by the Board of Trade to inform you that the German Consul General states in reply to a communication addressed to him from this Board that he agrees as to there being no necessity to procure plans of the time-balls in use at Edinburgh and Liverpool those at Deal being in all important points of construction exactly similar.

This correspondence suggests strongly that the Deal arrangement, using operation by electric telegraph from a remote observatory, was a key reference for development of time balls in Germany. Most had a diameter of 1.5 m and a drop height of 3 m, as at Deal.

The following notice from Berlin was published in *Lloyd's List* (Zeltz, 1875)

> Being in receipt of your esteemed letter of the 18th inst. I beg to inform you that the Government of the German Empire intends to establish four time-ball stations on the German coast at Neufahrwasser, Swinemunde, Cuxhaven and Bremerhaven. These stations being in telegraphic communication with the next observatories will give regular and certain time-signals to the passing vessels twice a day, through the fall of a ball visible at a long distance. The first fall of the ball states the middle of the time of the signal station, the second that of Greenwich. There is no doubt the said intended establishments will be of great importance to the shipping, and therefore I do not delay to give you notice of it.

Another notice indicated that time balls had been erected at the four locations by 20 October 1875 (*Shipping and Mercantile Gazette*, c1875). All were operated using electric telegraph signals. Some were positioned on towers that were constructed specifically for the time ball but others used existing buildings.

Count Münster, The German Ambassador, provided comprehensive notes on the German time balls which had been implemented by 1877 (Münster, 1877).

1.2 Time Signals in Admiralty Lists

The 1880 Admiralty list included seven time balls in Germany. These were located on the North Sea at Wilhelmshaven, Bremerhaven, Cuxhaven and Hamburg, and on the Baltic at Kiel, Swinemünde and Neufahrwasser, in order of longitude East. All were stated to have diameters of 5 feet (1.5 m). There was, however, an error in the 1880 Admiralty list, which gave the drop heights as 5 feet (rather than the correct 10 feet) at all locations except Kiel, where the drop height was unstated. Only one drop time was specified for each location in 1880. It was at noon local astronomical time, except at Hamburg where it was at noon GMT.

There were eight time balls in the 1898 list, with the addition of Bremen on the North Sea. The 1898 list gave the drop height as 10 feet at all locations except Kiel, where it was 11 feet (3.3 m) and Neufahrwasse, where it was only 7 feet (2.1 m). The 1898 list showed that the German time balls were then dropped twice, at 11 am and noon GMT (noon and 1 pm Mid-European time), except at Hamburg where the drop was at noon GMT only. The same signals were extant in 1908.

By 1915, time balls had also been introduced at Emden and Stettin, with a second time ball at Kiel, giving a total of 11 German time balls. The introduction of time lights can be traced through lists from 1908 onwards.

Most German time light signals were repeated every 6 hours at 5 am, 11 am, 5 pm and 11 pm GMT. The first signal of this type was at Hamburg, noted as a supplementary signal in 1908 and as a primary signal in Admiralty lists after 1911. The lights were illuminated 5 minutes before extinction as the time signal. Other sequences were used in Danzig. Many of these lights continued in operation until at least 1939 and were often operated in parallel with time balls.

2 North Sea Ports

2.1 Hamburg

Hamburg introduced a time ball service in 1875, as the principal North Sea port in Germany. The Admiralty lists for 1880 until 1939 included a black time ball on the Kaiser Quay. It was located at 53° 32′ 30″ N, 9° 58′ 57″ E and was dropped once at noon. It was over 50 m above the water, as shown in Fig. 12.2. It was complemented by time lights and controlled clocks.

2.1.1 Hamburg Time Lights and Controlled Clocks

Although they were not listed as primary signals until after 1911, time lights and controlled clocks in Hamburg were first mentioned under "Additional Details" in the 1908 Admiralty list. These details are transcribed below:

> A standard clock has been established, in a clock house, situated about 1,150 yards in a S. 67° E. (true) direction from the fixed light on the head of Kronprinz quay, Hamburg. This

Fig. 12.2 The Hamburg time ball. (Author's collection)

> clock beats in unison with the Observatory clock, by electricity, and keeps Mid-European time, or 1 hour fast on Greenwich mean time. It may be relied on to 1 second.
>
> An electric time signal has also been established in the tower of the electric centre of Kaiser Wilhelm harbour, Hamburg, situated about 70 yards in a due west (true) direction from the standard clock house. The signal consists of a horizontal beam of light, which is shown for 5 minutes (four times a day) before the hours of 6 a.m., noon, 6 p.m. and midnight, Mid-European time, and is suddenly extinguished at each of these hours precisely. Position, latitude 53° 31′ 46″ N., longitude 10° 58′ 23″ E.

This appears to be the first time that electric lights had been introduced as an alternative signal anywhere in the world. The 1911 list included the same details with the addition of "A similar signal is made from the tower of St Pauli pier." The electric lights at both locations were included as primary signals between 1915 and 1939. The Hamburg controlled clocks were not included as primary signals until after 1928. Two controlled clocks had been included by 1932. They were listed as at "Tower on St. Pauli landing stage" and "Clock-house on the Reiherdamm at Kuhwärder".

The 1932 list gave specific details about the time light and clock arrangement at the landing stage:

> The signal consists of incandescent lamps arranged in groups of 3 above the tower clock and between the 8 columns of the cupola of the tower in such a manner that the signal is visible from all sides.
>
> The lamps are lit 5 minutes before the time of the signal, and are extinguished at 6h 00m 00s, 12h 00m 00s, 18h 00m 00s and 0h 00m 00s Standard Time.
>
> The pendulum clock, whose face is visible from the tower gateway, is in electrical connection with the Observatory, and shows Middle European Time correct to 1.0 sec.

The same sequence of time signals was used from 1929 at all German locations having time lights.

2.2 Bremen

Figure 12.3 shows the time ball arrangement at Bremen which used an existing building, as at Hamburg, rather than a purpose-built column. It was located on the "Tower of Harbour Office at S.E end of Freehaven (*sic*)" (1908 list) at an elevation of 44 m above ground. It was replaced by time lights in 1929.

2.3 Bremerhaven

The following information for Bremerhaven illustrates the level of detail that was provided in 1877 (Münster, 1877). The latitude and longitude of the time ball were 53° 33′ N, 8° 34′ E from Greenwich. It was located "137 m Easterly from the

Fig. 12.3 Bremen time ball. (Courtesy Klaus Hülse collection)

lighthouse". The Bremerhaven time ball tower was very similar to the design for Cuxhaven. Figure 12.4 shows a fine engraving of the Bremerhaven time ball, from a drawing that was made soon after its introduction.

The ball was dropped twice each day, at noon Bremerhaven Time and noon Greenwich Time. The two drops were separated by 34 minutes, 16.5 seconds. Additional details are transcribed below.

> Ball 1.5 metres diameter, black, at full height 39.28 metres above high water, 36.93 metres above ground. Descends 3 metres. 10^m before, to half height: 3^m before, to whole height. If the fall not correct, then within 3^m a red ball 0.4 metres is raised on one of the wirecords, for 5^m, then a placard. If no descent, the red ball will be raised to half height and remain till the ball is dropped. If things are wrong, the red ball is kept at half height.

The policy for signalling errors was common to most German time balls. The time ball at Bremerhaven was replaced by time lights in 1929.

Fig. 12.4 Bremerhaven time ball. (Wood engraving by Rudolf Stieler, based on drawing by H. Marutsky, c1876)

2.4 Cuxhaven

The development of German time ball arrangements is attributed to Hugo Lentz, Leiter der Cuxhavener Wasserbauinspektion, who received a patent for his design. The first implementation was at Cuxhaven, which was operational in October 1875 (The time ball column, 2022).

Figure 12.5 shows the time ball at Cuxhaven and its time light replacement on the same tower in 1929. The ball was always described in Admiralty lists as black with a diameter of 5 feet, but the photograph suggests that the ball was not spherical and used open framework construction, at least for part of its existence. The list entry for 1932 states that the signal was from "an electric light over a ball-shaped beacon" which can be seen in the photograph.

Fig. 12.5 Cuxhaven time ball and time light. (Courtesy Klaus Hülse collection)

2.5 Emden

The time ball and time light at Emden were both established in 1912 on "Tower, north-east corner of outer harbour" at 53° 20′ 46″ N, 07° 11′ 38″ E. The combined time ball and light signal was made only once per day (1926 list) in contrast to other light signals that were usually made at 6 hour intervals:

> Ball hoisted half-way up at 10 minutes and close-up at 5 minutes before the signal, when a light at upper edge of ball is switched on. The light is extinguished and the ball dropped at Noon, Standard time.

The time ball was discontinued in 1928 but the light signal then continued at the usual 6 hour intervals until it at least 1939.

2.6 Wilhelmshaven

The time ball at Wilhelmshaven was located first on the "East Tower of the Observatory". Its location was given as 53° 31′ 54″ N, 08° 08′ 48″ E in 1880 and 1898. It was moved in 1905 to a new location on the "South West Jetty of 2nd

Fig. 12.6 First and second time balls at Wilhelmshaven. (Courtesy Klaus Hülse collection)

entrance to the harbour", at a higher elevation, retaining its black colour, 1.5 m diameter and 3 m drop height. Its location was given as 53° 30′ 59″ N, 08° 09′ 22″ E (Notice, 1905). It appears to have been moved again to the "Third harbour entrance" with another increase in elevation by 1932. Figure 12.6 shows the time ball at its first two locations. The support structure at the third location is believed to have been similar to that of its second. It was the only German time ball that was still listed in 1934, other than at Hamburg. There were no time lights at Wilhelmshaven.

3 Baltic Ports

3.1 Kiel

The first time ball at Kiel was located "At the Royal Observatory" at 54° 20′ 30″ N, 10° 08′ 52″ E according to the 1880 Admiralty list. It was dropped at noon local mean time (50 m 25 s ahead of GMT). These coordinates and drop time had been specified by Münster (1877). Its location had changed by 1898 to "On the roof of the Artillery Magazine on the Imperial Wharf. Kiel fiord" at 54° 19′ 25″ N, 10° 09′ 48″ E. It was then dropped at 11 am and noon GMT, 19 m 21 s behind local mean time. It was discontinued in 1919. The black ball had a diameter of 1.5 m and a drop of 3.3 m at both locations.

The last Kiel time ball was erected in 1911 on the "Tower of the Naval barracks" at Wik with co-ordinates 54° 21′ 40″ N, 10° 08′ 45″ E. It was listed until 1928. Unusually for Germany, it had a diameter of almost 2 m. The drop height was 3.3 m, as at the other Kiel locations. It was dropped twice daily, except Sundays and holidays, at 11 am and noon GMT.

The locations of the time balls are shown in Fig. 12.7. The last time ball was close to the Kiel Canal.

Fig. 12.7 Location of time balls at Kiel. (Google Maps)

3.1.1 Kiel and Mürwik Time Guns

A time gun at Kiel, fired at 11 am GMT, was listed from 1898 until 1911. It was then listed as being on a training ship (S.M.S *König Wilhelm*) at Mürwik Naval Academy from 1915 until 1922. The gun location was then given as 54° 49′ 00″ N, 9° 27′ 00″ E. It was fired at noon GMT (1 pm German standard time). The training ship was about 80 km NW of Kiel Observatory.

3.2 Stettin

The time ball at Stettin was established in 1912 and listed until 1928. It was replaced by time lights in 1929, like many other time balls in Germany. The time ball was located on the "Northern tower of Government buildings" at 53° 25′ 58″ N, 14° 34′ 15″ E. It had a drop of 12 feet (3.6 m). The 1929 time light was on the same building.

3.3 Swinemünde

The time ball at Swinemünde was one of the first to be established in Germany. It was located at 53° 54′ 36″ N, 14° 15′ 58″ E. "120 yards East of tower of new Navigation House" (1898 list). This navigation house had not been built in 1880, when the time ball was dropped at noon local mean time, but the elevation of the signal had not changed between 1880 and 1898. As at other locations in Germany it was dropped at 11 am and noon GMT in 1898. Unusually, it was dropped additionally at 10 am and 3 pm in 1908 and 1911. A later notice indicated that the signal had been discontinued but then re-established (Notice, 1914). The specified drop times were again 11 am and noon GMT. The arrangements remained the same until 1929, when the time ball was replaced by time lights.

4 Danzig

4.1 Neufahrwasser and Gdańsk

Neufahrwasser was a district of Gdańsk (in Germany before World War I, then in Danzig from 1920 to 1939, in Germany up to 1945 and now in Poland). The first time ball had been erected on the "Pilot's Watchhouse" in 1876. Details of the first time ball at Neufahrwasser had been published in *Lloyd's List* (Time ball at Neufahrwasser, 1876). The ball was painted black and had a diameter of 5 feet (1.5 m). It was 92 feet above high water and 65 feet above ground with a drop height of 10 feet. The 1876 notice indicated that the time ball service had started on 15 July 1876 and that the ball was dropped twice daily: at noon mean Neufahrwasser time and at noon GMT, 1 h 14 m 39.2 s later.

The entry in the 1880 Admiralty list gave similar details, except that the drop height was specified, incorrectly, as 5 feet. Also, only the first signal was included in the 1880 list, suggesting that the noon GMT signal at Neufahrwasser may have been discontinued, at least temporarily. Its location was given as 54° 24′ 24″ N, 18° 39′ 48″ E. This longitude is consistent with the local astronomical time in the 1876 notice, which was repeated in the 1880 list.

A new arrangement was erected in 1894 on the "Principal lighthouse". Figure 12.8 shows the lighthouse in 1894 (The time ball on Gdańsk Nowy Port Lighthouse, 2022). Its details were included in the 1898 Admiralty list. This time ball was dropped twice daily, at 11 am and noon GMT, corresponding to noon and 1 pm Mid-European time, as at other locations in Germany. It was 124 feet above high water and 100 feet above ground. The black ball had a diameter of 5 feet and a modified drop height of 7 feet. Its location was now 54° 24′ 30″ N, 18° 39′ 58″ E. The local astronomical time was 1 h 14 m 39.9 s ahead of GMT, the expected increase of 0.7 seconds for a 10 arc-seconds of change in longitude.

The history of the Gdańsk time balls and the construction of a modern working replica in 2010 has been described by Szychliński (2010). The history is also available online (The time ball, 2022). The new ball has the same open steel framework as the original with a diameter and weight of 1.6 m and 76 kg. The diameter, derived from original drawings, appears to have been slightly larger than the 1.5 m stated by the Admiralty in 1898 and subsequent lists.

Fig. 12.8 The second time ball at Neufahrwasser. (Latarnia-widziana-z-Westerplatte-1894)

4.1.1 Neufahrwasser Time Light

The Neufahrwasser time light was introduced in 1927 and relaced the time ball. It was on "Langfuhr Observatory Tower" at 54° 23′ N, 18° 37′ E. It was more complicated than others in Germany and Danzig. The signal was made at 8 am, noon, 4 pm and 7 pm GMT, following the sequence below (1932 and 1938 lists).

> Light at 2 minutes before time of signal shows three groups of 4 flashes each, corresponding to Morse letter Z, *every 20 seconds.*, thus … At 1 minute before time of signal a steady light is shown, which is extinguished at exact hour. If the signal is correct, 3 flashes of 2 seconds each will be shown immediately after the time signal. …

4.2 Gdynia

An additional time light was introduced at Gdynia in 1930. It was located on "Tower of Marine Observatory" at 54° 32′ N, 18° 33′ E and was extraordinarily powerful with high accuracy. The signal was repeated every hour, night and day. The following description is from the 1938 Admiralty list.

> Light is of 1,000,000 C.P. and is clearly visible during daylight at a distance of 12 miles. Light is switched on 57m 30s, 58m 30s, and 59m 30s and extinguished at 58m 0s, 59m 0s, and 00m 00s. The degree of accuracy is 0.2sec. …

The time lights at Gdynia and Neufahrwasser were in parallel operation. Their stated coordinates may contain errors as they appear to be implausibly close together.

Acknowledgements I am grateful to Paul Fuller for locating notices about the German time balls in the British Newspaper Archive.

References

Malcolm. (1873). *Letter from the Board of Trade to the Astronomer Royal*, dated 9 June 1873. RGO 6 617, File 3, Papers of George Airy, Cambridge University Library.

Münster, Count. (1877). Letter to the Earl of Derby dated 20 December 1877, with accompanying notes about German time balls. RGO 6 618, Papers of George Airy, Cambridge University Library.

Notice. (1875). *Shipping and Mercantile Gazette*, 16 November 1875, 6.

Notice. (1905). North Sea – Germany – Jade River – Wilhelmshaven – Position of time ball, *Nautical Magazine*, 47; *Nachrichten für Seefahrer*, 44, Berlin.

Notice. (1914). Baltic Sea – Germany – Swinemünde – time signal reestablished, *Nautical Magazine*, 18; *Nachrichten für Seefahrer*, No 12 (748), Berlin.

Szychliński, G. (2010). The Gdańsk Time Ball, *Horological Journal*, May 2010, 216–219.

The time ball column at the "Alte Liebe" (section of the harbour in Cuxhaven). Article describing time ball development in Germany. Accessed 2022. http://www.lentz-familie.de/r/zeitballe.html

The time ball on Gdańsk Nowy Port Lighthouse. (2022). http://www.latarnia.gda.pl/time-ball/

Time ball at Neufahrwasser, Danzig. (1876). Notices to Mariners, Prussian coast, published in *Lloyd's List*, 3 November 1875, 10.

Wilkes. (1873). *Letter from the Consul General of the German Empire to the Board of Trade*, dated 30 April 1873. RGO 6 617, papers of George Airy, Cambridge University Library.

Zeltz, H. (1875). Signal stations on the German coast, dated Berlin, 22 October 1875, published in *Lloyd's List*, 3 October 1876, 10.

Chapter 13
The Netherlands, Belgium and Scandinavia

Roger Kinns

Abstract Recent research into the history of time signals in the Netherlands and Belgium has exposed time signals that appear to have been unknown to the British Admiralty in their early years. Time discs, often called flaps or boards, became the preferred signal in the Netherlands after 1854, superseding time balls that were first dropped from guard ships at Hellevoetsluis and Nieuwe Diep in 1846. From a distance, discs rotating about a horizontal axis looked like a disappearing time ball. They were usually perforated to reduce aerodynamic drag and weight, allowing operation in high winds as well as rapid rotation to reduce signal ambiguity. Antwerp introduced a time ball in 1856, superseded by time discs in 1883.

Time balls and drums were used at many locations in Scandinavia and time lights were introduced in the Netherlands and Norway after 1919. Time guns were never listed. The first Scandinavian time ball was introduced at Copenhagen in 1868, replacing a flag signal.

1 Introduction

Recent research (Willems, 2023a, b, c, d, e) into the history of time signals in the Netherlands and Belgium has exposed time signals that appear to have been unknown to Airy (1861a). Airy's additional list of time balls that he knew to have been projected but not implemented included a time ball for Copenhagen, but not for any other locations in the Netherlands, Belgium or Scandinavia (Airy, 1861b). Time balls, drums and discs (often described as flaps) were later used at many locations in the region and time lights were introduced in the Netherlands and Norway after 1919. Time guns were never listed.

R. Kinns (✉)
The Maudslay Society, Pembroke College, Cambridge, England

R. Kinns (ed.), *The World of Visual Time Signals for Mariners*, Historical & Cultural Astronomy, https://doi.org/10.1007/978-3-031-57334-7_13

1.1 Time Discs

Time discs were invented by Professor Frederik Kaiser (1808–1872), Director of Leiden Observatory. Often called flaps, they became the preferred signal in the Netherlands during the 1850s (Mörzer Bruijns, 1993; Willems, 2023e). From a distance, the rotating discs looked like a disappearing time ball. They were usually perforated to reduce aerodynamic drag and weight, allowing operation in high winds as well as rapid rotation to reduce signal ambiguity.

Figure 13.1 shows the discs in raised and lowered locations on the Meteorological Institute (also described as the Port or Gate building) in Rotterdam. The discs were rotated from horizontal to vertical positions 5 minutes before the signal, which was the time of return to the horizontal rest position (Meyers Lexikon, 1902; illustration included by Hite, 2014).

1.2 Entries in Admiralty Lists

The range of signals for the Netherlands, Belgium and Scandinavia in Admiralty lists is shown in Table 13.1. Their locations are shown in Fig. 13.2. Willems (2023e) has established that some time discs in the Netherlands and Belgium were listed by the Admiralty after they had been withdrawn from service. It appears that notices of signal withdrawal had not always been received by the editors of Admiralty lists. There is also evidence that their earlier existence was not widely recognised outside the Netherlands.

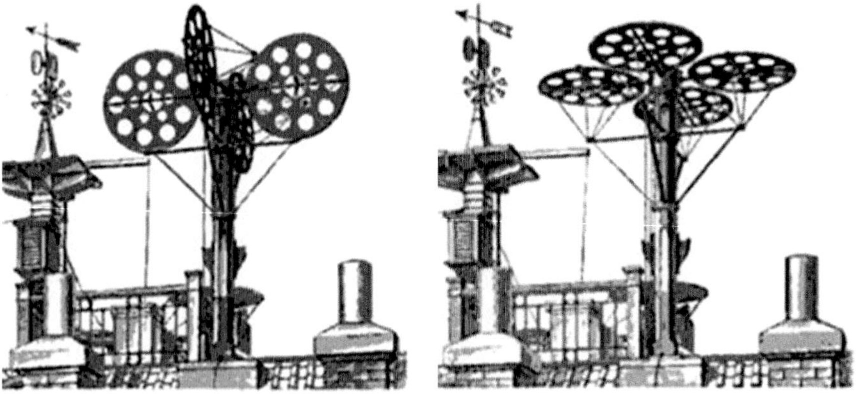

Fig. 13.1 Time discs on the Meteorological Institute. (Meyers Lexikon, 1902)

Table 13.1 Signals in the Netherlands, Belgium and Scandinavia

Country	Location	1880	1898	1904	1911	1915	1919	1922	1928	1934	1939	1947
Netherlands	Amsterdam		Discs (black)	Discs (black)	Discs (black)	Discs (black)	Discs (black)	Discs (black)	Discs (black)	Discs (black)	Discs (black)	
									Light	Light	Light	Light
	Hellevoetsluis	Discs	Discs	Discs (black)								
	Nieuwe Diep	Discs	Discs	Discs (black)	Discs (black)	Discs (black)	Discs (black)	Discs (black)	Discs (black)	Discs (black)	Discs (black)	
	Rotterdam: RDYC	Discs										
	Rotterdam: Gate building		Discs	Discs (black)	Discs (black)	Discs (black)	Discs (black)	Discs (black)	Discs (black)	Discs (black)	Discs (black)	
								Light	Light	Light	Light	Light
	Rotterdam: Water tower				TB (black)	Discs (black)	Discs (black)	Discs (black)	Discs (black)	Discs (black)	Discs (black)	
								Light	Light	Light	Light	Light
	Vlissingen	Discs	Discs	Discs (black)	Discs (black)	Discs (black)	Discs (black)	Discs (black)	Discs (black)			
Belgium	Antwerp		Discs	Discs	Discs	Discs	Discs	Discs	Discs	Discs	Discs	
Denmark	Copenhagen	TB	TB	TB	TB	TB	TB	TB	TB	TB	TB	
	Elsinor	TB	TB	TB	TB	TB	TB	TB	TB	TB	TB	
Sweden	Gothenburg	TB	TB	TB	TB	TB	TB	TB	TB	TB		
	Karlskrona		TB (black)	TB (black)	TB (brown)	TB (brown)	TB (brown)	TB (brown)	TB (brown)	TB (brown)		
	Malmö		TB (black)	TB (black)	TB (black)	TB (black)	TB (black)	TB (black)	TB (black)	TB (black)	TB (black)	
	Stockholm	TB (red)	TB (red)	TB (b&w)	TB (black)	TB (black)	TB (black)	TB (black)	TB (black)	TB (black)		
Norway	Bergen	TB	TB (r&w)	TB (r&w)	TB (red)	TB (red)	TB (red)	TB (red)	TB (red)			
										Light	Light	Light
	Carl Johansvaern							Light	Light			
	Oslo (Christiania)	TB (b&w)			TB (gold)							
			Drum (b&w)	Drum (b&w)	Drum (b&w)	Drum (b&w)	Drum (b&w)	Drum (b&w)	Drum (b&w)	Drum (b&w)	Drum (b&w)	
									Light	Light	Light	Light
	Trondheim	Drum (b&w)	Drum (b&w)	Drum (b&w)	Drum (b&w)							
						TB (b&w)	TB (b&w)	TB (b&w)	TB (b&w)			
Signal type totals		1880	1898	1904	1911	1915	1919	1922	1928	1934	1939	1947
Time ball, including collapsible		6	7	7	9	8	8	8	8	6	3	0
Gun, including powder flash		0	0	0	0	0	0	0	0	0	0	0
Lights		0	0	0	0	0	0	3	5	5	5	5
Other: disc, drum, flag, etc.		5	8	8	7	7	7	7	7	6	6	0

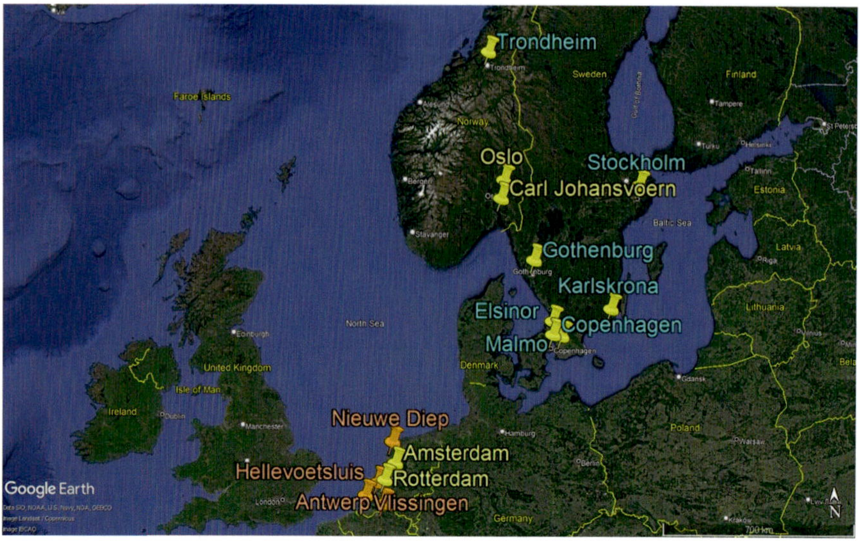

Fig. 13.2 The Netherlands, Belgium and Scandinavia. (Google Maps)

2 The Netherlands

The first time balls in mainland Europe are believed to have been in the Netherlands. They were dropped from the main topgallant yards on guardships at Hellevoetsluis and Nieuwe Diep in January1846 (Oprigting van Tijdballen, 1846; Cittert-Eymers, 1988; Willems, 2023b). These guardships were at principal naval establishments.

2.1 Hellevoetsluis, Nieuwe Diep and Vlissingen

2.1.1 Time Balls

The guardship time balls were both described in pilot guides for 1863 and 1878 (North Sea Pilot, 1863, 125, 143; 1878, 140, 167) but it now seems likely that they had been discontinued in the 1850s when time discs were first introduced. There is no later reference to them as operational signals in the Netherlands (Willems, 2023e). It appears that British authorities had not received timely notification of either time ball withdrawal, or the introduction of time discs. The entries in 1878 are still useful, because they describe the mode of operation.

> **Time Ball.** – In order to facilitate the rating of chronometers at Hellevoetsluis, the instant of mean noon is shown on board the guardship by the dropping of a time-ball from the main topgallant yard. The ball is hoisted half-way from the topsail yard at 5 minutes, and close up at 2 minutes before twelve o'clock, and at the instant of noon, Hellevoet mean time, the ball will be dropped from the topgallant down to the topsail yard.

As the officer counting the seconds guides the line himself, no allowance for loss of time is considered necessary, and the instant of mean noon is that on which the ball leaves the topgallant yard. The longitude of Hellevoetsluis, as fixed by triangulation, is 4° 8′ 0″ East of Greenwich.

Time-ball. – A time-ball is exhibited at the main top-gallant yard of the guard-ship lying in the Nieuwe diep, for the rating of chronometers. The ball is hoisted half up at 5 minutes, and close up at 2 minutes before 12 o'clock, and is dropped at the instant of noon, Nieuwe diep mean time; the moment of noon being that in which the ball leaves the topgallant yard. The longitude of the Nieuwe diep is 4° 46′ 41″ East of Greenwich.

2.1.2 Time Discs

Time discs were used for the first time at Vlissingen in 1854 (Willems, 2023c). They were then at the "Arsenal of the State Equipment Yard". The time discs at Vlissingen were relocated in 1875 to the "tower of the Weather Station next to the sea lock" (Willems, 2023c). The location was described as "Stone tower west side of great sluice" in Admiralty lists. Figure 13.3 shows the installation at its second location in Vlissingen, with discs in their rest position, highlighted using a red arrow. Model time disc installations at the *Amsterdam Rijksmuseum* and other Vlissingen photographs are described by Willems.

Time discs were erected on shore buildings at Nieuwe Diep and Hellevoetsluis in 1855 and 1856 respectively (Willems, 2023e). These discs were introduced because of concerns about the accuracy of the guardship time balls, which were not supported by an astronomical observatory and used sextant measurements to regulate

Fig. 13.3 Vlissingen time discs. (Courtesy Klaus Hülse collection)

an onboard chronometer. The discs were regulated initially using precise astronomical clocks and then by electric telegraph from Leiden Observatory after 1859 (Mörzer Bruijns, 1993; Willems, 2023e).

Time discs at Nieuwe Diep and Hellevoetsluis were not mentioned in the 1863 or 1878 pilot guides, but they replaced the guardship time balls entries in the 1887 edition (North Sea Pilot, 1887, 180–181). The term "time ball" was often used loosely to describe time discs and other devices, as in the 1887 entries which are transcribed below.

> **Time Ball.** – In order to facilitate the rating of chronometers at Hellevoetsluis, the instant of mean noon is shown from the roof of the chief office of Marine Establishment, by the dropping of four discs, at a height of 60 feet above the ground. The discs are placed perpendicularly at 5 minutes before the signal, and they fall into a horizontal position at the instant of noon, Hellevoetsluis mean time. Equivalent to 23h. 43m. 39.3s., Greenwich mean time.
>
> **Time-ball.** – Four black circular discs from the roof of the chief marine establishment at Nieuwe diep, 57 feet above the ground. The discs are placed perpendicularly at 5 minutes before noon, and fall into a horizontal position at noon, Nieuwe diep mean time. Equivalent to 23h. 40m. 53.6s. Greenwich mean time.

Surprisingly, there was no mention of the time discs at Vlissingen in the 1863, 1878 or 1887 pilot guides.

2.1.3 Entries in Admiralty Lists

Discs at Hellevoetsluis, Vlissingen and Nieuwe Diep (Willemsoord) were listed by the Admiralty in 1880. The same coordinates and signal descriptions were given in Admiralty lists for all three installations up to their dates of withdrawal. The discs at Hellevoetsluis were not listed after 1904 and were withdrawn in 1905 (Willems, 2023e). The time discs at Vlissingen were still listed in 1928, but not in 1932. They had been withdrawn in 1927 (Willems, 2023e). The discs at Nieuwe Diep were still listed in 1939 but are believed to have been withdrawn much earlier.

2.2 Rotterdam

Rotterdam time signals were listed for three locations in Rotterdam, as shown in Fig. 13.4. The assumed GPS latitude and longitude are those specified in Admiralty lists, without correction. The first disc signal was replaced in 1890 by another at a different location. A further disc signal was added in 1909.

In 1880, the Rotterdam discs were listed as being on the "Tower of Royal Dutch Yacht Club". This signal operated from 1879 to 1890 (Willems, 2023e). There were four black discs with diameters of 2 feet 8 inches (0.8 m), 27 m above ground. The latitude and longitude were given as 51° 54′ 30″ N, 4° 28′ 51″ E.

In 1898 and later lists, the time signal location was given as the "Gate building" at Feijenoord with coordinates 51° 54′ 39″ N, 4° 29′ 47″ E. The discs had been

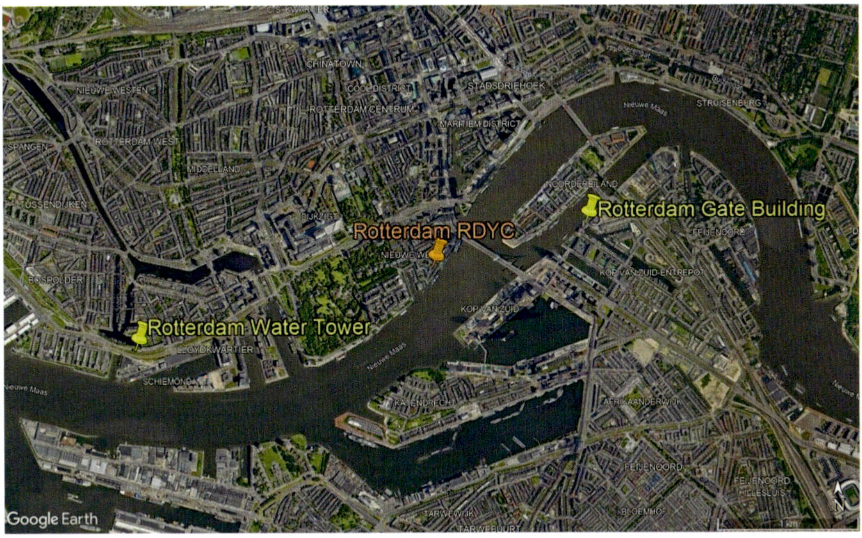

Fig. 13.4 Rotterdam signal locations. (Google Maps)

Fig. 13.5 Time discs on the Rotterdam water tower. (https://nl.wikipedia.org/wiki/Watertoren_Rotterdam_Delfshaven)

erected in 1890. The disc diameters and elevation remained the same throughout. The Gate building location was on the south bank of the Nieuwe Maas, about 1.1 km ENE of the first location on the north bank.

The signal on the Rotterdam water tower at Delfshaven had been erected in 1909. Unusually for the Netherlands, it was listed by the Admiralty as a time ball in 1911. The listing as a time ball appears to have been incorrect. Local announcements (Willems, 2023e) indicate that the signal had been time discs from the outset. The description had been corrected to four discs in the 1915 list. Figure 13.5 shows the installation on the Rotterdam water tower which was listed until 1939. Its location was given as 51° 54′ 13″ N, 4° 27′ 01″ E, on the north bank of the river, about 3.3 km WSW of the Gate building on the south bank.

The discs on the Gate Building and Water Tower were complemented by time lights in 1921. The Rotterdam discs were listed by the Admiralty as substitute signals for use in case of time light failure from 1921 until at least 1939. The lights were still listed in 1947.

2.3 Amsterdam

Time discs at Amsterdam had been erected in 1881 on the tower of the Navigation School (Willems, 2023e) and relocated in 1886 to "Public Buildings, Commercial Quay" (Admiralty lists). Figure 13.6 shows the implementation which was included in lists from 1898 to 1939. The four black discs had diameters of 1.0 m, 31 m above ground.

Time lights were introduced at Amsterdam in 1927. They were illuminated 5 minutes before extinction as the signal. The lamps in Amsterdam were on towers at three harbour locations, one close to the discs and the others on separate buildings (Willems, 2023e). According to the Admiralty, the discs were retained for use in

Fig. 13.6 Amsterdam time discs. (Courtesy Klaus Hülse collection)

case of light failure until at least 1939. As at Rotterdam, only the time lights were listed in 1947.

3 Belgium

The Belgian government undertook four initiatives in the nineteenth century to help shipmasters rate their chromometers at Antwerp. Early time measurements were made locally in Antwerp, replaced later by electric telegraph signals from The Royal Observatory at Brussels and then at Uccle (Ukkel) from 1890.

3.1 Antwerp

The Antwerp initiatives are shown in the list below (Willems, 2023e).

1. Local time measurement using a portable transit instrument made by Troughton and Simms at the Pavillon Astronomique.
2. Local time measurement using the portable transit instrument, with the addition of a time ball at the Observatoire Maritime in 1856.
3. Time discs at the Hanseatic House from 1883 to 1893, using time measurements by the Royal Observatory at Brussels (moved to nearby Uccle in 1890).
4. Time discs at the Pilot Station from 1895, using time measurements by the Royal Observatory at Uccle.

There was a disastrous fire at the Hanseatic House in 1893 which destroyed the time signal, but the decision to erect a new building for harbour services on the site of the Old Zelande Corn Market had been made in 1890 (Willems, 2023d).

3.1.1 Admiralty List Entries for Antwerp

The 1856 Antwerp time ball was not included in the 1880 Admiralty list. The only known photograph of it and the associated maritime observatory, taken in 1875, is described by Willems (2023b). The discs on the Hanseatic House were introduced in 1883 and are likely to have been included in the 1888 and 1892 lists.

The new disc signal for Antwerp was included in Admiralty lists from 1898 onwards. The signal was made at 1 pm GMT and became operational in December 1895 (Willems, 2023d). It consisted of four circular discs, as at many locations in the Netherlands. It was described by the Admiralty in 1898 as being on the tower of "the Pilot House", known in Belgium as Het Loodswezen (The Pilot Station). A blue flag was raised when the signal was not available.

From February 1901 onwards the signal was made at both 1 pm and 1.15 pm GMT (Willems, 2023d). A ball above the discs was raised to indicate an error and

Fig. 13.7 Antwerp time discs. (Author's collection)

was not itself the time signal: "should the signal fail, or be inaccurate, a ball will be immediately hoisted" (Admiralty lists). Figure 13.7 shows the location of the time discs on the tower in about 1906.

The location was described as the tower of the Government building in the Zelande Corn Market in 1908 and later lists, but there was no change in the listed latitude and longitude. The change was in the description of the building. The disc signal was dismantled in 1927 and should not have been included in later Admiralty lists. Another, larger ball was installed after the time discs had been removed in 1927, but it was never used as a time signal (Willems, 2023d).

4 Denmark

Copenhagen is believed to have used a flag signal in the eighteenth century that was hauled down when the sun reached its highest point, later changed to a flag signal indicating mean time for chronometer rating.

Time balls were listed by the Admiralty for Copenhagen and Elsinore from 1880 to 1939 but had been withdrawn by 1947. Denmark provided telegraph time signals at a large number of locations around the coast which were specified in Admiralty lists by 1898 but could only be accessed by going ashore.

4.1 Copenhagen

Airy had been approached by the astronomer in Copenhagen in 1857, who asked for details of the time ball at Greenwich (Schjellerap, 1857). A flag was then being used as the Copenhagen time signal, but there was a desire to change this to a "Ball signal, which in every part of the world is known as excellent and only of the seaman respected". Airy sent a detailed reply, with paragraphs covering time ball construction and operation at Greenwich and four other locations (Airy, 1857). The other locations were in the Strand and at Cornhill in London, Deal and Edinburgh. Airy also included sketches showing mast construction and mechanisms. The letter ended with an invitation to London. A time ball in Copenhagen replaced the flag signal in 1868.

The Copenhagen time ball was located on the Nikolai Tower from July 1868 until 1908 (Graham, 2016). Its location had been changed to the "Silo storehouse, Frihavn" before 1911. The two locations are shown in Figs. 13.8 and 13.9. Time balls at both Copenhagen locations were described as "wicker balls" with a

Fig. 13.8 Copenhagen Nikolai Tower. (Courtesy Klaus Hülse collection)

Fig. 13.9 Copenhagen Frihavn. (Courtesy Klaus Hülse collection)

diameter of 1.5 m and a drop height of 4.5 m. Figure 13.10 suggests that they were more massive than early time balls at the Cape of Good Hope, for example, which used traditional wickerwork construction to reduce cost and weight (Kinns, 2021). The ball diameter of 1.5 m was that suggested by Airy in 1857 but the ball construction was different and the drop height had been increased from 3 m to 4.5 m.

4.2 *Elsinore*

The wicker time ball at Elsinore (Helsingör) was at the "entrance to harbour", but its diameter and drop height were not stated in the 1880 Admiralty list. Its latitude and longitude were given as 56° 2′ 4″ N., 12° 37′ 24″ E. in 1880 and remained the same thereafter. More detail was given in the 1898 and later lists. It was dropped "from the mast on the Quarantine and Pilot House on the south harbour mole" and had a large drop height of 12 m, according to the lists between 1898 and 1908. It was raised half-way at 5 minutes before the signal, close-up at 2 minutes before the signal and then dropped at noon GMT (1 pm Mid-European time).

Fig. 13.10 Copenhagen time balls. (Courtesy Klaus Hülse collection)

The drop height had been reduced to 3 m with the same elevation, by 1911. This appears likely to have coincided with a change of arrangement from a ball dropped from a yard-arm to a ball sliding down a mast above the building. The diameter of the wicker ball was given as 3.5 feet (1.05 m) in 1911. The diameter was stated to be 5 feet (1.5 m) with the same drop height of 3 m between 1915 and 1941.

Figure 13.11 shows a postcard used in 1905, so the photograph must show the early arrangement. Figure 13.12 shows the second installation on the Quarantine and Pilot House that had a large wind direction indicator on one wall at the time of the photograph. The mast is notably more robust than that shown in Fig. 13.11, with prominent cross-trees.

5 Sweden

Sweden had time balls at Göteborg (Gothenburg), Karlskrona, Malmö and Stockholm, but only the Malmö time ball remained in 1939 and all had been withdrawn by 1947. They varied in colour with time and between locations. Time balls at Göteborg and Stockholm were listed in 1880. Those at Karlskrona and Malmö were included in the 1898 list. All were dropped at 1 pm Swedish mean time (noon GMT). There was, however, an apparently unique procedure at Gothenburg, described below.

Fig. 13.11 Elsinor first time ball. (Courtesy Klaus Hülse collection)

5.1 Gothenburg

The time ball at Gothenburg was at the Navigation School and was dropped on Tuesdays, Thursdays and Saturdays only. It was listed from 1880 to 1934 with the same details throughout. The time ball location was given as 57° 42′ 34″ N., 11° 58′ 0″ E. in all Admiralty lists. It was 42 m above the water and 21 m above ground, with a diameter of 1.2 m and a drop height of 2.6 m. However, the Navigation School building and time ball tower were rebuilt in 1915 and the appearance of the building then changed considerably. The original building is shown in Fig. 13.13. The later building is shown in Fig. 13.14.

The list entries from 1880 to 1932 included the following procedure for time ball operation at Gothenburg (1908 list):

> Ball hoisted close up at 5 minutes before signal.
> Ball lowered half way at 2 minutes before signal.
> Ball dropped at noon Greenwich mean time, or at 1h 0m 0s Swedish mean time.

This was an extraordinary departure from the common practice of raising the ball in two stages to its maximum height and then dropping the ball as the signal, which was the procedure used at Karlskrona and Stockholm. It seems unlikely to have been a transcription error, given the obvious departure from usual practice and its repetition for half a century.

Fig. 13.12 Elsinor second time ball. (Courtesy Klaus Hülse collection)

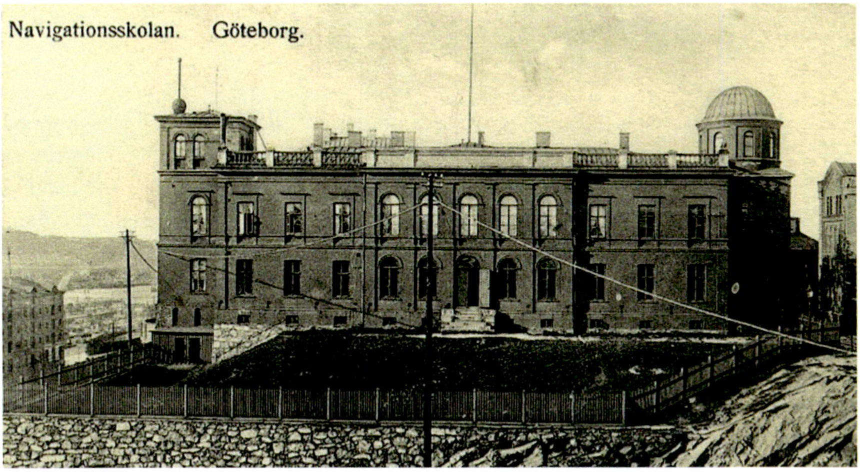

Fig. 13.13 Gothenburg time ball on first building. (Courtesy Klaus Hülse collection)

Fig. 13.14 Gothenburg time ball on new building. (Courtesy Klaus Hülse collection)

5.2 Stockholm

The time ball at Stockholm was also at the Navigation School there and was dropped daily. It had a diameter of 0.9 m and a drop height of 2.4 m. It was painted red in 1880 and 1898, but the colour had changed to black with a horizontal white band in the 1904 and 1908 lists. Figure 13.15 shows the time ball with the second colour scheme. The time ball in lists from 1911 until 1934 had an increased diameter of 1.5 m, a larger drop height of 3.3 m and an increased elevation. The same time ball latitude and longitude was specified throughout. Figure 13.16 shows the later arrangement.

Fig. 13.15 Stockholm first time ball. (Courtesy Klaus Hülse collection)

5.3 Karlskrona

The Karlskrona time ball was located on a "tower in the Dockyard". It had a diameter of a 0.9 m and a drop height of 3 m. It was dropped on Fridays only "by wire from Stockholm Observatory". It was listed as black in 1898 and 1904 and brown from 1908 until 1934.

5.4 Malmö

The Malmö time ball at the Navigation School was black with a diameter of 1.2 m and a drop height that changed from 3.9 m in 1898 to 3 m in 1908. It was dropped daily. The ball was raised to its maximum height without resting half-way at Malmö. Until 15 October 1916, the time ball was on a tower at the Navigation School. A new time ball was then introduced at a wharf location, but the earlier time ball was kept in reserve in case of signal failure (Notices, 1916). Subsequent Admiralty lists gave the new location as the "steamer pier".

Figure 13.17 shows the second time ball at Malmo, soon after its installation in 1916. The new tower had an open frame structure, which allowed easy access to the time ball operating system. Unlike the other time balls in Sweden, it was still listed in 1939.

Fig. 13.16 Stockholm second time ball. (Courtesy Klaus Hülse collection)

6 Norway

Norway used a variety of time signals, including balls, drums and lights but not guns. They were located at Bergen, Carl Johansvaern (now Karljohansvern), Christiana (Oslo) and Trondheim. Time lights were still listed at Bergen and Oslo in 1947.

6.1 Bergen

The colour of the time ball at Bergen changed over the years. Its colour was not stated in 1880, but it was listed as having red and white stripes in 1898, 1904 and 1908 and then as dark red in lists from 1911 to 1928. It was listed as a hollow iron ball with a diameter of 0.6 m and drop height of 3.7 m until 1908. Its location was given as the NE corner of the Observatory. The diameter of the dark red ball was unstated but it had a smaller drop height of 1.8 m. The ball was dropped on Wednesdays and Saturdays only at noon local time in 1880 and later at noon Mid-European time (11 am GMT). It was replaced by a time light in 1929.

Fig. 13.17 Malmö second time ball in 1916. (Courtesy, Malmö City Archives/archival collections/00570, harbor administration ID: 10016)

6.2 Carl Johansvaern

A time light was introduced in 1919 at Carl Johansvaern, then the home of the Royal Norwegian Navy. It ceased operation when a time light was introduced at Bergen in 1929. The Bergen light was illuminated 5 minutes before extinction at noon.

6.3 Christiana (Oslo)

Admiralty lists initially specified Christiana (often Kristiana in Norway) as the time ball location. Officially, the city name was changed to Oslo on 1 January 1925, but "Oslo (Christiana)" was listed as the location until the early 1930s.

Christiana had a time ball in 1880 with a diameter of 0.8 m and a drop height of 4.5 m on the roof of the Observatory. This was replaced by a drum which was listed from 1898 until at least 1939. The drum diameter was given as 3 feet 8 inches (1.05 m) up to 1911, but later as 5 feet (1.5 m), still with a drop height of 4.5 m. The ball and the drum were both black and white and were dropped on Wednesdays and Saturdays only at noon. Lists after 1922 included the comment "Discontinued (1923)" when a time light had been introduced. The fact that it was still listed after 1923 suggests that it was retained as a reserve signal in case of light failure.

A time light was erected on a mast at the nearby Navigation School in 1922 and operated in parallel with the drum. The signal was made every 4 hours, starting at 4 am. The light was switched on about 5 minutes before extinction at the signal time which was stated to have an accuracy of 0.2 second. The Oslo light was still extant in 1947 when the drum signal was no longer listed.

An additional time ball was listed in 1911 but appears to have had a short life. It was located on the "Tower of Glass warehouse in the market place", shown in Fig. 13.18. The ball was made of gilded metal with a diameter of 0.7 m and a drop height of 2.4 m. It was dropped electrically from the Observatory every day at 1 pm (noon GMT), except Sundays and holidays. It was no longer listed in 1915.

6.4 Trondheim

Trondheim (Trondhjem) had a cylindrical canvas drum signal on the roof of the Observatory in 1880, with a diameter of 1.5 m and a drop height of 3.3 m. It was painted black and white and was hoisted 15 minutes before the signal at noon local

Fig. 13.18 Christiana Glass Market time ball. (Courtesy Klaus Hülse collection)

time. The drop time had been changed to noon Mid-European time (11 am GMT) by 1898. Its location was 63° 25′ 48″ N, 10° 22′ 5″ E. The 1898 list onwards included the comment "The Observatory is not readily visible". The drum was replaced by a time ball from 1914 until 1928. This time ball was black with a white cross and was located on "Flagstaff north of custom house" at 63° 26′ 27″ N, 10° 24′ 30″ E.

Acknowledgement I am most grateful to Joris Willems, who is a Belgian river pilot on the River Scheldt between Antwerpen (Antwerp) and Vlissingen (Flushing), for information about time signals in Belgium and the Netherlands. He published papers in the *Bulletin of the Dutch and Flemish Sundial Society*, which were republished in *Ruimschoots* during 2023. They contain a wealth of information and illustrations concerning signals at Antwerp and Vlissingen. Generously, he has also shared the results of his wider research into signals in the Netherlands, Belgium and Indonesia that is presently unpublished.

References

Airy, G. B. (1857, February 10). Letter to H.C.F.C. Schjellerap. RGO 6.145, Section 8, Leaf 90, Papers of George Airy, Cambridge University Library.

Airy, G. B. (1861a, May). *List of Time Balls in Actual Operation*. RGO 6 615, Item 2, Papers of George Airy, Cambridge University Library.

Airy, G. B. (1861b). *List of Places at Which Time Balls have been Projected, the Erection of None of Which is Known to have been Proceeded with*. RGO 6/615, Item 3, Papers of George Airy, Cambridge University Library.

Cittert-Eymers, J. G. (1988). Tijdseinen, tijdballen en tijdkleppen. *Bulletin van de Nederlandse Zonnewijzerkring*, (2), 8. [In Dutch].

Graham, J. (2016). Pre-20th century Denmark: Time immaterial since time immemorial. https://cphpost.dk/2016-11-11/history/more-than-one-way-to-tell-the-time/

Hite, L. (2014). How Time balls Worked, featuring the Cincinnati Observatory, Birthplace of American Astronomy. http://www.voamuseum.org/media/1355/how_time_balls_worked

Kinns, R. (2021). Time signals for mariners in South Africa. *Journal of Astronomical History and Heritage, 24*(2), 285–314.

Meyers Lexikon. (1902). *Meyers Konversations-Lexikon* or *Meyers Lexikon* was a major encyclopedia in the German language.

Mörzer Bruijns, W. F. J. (1993). The astronomical clocks of Andreas Hohwü. In R. G. W. Anderson, J. A. Bennett, & W. F. Ryan (Eds.), *Making instruments count: Essays on historical scientific instruments* (pp. 454–461). Variorum.

North Sea Pilot, Part IV, first edition. (1863). *Rivers Thames and Medway, and the shores of the North Sea from Calais to the Skaw* (1st ed.). Printed for the Hydrographic Office, Admiralty, London

North Sea Pilot, Part IV, third edition. (1878). *Rivers Thames and Medway, and the shores of the North Sea from Calais to the Skaw* (1st ed.). Printed for the Hydrographic Office, Admiralty, London.

North Sea Pilot, Part IV, fourth edition. (1887). *Rivers Thames and Medway, and the shores of the North Sea from Calais to the Skaw* (1st ed.). Printed for the Hydrographic Office, Admiralty, London.

Notices. (1916, July-December). No. 3399, Sweden – The Sound – Malmo – Time ball moved, *Underrättelser för sjöfarende* No. 41, Stockholm.

Oprigting van Tijdballen aan het Nieuwe Diep en te Hellevoetsluis. (1846, January 30). Berigt aan de Zeevarenden. In *Nederlandsche Staatscourant*. [In Dutch].

Schjellerap, H. C. F. C. (Observ. Astronom., Copenhagen). (1857). Letter to G.B. Airy, 2 February 1857. RGO 6.145, Section 8, Leaf 89, Papers of George Airy, Cambridge University Library.

Willems, J. (2023a). Greenwich aan de Schelde: Tijdmeting in de Antwerpse haven gedurende de 19de eeuw, *Ruimschoots, driemaandelijks nautisch tijdschrift* (quarterly nautical magazine), *2023*(1), 4–10. [In Dutch].

Willems, J. (2023b). Greenwich aan de Schelde: Observatoire Maritime d'Anvers (1855–ca 1875), *Ruimschoots, driemaandelijks nautisch tijdschrift* (quarterly nautical magazine), *2023*(2), 10–15. [In Dutch].

Willems, J. (2023c). Greenwich aan de Schelde: De tijdkleppen van het Hanzehuis (1883–1893), *Ruimschoots, driemaandelijks nautisch tijdschrift* (quarterly nautical magazine), *2023*(3), 8–13. [In Dutch].

Willems, J. (2023d). Greenwich aan de Schelde: De tijdkleppen van het Loodswezen (1885–1914), *Ruimschoots, driemaandelijks nautisch tijdschrift* (quarterly nautical magazine), *2023*(4), 14–19. [In Dutch].

Willems, J. (2023e). Presently unpublished research into the history of time signals in The Netherlands, Belgium and Indonesia.

Chapter 14
Russia, Finland and Latvia

Roger Kinns

Abstract Signals in Russia ranged from the Baltic and Black Seas to the Arctic port of Archangel and to Vladivostok, on the Sea of Japan. The observatory at Polkowa near St Petersburg was an important reference for time definition from the 1840s onwards. Time balls and guns were used widely, with later addition of lights at Sevastopol and Archangel. Finland and Latvia were in Russia for part of the era of visual time signals.

Time balls in Finland were at Helsingfors (Helsinki), complemented by a time gun, and at Abo and Uleaborg. Latvia had time signals at Riga and Libau. Time lights at Riga were introduced in 1927, a nineteenth century time ball having been discontinued in 1921. A time ball and time gun operated at Libau between 1913 and 1921.

Remarkably, five time balls, five guns and two lights were still listed for Russia, Finland and Latvia in 1947.

1 Introduction

Signals in Russia ranged from the Baltic and Black Seas to the Arctic port of Archangel and to Vladivostok, on the Sea of Japan. These signals are shown in Table 14.1, together with those for Finland and Latvia which were part of Russia for much of the period from 1880 to 1947. Ukraine, whose coastline includes many Baltic ports, was listed as part of Russia during the visual time signals era. The locations of the signals in Europe are shown in Fig. 14.1. Vladivostock is on the Sea of Japan. Its location is shown on the map of signals for China, Japan and Korea (see Chap. 21).

Remarkably, five time balls, five guns and two lights were still listed for Russia, Finland and Latvia in 1947.

R. Kinns (✉)
The Maudslay Society, Pembroke College, Cambridge, England

R. Kinns (ed.), *The World of Visual Time Signals for Mariners*, Historical & Cultural Astronomy, https://doi.org/10.1007/978-3-031-57334-7_14

Table 14.1 Signals in Russia, Finland and Latvia

Country	Location	1880	1898	1904	1911	1915	1919	1922	1928	1934	1939	1947
Russia (Baltic)	Kronstadt	TB	TB	TB	TB	TB	TB	TB	TB	TB	TB (black)	TB (black)
					Gun	Gun	Gun	Gun	Gun	Gun	Gun	Gun
	St. Petersburg (Leningrad, Petrograd)	Gun	Gun	Gun	Gun	Gun	Gun	Gun	Gun	Gun	Gun	Gun
									TB	TB	TB	TB
Russia (Arctic)	Archangel								TB (black)	TB (black)	TB (black)	TB (black)
											Light	Light
Russia (Black Sea)	Theodosia								Gun	Gun	Gun	Gun
	Nikolaev	TB (black)	TB (black)	TB (black)	TB (black)	TB (black)	TB (black)	TB (black	TB (black)	TB (black	TB (black)	TB (black)
			Gun	Gun	Gun	Gun	Gun	Gun	Gun	Gun	Gun	Gun
	Odessa			TB (black)	TB (black)	TB (black)	TB (black)	TB (black	TB (black)	TB (black)		
	Sevastopol C-in-C house				TB	TB	TB	TB				
	Sevastopol Observatory				TB (black)	TB (black)	TB (black)	TB (black	TB (black)			
										Light	Light	Light
	Sevastopol								Gun	Gun		
Russia (Asia)	Vladivostok		TB	TB	TB	TB (red)	TB (red)	TB (red)	TB (red)	TB (red)	TB (red)	TB (red)
			Gun	Gun	Gun	Gun	Gun	Gun	Gun	Gun	Gun	Gun
Finland (was Russia)	Helsingfors (now Helsinki)		TB (black)	TB (black)	TB (black)	TB (black)	TB (black)	TB (black)	TB (black)	TB (black)	TB (black)	
			Gun	Gun	Gun	Gun	Gun	Gun				
	Abo		TB (black)	TB (black)	TB (black)	TB (black)	TB (black)	TB (black)	TB (black)	TB (black)	TB (black)	
	Uleaborg		TB (black)	TB (black)	TB (black)	TB (black)	TB (black)	TB (black)				
Latvia (Baltic) (was Russia)	Libau					TB	TB					
						Gun	Gun					
	Riga		TB (black)	TB (black)	TB (black)	TB (black)	TB (black)					
										Light	Light	Light
Signal type totals		1880	1898	1904	1911	1915	1919	1922	1928	1934	1939	1947
Time ball, including collapsible		2	7	8	10	11	11	9	9	8	7	5
Gun, including powder flash		1	4	4	5	6	6	5	6	6	5	5
Lights		0	0	0	0	0	0	0	0	2	3	3
Other: disc, drum, flag, etc.		0	0	0	0	0	0	0	0	0	0	0

2 Russia

In the early years, noon at Pulkowa Observatory was the signal time at several locations. This was the observatory, near St. Petersburg, that featured in 1845 and 1846 exchanges of chronometers to determine its longitude to high accuracy (Howse, 1997, 117). These exchanges have been described in Chap. 2.

Fig. 14.1 European Russia, Finland, Latvia and Ukraine. (Google Maps)

2.1 *Russia: Baltic Sea*

2.1.1 Kronstadt

The time ball service at Kronstadt was announced in the *Nautical Magazine* (Notice, 1865). The notice is reproduced below, showing that the time ball was then dropped at 1 minute before noon, at noon and at 1 minute after noon at Greenwich, after having been dropped at noon local time almost 2 hours previously.

> The Imperial Ministry of Marine at St Petersburg has given notice, that a Time-ball is dropped from the mast-head of the Naval Telegraph at Kronstat (sic) during the navigation, Sundays excepted.
>
> To call attention, the ball is hoisted midway up the mast about five minutes before noon; and entirely up at half a minute before each time it is to be dropped.
>
> The ball falls successively at one minute before noon; at noon; and at one minute after noon of Greenwich mean time.
>
> The ball will also be dropped at mean noon of Kronstat (sic) every day during the year; being hoisted to the mast-head five minutes before that time.
>
> The position of the Telegraph is given as lat. 59° 59′ 23.9″ N, long. 1h. 59m. 3.3s. E. of Greenwich.

Another notice in the *Nautical Magazine* (Notice, 1867), showed that the signal times had been simplified to one drop at noon, local mean time, and one at noon, GMT. The spelling of Kronstadt was then given as Cronstadt. The stated location of the signal remained the same in Admiralty lists from 1880 onwards.

Fig. 14.2 The Marine Engineering School at Kronstadt. (Courtesy Klaus Hülse collection)

The time ball at Kronstadt was included in the 1880 list. The ball was made of wickerwork with a diameter of 3 feet (0.9 m) and drop height of 9.5 m. It was located on the "Mast of the Submarine Telegraph Office" and was dropped at noon, Kronstadt mean time. A time gun at Kronstadt was added in the 1911 list, fired from "the wall of the Commercial harbour", also at noon. The listed time ball diameter had changed to 5 feet (1.5 m) in 1919, without change to other parameters, and then remained the same. The description of its location was expanded to "Mast of the submarine telegraph office, on marine engineering school". Figure 14.2 shows the Marine Engineering School. The time ball is not obvious and the date of the photograph is uncertain.

The signal time had changed to noon Standard time, 2 hours ahead of GMT, by 1926. It had changed again by 1938, to be 3 hours ahead of GMT.

2.1.2 St Petersburg

The time gun at St Petersburg was included in the 1880 list. St Petersburg was renamed Petrograd in 1914 and then Leningrad from 1924 to 1991, before reverting to the original name. Curiously, the Admiralty list entry was still for Petrograd in 1932. The St Petersburg gun at "Fort Peter and Paul" was initially fired at noon (Pulkowa) mean time. A time ball on "Western side of Nevaki gate" was added in 1923 and dropped at noon, which had changed from Pulkowa time to 2 hours ahead of GMT by 1926. Unlike Kronstadt, the time zone remained the same thereafter.

2.2 *Russia: Black Sea*

2.2.1 Nikolaev

A black time ball at Nikolaev (Nicolaev) Observatory was listed in 1880. It had a small diameter of 0.6 m and a drop height of 3 m and was dropped at noon. The same ball was listed in 1898, together with a time gun that was also fired at noon. Both were still extant in 1947, although noon had been changed to 2 hours ahead of GMT by 1926. Figure 14.3 shows the Observatory.

2.2.2 Odessa

By 1904, there was also a black time ball at Odessa. It had a diameter of 1.2 m and was dropped by hand at local noon. It was located on "Mast on the office of the Russian Steam Navigation Company". It was still extant in 1934, when it was dropped at noon, then 2 hours ahead of GMT.

Two balls at Sevastopol were listed in 1908. One with an unstated colour was on a "Mast near Commander-in-Chief's house. The other, painted black, was on"Tower of the Observatory on Cape Paulovski". Initially, both were dropped at noon, Pulkowa Observatory mean time. Only the Observatory time ball was listed in 1926. It was then dropped at noon, 2 hours ahead of GMT as at other Black Sea locations, accompanied by firing of a time gun which had been introduced in 1924. The time

Fig. 14.3 Nikolaev Observatory. (Courtesy Klaus Hülse collection)

ball was replaced by a light in 1930, which was illuminated 5 minutes before noon and flashed during the last minute before extinction. The light was still extant in 1947, but the time gun was no longer listed in 1938.

2.2.3 Theodosia

A time gun at Theodosia (Feodosiya) was introduced in 1926 and was still extant in 1947. If was fired at noon on Thursdays only.

2.3 Russia: Arctic

2.3.1 Archangel

A black time ball was introduced at the far northern port of Archangel in 1922. It was on the "Mast of Naval depot" and was dropped at noon, 2 hours ahead of GMT. A time light was introduced in 1934 as the primary signal located on "Turret of Ubekoserer building". It was initiated 5 minutes before noon with a code preparatory light signal for 30 seconds, followed by a fixed light for 30 seconds, repeated until 1 minute before the signal. The white light was then illuminated for 45 seconds before being flashed three times before extinction as the noon time signal. A black time ball was specified as the signal should the light be out of order. The same signals were specified in 1947.

2.4 Russia: Sea of Japan

2.4.1 Vladivostok

By 1898, a time ball and gun had been introduced at Vladivostok. The ball was dropped from a mast on the Harbour Board office and a gun in front of the office was fired at noon local time, 8 h 47 m 33 s ahead of GMT. The same times were given in 1911, with a note that the signals were unreliable in 1909. By 1919, it was listed as being on a mast at the Observatory. Its elevation had changed from 50 m to 27 m above the sea. Unusually for Russia, the second Vladivostok time ball was painted red. The gun had also been relocated to a platform below the mast. The ball had a diameter of 1.05 m and was dropped automatically at noon. The time was the same as before. By 1927, the signals were still at noon Standard time, changed to be 9 hours ahead of GMT.

3 Finland

Finland had been part of Russia up to 1917, when it declared independence. Independence was recognised by Russia at the end of that year and by most other countries after the end of World War I. The Admiralty list for 1919, prepared in 1918, still showed Finnish ports as being in Russia.

By 1898 there were time balls at Helsingfors (now Helsinki), Abo and Uleaborg as well as a gun to complement the Helsingfors time ball.

3.1 Helsingfors

The black time ball at Helsingfors was on the "Central tower of Astronomical Observatory. It had a diameter of 1.05m and was dropped at noon Observatory mean time, when the gun on Sketadden Island was fired. By 1908, an "explosive signal" was fired from the Observatory in place of the gun. The explosive signal was no longer listed in 1926. The ball was then dropped at noon, Standard time, 2 hours ahead of GMT.

3.2 Abo

The black time ball at Abo had a large diameter of 1.8 m and was located at the "Navigating School". The signal was made on Mondays only at noon, later 2 hours ahead of GMT The signal was still listed in 1939.

3.3 Uleaborg

The black time ball at Uleaborg had a diameter of 1.05 m. Also at the "Navigating School", it was dropped at noon on Mondays only. It was no longer listed in 1926.

4 Latvia

Latvia declared independence from Russia at the end of World War I, which was lost again during World War II and in the period up to 1991. The Latvian ports of Riga and Libau were listed as being in Russia up to 1919.

4.1 Riga

A time ball at Riga was listed between 1898 and 1919. The black time ball at Riga was located on a mast at the Sailors' Home. It is shown in Fig. 14.4. It had a diameter of 1.5 m and a drop height of 4.5 m. In 1898 it was dropped at 1 pm local mean time. In 1908, it was dropped at 1 pm Pulkowa Observatory mean time, 2 h 1 m 18 s ahead of GMT and 24 m 57 s behind local mean time. Clocks in Riga were set to Pulkowa Observatory time, which was Standard time in 1919. The same drop time was listed in 1919. The signal was discontinued in 1921.

Time lights at Riga were introduced in 1927 and were still listed in 1947. The Admiralty list for 1934 indicated that there were 8 lamps located on a chimney near the Sailors' Home. A proportion of them were illuminated for about I minute before extinction as the signal every "full hour" of GMT.

4.2 Libau

Libau was also shown as Port Emperor Alexander III in Admiralty lists. A time ball and gun at Libau were listed in 1915 and 1919. They had been introduced in 1913. The time ball was on the "Flagstaff of Marine telegraph station". The gun was fired

Fig. 14.4 Riga time ball on the Sailors' Home. (Courtesy Klaus Hülse Collection)

when the ball dropped at noon, Standard time, rather than 1 pm as at Riga. The signals were discontinued in 1921.

References

Howse, D. (1997). *Greenwich time and the longitude*. Philip Wilson Publishers Ltd.
Notice. (1865). Announcement concerning the Kronstadt time ball, issued by the Imperial Ministry of Marine at St. Petersburg. *Nautical Notices in the Nautical Magazine*, August 1865, 440.
Notice. (1867). Time ball at Cronstadt. *Nautical Notices in the Nautical Magazine*, December 1867, 704.

Chapter 15
The Mediterranean

Roger Kinns

Abstract The signals in the Mediterranean are described, excluding the signals at Toulon and Gibraltar which have been described in Chap. 11. They include the signal at Beirut, at the eastern end of the Mediterranean, and the Egyptian signals at Alexandria and Port Said on the coast of North Africa. Several ports before the First World War had been in the Austro-Hungarian Empire.

Many time signals were distributed around the coasts of Italy, Sicily and the Adriatic. Combinations of time balls and guns were used widely. Several unusual signals are described in detail, including an extraordinary shutter signal at Pola, then in Austria. Early time lights at Genoa, Trieste and Venice were followed by another at Taranto in 1933. Some were extinguished at 5 minutes past the hour.

There were early signals in Malta and Croatia, followed by others in Greece and Albania, with a time gun at Durazzo that appeared in the 1947 list.

1 Introduction

The signals in the Mediterranean are described in this chapter, excluding the signals at Toulon in France and Gibraltar in the western Mediterranean, which have been described in Chap. 11. It includes the signal at Beirut, at the eastern end of the Mediterranean, and the Egyptian signals at Alexandria and Port Said on the Mediterranean coast of North Africa.

Table 15.1 shows the evolution of signals in Italy and Sicily, derived from Admiralty lists between 1880 and 1947. Table 15.2 shows their evolution in the eastern Mediterranean and Adriatic, and North Africa, together with Greece and Malta. Figure 15.1 shows their locations. The time light at Taranto is highlighted in red, reflecting the colour of the light itself.

R. Kinns (✉)
The Maudslay Society, Pembroke College, Cambridge, England

R. Kinns (ed.), *The World of Visual Time Signals for Mariners*, Historical & Cultural Astronomy, https://doi.org/10.1007/978-3-031-57334-7_15

Table 15.1 Signals in Italy and Sicily

Country	Location	1880	1898	1904	1911	1915	1919	1922	1928	1934	1939	1947
Italy (including former Austria)	Ancona					TB	TB	TB	TB	TB	TB	
						Gun	Gun	Gun	Gun	Gun	Gun	
	Brindisi					TB	TB	TB	TB			
	Genoa	Gun	Gun	Gun	Gun	Gun	Gun	Gun	Gun	Gun	Gun	
					TB (black)							
						Light	Light	Light	Light	Light	Light	
	Naples			TB	TB	TB	TB	TB				
				Gun	Gun							
	Spezia		Gun	Gun	Gun	Gun	Gun	Gun	Gun	Gun	Gun	
	Taranto: Arsenal				Cones	Cones	Cones	Cones	Cones	Cones		
	Taranto: Fort				Cones	Cones	Cones	Cones	Cones	Cones		
	Taranto		Gun	Gun	Gun	Gun	Gun	Gun	Gun	Gun		
				TB						Light (red)		
	Trieste	TB (black)	TB (black)	TB (black)	TB (black)	TB (black)	TB (black)	TB (black)				
		Gun	Gun	Gun	Gun	Gun	Gun	Gun	Gun	Gun	Gun	
									Light	Light	Light	
	Venice				TB (red)							
	Venice: San Georgio					Light	Light	Light	Light	Light	Light	
	Venice: Naval Station					Light	Light	Light	Light	Light	Light	
Sicily	Catania			TB (black)	TB (black)	TB (black)	TB (black)	TB (black)	TB (black)	TB (black)		TB (black)
				Gun	Gun	Gun	Gun	Gun	Gun	Gun		
	Messina			Disc (b&w)								
						TB	TB	TB	TB	TB	TB	
				Gun		Gun	Gun	Gun	Gun	Gun	Gun	
Signal type totals		1880	1898	1904	1911	1915	1919	1922	1928	1934	1939	1947
Time ball, including collapsible		1	1	4	5	6	6	6	4	3	2	1
Gun, including powder flash		2	4	7	6	7	7	7	7	7	5	0
Lights		0	0	0	0	3	3	3	4	5	4	0
Other: disc, drum, flag, etc.		0	0	1	2	2	2	2	2	2	0	0

2 Italy

Time signals were distributed around the coast of Italy including Trieste which had been in Austria before the First World War. Time in Italy and Austria was based on GMT from 1893 onwards (Howse, 1997). Mid-European time was 1 hour ahead of GMT.

Table 15.2 Signals in the Eastern Mediterranean and Adriatic, and North Africa

Country	Location	1880	1898	1904	1911	1915	1919	1922	1928	1934	1939	1947
Croatia (including former Austria)	Fiume	TB (black)	TB (black)	TB (black)	TB (black)	TB (black)	TB (black)	TB (black)	TB (black)	TB (black)	TB (black)	
		Gun	Gun	Gun	Gun	Gun	Gun	Gun	Gun	Gun	Gun	
	Lussin Piccolo		Discs (black)	TB (black)	TB (black)	TB (black)	TB (black)	TB (black)	TB (black)			
	Pola	TB (black)	TB (black)	TB (black)								
					Shutter	Shutter	Shutter	Shutter				
		Gun	Gun	Gun	Gun	Gun	Gun	Gun				
	Sibenik (Sibenico)					Gun	Gun	Gun	Gun	Gun	Gun	
										Flag		
Albania	Durazzo											Gun
Greece	Piraeus							TB (black)	TB (black)	TB (black)	TB (black)	TB (black)
								Gun	Gun	Gun	Gun	Gun
	Salonika								Gun	Gun	Gun	
Malta	Valetta Palace Tower	TB (black)	TB (black)	TB (black)	TB (black)	TB (black)	TB (black)	TB (black)	TB (black)	TB (black)		
	Valetta Custom House		TB (red)	TB (red)								
	Valetta				Gun	Gun	Gun	Gun				
Lebanon (was Syria)	Beirut					TB	TB	TB	TB	TB	TB	TB
Egypt	Alexandria		TB (black)	TB (black)	TB (black)	TB (black)	TB (black)	TB (black)	TB (black)	TB (black)	TB (black)	TB (black)
			Gun	Gun	Gun	Gun	Gun	Gun	Gun	Gun	Gun	Gun
	Port Said		TB (black)	TB (black)	TB (black)	TB (black)	TB (black)	TB (black)	TB (black)	TB (black)	TB (black)	TB (black)
Signal type totals		1880	1898	1904	1911	1915	1919	1922	1928	1934	1939	1947
Time ball, including collapsible		3	6	7	5	6	6	7	7	6	5	4
Gun, including powder flash		2	3	3	4	5	5	6	5	5	5	3
Lights		0	0	0	0	0	0	0	0	0	0	0
Other: disc, drum, flag, etc.		0	1	0	1	1	1	1	0	1	0	0

A time ball and gun at Trieste were listed in 1880, together with a gun at Genoa. Several unusual signals are described in detail with a brief description of other Italian signals. Unusually, time lights at Genoa, Trieste and Venice were illuminated on the hour and extinguished as the signal at exactly 5 minutes past the hour. The light signal at Taranto, introduced in 1933, followed the usual practice of extinction on the hour.

Fig. 15.1 Signals in the Mediterranean. (Google Maps)

2.1 Genoa

The time gun at Genoa was at Fort Castellaccio, on a hill northeast of the port. It was a precision signal. By 1880, a preparatory flag was hoisted on the Hydrographic Office 5 minutes before the gun was fired. It was hauled down 30 seconds before the gun was fired electrically from the Office at noon Genoa mean time. The same procedure was used in 1898 but the gun was then fired at noon, Mid-European time, 1 hour ahead of GMT. A black time ball at Genoa was included in the 1908 and 1911 Admiralty lists. The apparatus there is known to have used the Devonport principle (Lewis, 1910). It was at the "Semaphore of S. Beningo" and was dropped three times daily, at 10 am, noon and 3 pm. The gun was fired at noon only. The time ball was not included in the 1915 or later lists, but the gun was still listed in 1939.

Electric lights at Genoa were introduced in 1914 and replaced the time ball at San Beningo. The lights were switched on 5 minutes before extinction as the signal at 10.05 am, 12.05 pm and 3.05 pm. The original announcement is shown below (Notice, 1914).

> The Italian Government has given notice that the time ball at the semaphore station on Cape San Benigno, Genoa, Italy, will be replaced by a flash light signal consisting of a group of 8 flashes of 1,000 candlepower each, visible about 4½ miles. The flashes will begin at 22h, 0h, and 3h and will cease at 22h 05m, 0h 05m and 3h. 05m, mean Central European time, corresponding to 21h, 23h and 2h, and 21h 05m, 23h 05m and 2h 05m, Greenwich Mean Time.

The same lights were listed from 1932 to 1939 as at the "Semaphore at Cape Faro", apparently 7 seconds of arc (about 200 m) further south than before.

2.2 *Trieste*

The Trieste time ball was included in the 1880 list. It had a diameter of 0.9 m with a drop height of 7 m. It was located on the NW side of the lighthouse and was dropped at noon, Trieste mean time. A gun was also fired at noon. The same arrangements were listed in 1898, except that noon had changed to Mid-European time. The ball was at the same location in 1908, but the drop height had been reduced to 4 m.

Electric time lights at Trieste had been introduced by 1926 and replaced the time ball. They were located on a "Framework iron column on top of barracks at end of Fratelli Bandiera mole" and switched on 5 minutes before extinction at 12.05 pm. The time gun continued to be fired at noon but was noted as unreliable from 1926 to 1939.

2.3 *Taranto*

2.3.1 Time Gun

The first signal at Taranto was a time gun at St. Angelo Castle, which had been introduced by 1898 and fired at noon. A time ball located on a signal staff at the entrance to the Arsenal was also listed in 1904 and 1908 and dropped at noon. It was no longer listed in 1911 when the signal was made using two black cones at two locations: at the entrance to the Arsenal and at Fort St. Angelo. They were dropped electrically at noon.

2.3.2 Time Light

A red time light at Taranto was added in 1933 and also included in the 1934 list. It was at "Extremity of arm extending from signal mast at Fort St. Angelo semaphore station". It was switched on 3 minutes before extinction at noon.

The Taranto signals had been withdrawn by 1938.

2.4 *Venice*

2.4.1 Time Ball

A red time ball at Patriarcale Observatory in Venice was included in the 1908 and 1911 lists and dropped at noon. It was hoisted 2 minutes before the signal at noon Mid-European time, 1 hour ahead of GMT.

2.4.2 Time Lights

The time ball was discontinued in favour of time lights at two locations in 1913. These were on the "Eastern turret of San Georgio" and at the "Naval Station, highest part of Sylos Factory". They were switched on at noon and extinguished as the signal at 12.05 pm.

2.5 Other Locations in Italy

2.5.1 Naples

A time ball at Naples was listed between 1904 and 1922, accompanied initially by a time gun that was not listed in 1915 onwards. The ball was on a flagstaff on Maschio Angioino. The gun was on the SE bastion of St. Elmo Castle.

2.5.2 Ancona

A time ball and gun were used at Ancona from 1913. The ball was located at the "Semaphore on Mt. Cappuccini". They were still listed in 1939.

2.5.3 Brindisi

A time ball at Brindisi was introduced in 1914 and listed until 1928, but there was no time gun there. It was on the eastern mast of the wireless telegraph station.

2.5.4 Spezia

Spezia used a gun on a "Bastion at root of Lagora mole", fired at noon. It was listed in 1898 and was still extant in 1939.

2.6 Sicily

Listed time signals in Sicily were at Messina and Catania, with first Admiralty list entries in 1904.

2.6.1 Messina

In 1904, Messina used a black and white disc signal, accompanied by a gun. A time ball had been introduced by 1908. The ball was chequered red and white and was located on the tower of the Observatory. Neither signal was listed in 1911. A different time ball at Messina was operational in 1912 and a gun at Fort Gonzanga was again in use. The second ball with an unstated colour was located at the northeast end of Fort San Salvadore. The time ball and gun were still listed in 1939.

2.6.2 Catania

Catania used a black time ball accompanied by time gun which were still listed in 1934, but not in 1939. The ball was located on the roof of the Benedictine convent, shown in Fig. 15.2. A new time ball had been re-established at the Astronomical Observatory by 1947.

2.7 *Eastern Adriatic*

The eastern shore of the Adriatic had been part of the Austro-Hungarian Empire (listed as Austria) and then of Italy after the First World War. Many locations are now in Croatia.

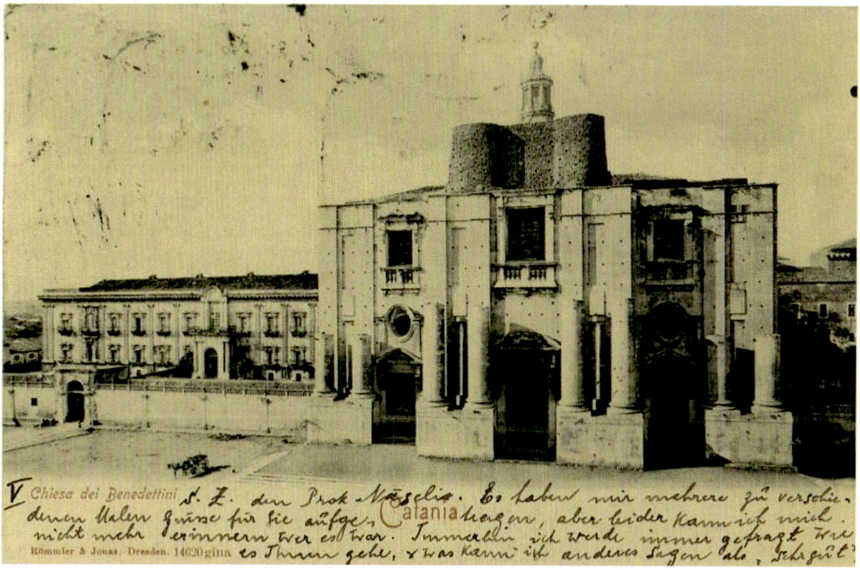

Fig. 15.2 Catania time ball at Benedictine convent. (Courtesy Klaus Hülse Collection)

2.7.1 Fiume

There was a black time ball and gun at Fiume in 1880 which were still listed in 1939. Initially, the black time ball was located on a staff at the inner end of the breakwater. By 1908, it had been relocated to "An iron framework on the top of a low square tower" near the western end of the Maria Teresa mole.

2.7.2 Pola

There was a black time ball at Pola (Pula) in 1880, located on the SW bastion of the Harbour Castle which was changed to a 6 feet square shutter signal in 1910, on the roof of the "Imperial Hydrographic office". The signal was made at 11 am when required. It was made at noon daily, accompanied by firing of a gun at the Harbour Castle. Its extraordinary operating procedure is transcribed below (1919 list).

> The signal at 23h 00m 00s, Standard time is made only when required by vessels of the Austrian Imperial Navy.
>
> The shutter is closed at 22h 45m, and opened by hand, in such a manner that the sky can be seen through the frame of the apparatus, at 23h 00m 00s, Standard time. Ten or more signals will follow the first at intervals of one minute, the shutter remaining open for 10 seconds after each signal. When the series is completed, the shutter will be rapidly opened and closed a number of times.
>
> The signal at Noon, Standard time, is made daily.
>
> The shutter is closed at 5 minutes before the signal, and opened, by hand, at Noon, Standard time.

The Pola signals had been discontinued before 1927.

2.7.3 Lussin Piccolo

The signal at Lussin Piccolo was also unusual. It was listed in 1898 as two black circular discs which were rotated from the vertical to the horizontal position as the signal, so appears to have been similar to signals used at various locations in the Netherlands. The discs were placed vertically "a short time before signal". They rotated to the horizontal at 11 am GMT. Their location was described as "S.W. extreme of the Quay, fronting the public square". The signal description had changed in the 1904 list to "Time ball also Two Black Circular Discs" at the same location. The ball had a drop height of 3 m. This changed again to "Ball, formed by two black vertical discs" in the 1919 and later lists, still at the same location. This later description was probably a clarification of the time ball design, which had been introduced by 1904. The ball was dropped electrically from the Imperial Nautical School.

Either a single time gun or a time gun volley at Fort Santa Anna in Sibenik (Sebenico) was listed between 1915 and 1939. It was fired at noon. An accompanying flag signal was listed for the first time in 1934. It was raised 5 minutes before noon and hauled down when the gun was fired but had been withdrawn by 1937.

2.8 *Albania*

A late introduction was a noon gun at "Torrioni di Durazzo" in Albania. It was introduced in 1939, listed in 1940 and was still extant in 1947.

2.9 *Greece*

2.9.1 Piraeus

A black time ball and a gun at Piraeus were introduced in 1920 and were still listed in 1947. The ball was located on a mast on the lighthouse at the south side of the harbour entrance. It was dropped at 8 am and noon, 2 hours ahead of GMT.

2.9.2 Salonika

Another time gun on a white tower at Salonika had been introduced in 1926. It was fired at noon, Standard time, 2 hours ahead of GMT. It was still listed in 1939, but not in 1947.

3 Malta

A black time ball at the Auberge de Castille in Valetta was listed from 1880 to 1934. It had a small diameter of 0.9 m but a large drop height of 6.5 m at an elevation of over 100 m. The drop height had been further increased to 15 m in 1934. A red time ball at the Custom House was also listed in 1898 and 1904. It had been operational before the end of 1881 and had originally been painted black (Mediterranean-Malta, 1881). It had been withdrawn by 1908. Its diameter, drop height and elevation were 1.4 m, 4.2 m and 18 m. A time gun fired from the Upper Barracca Saluting Battery was listed from 1911 to 1922, but not in earlier lists despite the known existence of a time gun at Valetta before 1880.

4 Syria

Beirut, initially in Syria and now the capital of an independent Lebanon, had a time ball in 1914 that was still listed in 1947. It was located on "Tower on main building of American College".

The ball was dropped at 8 am local mean time, 2 h 21 m 53 s ahead of GMT, having been raised to full height about 5 minutes previously. The same drop time was retained until 1947, when most locations had long changed to Standard times based on time zones.

5 Egypt

Egypt had time balls at Alexandria and Port Said. Both were painted black, established before 1898 and still listed in 1947. They were operated electrically from the observatory in Cairo about 200 km from either time ball. The Alexandria and Port Said time balls, as well as an inland time ball at Cairo, used the Devonport principle (Lewis, 1910).

In 1898, official time throughout Egypt was mean time at the Great Pyramid of Gizeh, 2 h 4 m 30.5 s ahead of GMT. This had been changed by 1908 to Egyptian Standard time, 2 hours ahead of GMT.

5.1 Alexandria

The Alexandria time ball was at the Signal Station, Fort Napoleon (Caffarelli). It is shown in Fig. 15.3. It was 4.5 m above ground and 35 m above the sea. The ball was dropped twice daily throughout its life. The drop times were initially at noon, Gizeh mean time and noon Alexandria mean time, which was 1 h 59 m 43 s ahead of GMT, giving an interval of 4 m 57.5 s between drop times. These times had been changed by 1908 to 10 am GMT and 11 h 0 m 27 s GMT, so the interval was then 1 h 0 m 27 s. These drop times were retained subsequently. A time gun at the fort was fired at the time of the first drop only.

5.2 Port Said

The Port Said time ball was on a lighthouse tower, giving an elevation of 60 m above sea level near the Mediterranean entrance to the Suez Canal. The lighthouse was completed in 1869, shortly before the canal was opened. No time ball is shown in an 1882 engraving (Lighthouse, 1882). A time ball apparatus was ordered from London in 1896 (Gillett & Johnson, 1896). It is likely have been erected in 1896 or 1897, not long before preparation of the 1898 Admiralty list. The ball had a diameter of 4½ feet (1.35 m) and a drop height of 3 m, corresponding to 2.2 ball diameters. It was supported on an "Iron lattice work mast 30 feet high on top of high lighthouse" and was dropped three times each day from Cairo. Figure 15.4 shows the time ball during its period of operation.

Fig. 15.3 Alexandria time ball. (Courtesy Klaus Hülse Collection)

Fig. 15.4 Port Said time ball. (Author's collection)

In 1898, the time ball at Port Said was dropped at 8 am, noon and 4 pm Port Said mean time, which was 2 h 9 m 15 s ahead of GMT. By 1908, the drop times had been changed to 8 am, noon and 4 pm Standard time and remained the same thereafter.

References

Gillett, & Johnston. (1896). *Ledger entry* showing an order for Port Said on 1 August 1896.

Howse, D. (1997). *Greenwich time and the longitude*. Philip Wilson Publishers Ltd.

Lewis, T. (1910). *Notes for the Astronomer Royal dated 21–25 April 1910*. Royal Greenwich Observatory Archives RGO 7 257, Papers of William Christie, Cambridge University Library.

Lighthouse and breakwater at the entrance of the Suez Canal, Port Said, (1882). *The Illustrated London News*, 22 July 1882.

Mediterranean-Malta. (1881). Notice published in *Lloyd's List,* 17 December 1881, Erection of a time ball on top of the Port Office, Great Harbour, Valetta.

Notice. (1914). Mediterranean Sea- Italy – Genoa – Time signal to be changed. *Nautical Magazine*, 24.

Chapter 16
South Africa

Roger Kinns

Abstract South Africa used an extraordinary range of time signals, including the first shuttered lamp in 1823, one of the earliest time balls and a wide range of subsequent types. A time gun, later a precision signal, was introduced in 1807. The 1836 time ball at the Cape Observatory was constructed locally and supplemented by a repeater ball in 1853. The Observatory ball was replaced in 1863 and a time ball at Alred Docks became the principal signal for Table Bay in 1873. A time ball at Simon's Bay was added in 1857, later replaced by a disc and then by another time ball in 1898. Additional time signals were provided at Port Elizabeth, Port Alfred and East London. The first 1873 time ball at the Docks was replaced in 1894. A time ball at Durban in Natal was established in 1883 and relocated in 1904. All were ultimately controlled from the Cape Observatory. The Docks time ball was restored in 1997.

1 Introduction

This chapter covers what is now the Republic of South Africa and is based on a recent paper by the author (Kinns, 2021). It builds on extensive previous research into the development of time signals in South Africa, especially those controlled by the Cape Observatory. Apparent contradictions between different sources have been resolved as far as possible.

The earliest known precision time signal for mariners was a shuttered lamp in about 1823 at the Cape of Good Hope. The Cape Observatory offered one of the earliest time ball services from 1836. Later time balls were distributed around the coast from Cape Town to Durban in Natal.

Table 16.1 shows entries for South Africa in Admiralty lists between 1880 and 1947. Figure 16.1 shows their locations.

R. Kinns (✉)
The Maudslay Society, Pembroke College, Cambridge, England

R. Kinns (ed.), *The World of Visual Time Signals for Mariners*, Historical & Cultural Astronomy, https://doi.org/10.1007/978-3-031-57334-7_16

Table 16.1 Signals in South Africa

Area	Location	1880	1898	1904	1911	1915	1919	1922	1928	1934	1939	1947
South Africa	Cape Town: Alfred Docks	TB	TB	TB	TB	TB	TB	TB	TB	TB		
	Cape Town	Gun	Gun	Gun	Gun	Gun	Gun	Gun	Gun	Gun	Gun	Gun
	Durban: First location		TB	TB								
	Durban: Bluff lighthouse				TB	TB	TB	TB	TB			
	East London		TB	TB	TB	TB	TB	TB	TB	TB	TB	
	Port Alfred	TB	TB	TB								
	Port Elizabeth	Disc (black)	TB (black)	TB (black)	TB (black)	TB (black)	Disc (black)	Disc (black)	Disc (black)			
	Simon's Bay	Disc	Disc									
				TB (white)	TB (b&w)	TB (b&w)	TB (b&w)	TB (b&w)	TB (b&w)			
Signal type totals		1880	1898	1904	1911	1915	1919	1922	1928	1934	1939	1947
Time ball, including collapsible		2	5	6	5	5	4	4	4	2	1	0
Gun, including powder flash		1	1	1	1	1	1	1	1	1	1	1
Lights		0	0	0	0	0	0	0	0	0	0	0
Other: disc, drum, flag, etc.		2	1	0	0	0	1	1	1	0	0	0

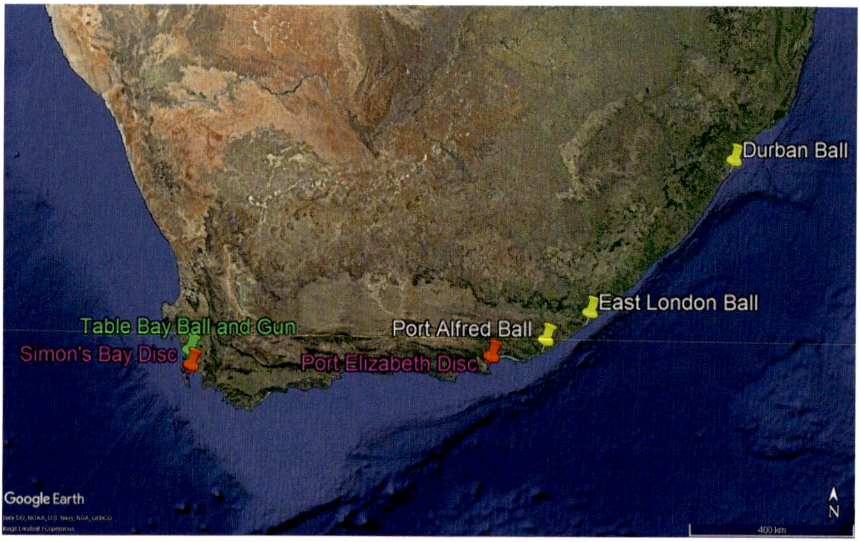

Fig. 16.1 Signals in South Africa. (Google Maps)

2 Signals Under Cape Observatory Control

Table 16.2 showing time signal development is derived from a summary by Evans (1993) for the period up to 1934, with references to research by Harding (1971), Warner (1979) and Bisset (1984). Entries are those by Evans with minor adjustments and inserted clarification in italics within square brackets. Entries that are not directly related to time signals for mariners have been excluded.

Table 16.2 Development of Time Signals controlled by the Cape Observatory. (From Evans, 1993)

Date	Event	Reference
1807	Noon Gun fired from Imhoff Battery, Cape Town Castle. Used to rate ships chronometers.	Bisset, 1984.
1821	Instruments for time determination erected at the Observatory. [*A temporary hut was used to house portable instruments that included a transit telescope. This allowed astronomical observations prior to erection of permanent observatory buildings. The new observatory was operational in 1829.*]	Harding, 1971
4-Jan-1833	Flash pistol and powder magazine purchased for visually signalling time.	Harding, 1971
30-Sep-1836	Time-Ball erected to S E of the Observatory	Warner, 1979
Oct-1853	Observatory time ball not visible from whole of Table Bay. Repeating time-ball on Lion's Rump	Warner, 1979
1857	Time-Ball in Simons Town. A portable transit instrument determined the time at which to drop the time ball.	Warner, 1979
21-May-1860	Observatory time-ball hidden by trees, re-located N.	Warner, 1979
Sep-1861	Electric release of 3 time-balls from Observatory.	Warner, 1979
Jun-1863	Observatory time-ball once again moved.	Warner, 1979
1864	Gun fired electrically from Royal Observatory.	Bisset, 1984
1873	Return signal from Port Elizabeth time ball 0.3 to 0.6 sec after trigger signal sent. [*the return telegraph signal indicated that the ball had actually been released*]	Royal Observatory, 1873
1877	Time distribution by telegraph.	Royal Observatory, 1877
1878	Noon ball dropped at docks for shipping. 13:00:00: Time Balls dropped at: - Observatory, Simons Town, Port Elizabeth, Kimberley.	Royal Observatory, 1878
21-Jul-1889	Electric time signals: Noon signal drops time ball at docks. 13:00:00: Time Ball dropped at Simon's Bay; Time Ball dropped at Port Elizabeth; Time Ball dropped at East London; Gun fired at Imhoff battery. Government telegraphic system distributed a time signal for use in the Cape colony, Orange Free State and Transvaal.	Report of H.M. Astronomer at the Cape of Good Hope for the period 26 May 1879 to 21 July 1889
7-Feb-1892	Arrangements made for changing civil time of colony. Previously Observatory mean time was used for telegraphic purposes throughout the colony and different railway systems used local time of their main station. At junction of E and W railway systems it was decided to adopt 22.5 degrees E as longitude for all time purposes. This worked so well that that the uniform time of the Cape Colony was adopted by both the Transvaal and the Orange Free State.	Report of HMA, 1889 to 1892
8-Feb-1892	Time signals at Observatory mean time noon and 1pm discontinued in favour of a single signal at Greenwich mean noon which corresponds to 13 30 00 of Uniform South African time.	
1893	Daily signals from Observatory at GMT noon: Disc dropped at Simon's Bay for shipping. Gun fired at Imhoff Battery for shipping and town. Time ball dropped at docks in Cape Town. Disc dropped at Lighthouse in Port Elizabeth. Ball dropped at East London. Ball dropped at Port Alfred. Signal to all telegraph offices in Cape, Orange Free State and Transvaal. A clock in the Harbour Tower at the Cape Town docks is electrically controlled from the observatory.	Report of HMA, 1893
1894	Daily signals from the Observatory at GMT noon: - Same as 1893. New Time-Ball erected in a conspicuous position near Resident Engineer's Office of the Cape Town Docks.	Report of HMA, 1894
1895	Daily signals from the Observatory at GMT noon: - Same as 1894.	Report of HMA, 1895
20-Mar-1895	Two return signals from Cape Town time ball at the instant the ball begins to fall at the bottom of its drop. [*The return telegraph signals indicated that the ball had been released and arrived at its rest position*]	

(continued)

Table 16.2 (continued)

Date	Event	Reference
1896	Daily signals from the Observatory at GMT noon: - Same as 1895. Simon's Bay disc failed often from old age.	Report of HMA, 1896
1897	Daily signals from the Observatory at GMT noon: - Same as 1896. Simon's Bay disc dropped up to 11 Dec 1897	Report of HMA, 1897
1-Jan-1898	Simon's Bay disc being replaced by time-ball.	
2-Feb-1898 to 9-Feb-1898	Cape Town gun under repair.	Report of HMA, 1898
14-Apr-1898	Simon's Bay time-ball operational.	
1898	Daily signals from the Observatory at GMT noon: - Same as 1897 except for the Port Alfred ball which was discontinued. Telegraphic signals also go to Rhodesia [*now Zimbabwe*].	
18-Dec-1900	Cape Town time-ball position: - 74.83 ft above l.w.o.s.t. 30 ft above ground level (ball drops 10 ft) Latitude 33° 54' 24"S. Longitude 18° 25' 10.4"E	SAAO Archives No A0035: Chronometers and Time Signals 1900-1906.
1901	Daily signals from the Observatory at GMT noon: - Same as 1898. Failures at Port Elizabeth and East London partly due to interruption of the lines by the Boers.	Report of HMA, 1901
1-Nov-1901	Hourly signal sent to Wynberg station - set clocks every hour.	
1902	Daily signals from the Observatory at GMT noon: - Same as 1901.	Report of HMA, 1902
4-Aug-1902	Noon Gun to Lion Battery on Signal Hill. Gun now fired from Signal Hill battery.	
28-Feb-1903 Midnight	The whole of South Africa adopts longitude 30 degrees E for time determination. That is 2 hrs ahead of Greenwich time.	Government Gazette, February 1903
1904	Second floor added to time-ball building.	Government Gazette 27 August 1982
1914	Daily signals from the Observatory at 10 00 00 GMT: Ball dropped at Simon's Bay for shipping. Gun fired from battery on Signal Hill. Time ball dropped at docks in Cape Town. Disc dropped at Lighthouse in Port Elizabeth. Ball dropped at East London.	Report of HMA, 1914
1-Sep-1914	Daily transmission of wireless time signal for shipping on Union Government wireless station at Slangkop.	

Table 16.2 (continued)

Date	Event	Reference
06-Feb-1925	Broadcast of time but Noon Gun retained	Bisset, 1984
1925	Daily signals from the Observatory at 10 00 00 GMT: Ball dropped at Simonstown dockyard for shipping. Gun fired from battery on Signal Hill. Time ball dropped at docks in Cape Town. Disc dropped at Lighthouse in Port Elizabeth. Ball dropped at East London. Time signal sent to telegraphic system and railway services. A clock in the Harbour Tower at the Cape Town docks is electrically controlled from the observatory. Daily transmission of wireless time signal for shipping on Union Government wireless station at Slangkop.	Report of HMA, 1925
6-Feb-1925	Time signal from the Observatory consisting of 4 dashes, the commencement of the last being on the hour, was transmitted by Local Broadcasting Association 3 times per day.	Report of HMA, 1925
1926	Daily signals from the Observatory at 10 00 00 GMT: Ball dropped at Simonstown dockyard for shipping. Gun fired from battery on Signal Hill. Time ball dropped at docks in Cape Town. Disc dropped at Lighthouse in Port Elizabeth. Ball dropped at East London. Time signal sent to telegraphic system and railway services. A clock in the Harbour Tower at the Cape Town docks is electricall y controlled from the observatory. Daily transmission of wireless time signal for shipping on Union Government wireless station at Slangkop.	Report of HMA, 1926
1927	Daily signals from the Observatory at 10 00 00 GMT: - Same as 1926.	Report of HMA, 1927
1928	Daily signals from the Observatory at 10 00 00 GMT: - Same as 1927.	Report of HMA, 1928
Aug-1928	Automatic control of clock in Harbour Tower discontinued as Harbour Administration considered it unnecessary.	
15-Sep-1928	Belin apparatus to send international (ONOGO) time signals installed. Transmission from Slangkop once a day. Local Broadcasting Association transmits last group of ONOGO signal (6-pips) at 1500 and 1800 GMT. Beginning of 6th pip is on the hour.	
1929	Daily signals from the Observatory at 10 00 00 GMT: - Same as 1928.	Report of HMA, 1929
8-Feb-1929	Simonstown time-ball dismantled.	
30-Sep-1930	Disc in Port Elizabeth dismantled.	Report of HMA, 1930
1933	Daily signals from the Observatory at 10 00 00 GMT: Gun fired from battery on Signal Hill. Time ball dropped at docks in Cape Town. Time signal sent to telegraphic system and railway services. Time signal also received at meteorological stations in Bulawayo and Salisbury. Daily transmission of wireless time signal for shipping on Union Government wireless station at Slangkop.	Report of HMA, 1933
31-Oct-1933	6-pip signal transmitted by African Broadcasting Company changed from 1700 and 2000 SAST to 1300, 1700, 1800, 1900, 2000, 2100 SAST.	
1934	Daily signals from the Observatory at 12 00 00 SAST: Gun fired from battery on Signal Hill.	Report of HMA, 1934
1-Feb-1934	Time ball at docks in Cape Town discontinued. Time signal sent to telegraphic system and railway services. Time signal also received at meteorological stations in Bulawayo and Salisbury. Daily transmission of wireless time signal for shipping on Union Government wireless station at Slangkop.	
7-Sep-1934	Removal of the Time Ball at the Docks. The ball was made in 1932 from thin steel plate.	The *Cape Times*, 7 September 1934

The paper by Harding has not been found, but the entries have been checked using other sources. Evans referred to reports by HM Astronomer at the Cape of Good Hope, which are shown in the table as reports of HMA with the year to which they refer. There are a few references to time signals at the Royal Observatory in *Monthly Notices of the Royal Astronomical Society*. These are included separately in the main reference list at the end of this paper. Additional references are cited in the form used by Evans.

2.1 Noon Gun in Cape Town

By 1807 a noon gun was fired from the Imhoff Battery on the seaward side of the Castle in Cape Town. It was one of the means by which ships in Table Bay "could determine the error and rate of their chronometers" (Warner, 1979, 47). The accuracy of the gun signal is likely to have been modest in its early years of operation, as it was not controlled from an astronomical observatory. It would have had high precision from 1864 onwards, when it was fired electrically from the Observatory.

The history of time guns in Cape Town was explored by Bisset (1984), using Army records. There were various changes to the location and time of the gun signal during the nineteenth and twentieth centuries which have been summarised by Evans (1993).

In February 1892, time signals at noon and 1 pm were discontinued in favour of a single signal at noon, Greenwich mean time. The gun was moved from the Imhoff Battery to the Lion Battery on Signal Hill in August 1902. From midnight on 28 February 1903, the whole of South Africa adopted a time zone based on longitude 30° E, two hours ahead of Greenwich. The gun was then fired at noon, South African Standard Time. The gun signal continued "at the request of the Municipality of Cape Town" after time balls and time discs had been withdrawn (Bisset, 1984, 67; Warner, 1979, 115).

The noon gun is still operating in 2023 and is a popular tourist attraction. Figure 16.2 shows the gun being fired on 4 November 1993 for the first time after the circuits had been modernised. The gun had been fired automatically since 1864,

Fig. 16.2 The Noon Gun being fired on Signal Hill in 1993. (Photograph by Willie Koorts)

but the Lion Battery site on Signal Hill was not open to the public until 1997. Evans (1993) noted that the report of the gun can be heard at the Observatory about 18.5 s after detonation.

The guns are 18-pounder muzzle loaders bearing the monogram of King George III, cast in the Bahamas (Bisset, 1984, 69). Several guns of this type have been used for signalling time and two were later operated in parallel, with one acting in reserve should the other fail to fire.

2.2 *The First Directors of the Cape Observatory*

In *A history and description of the Royal observatory, Cape of Good Hope*, Gill (1913) gives biographies of the Directors who preceded him. The following short summaries are derived largely from that book. Sir David Gill (1843–1914) is highly regarded for his pioneering work in astronomy. He was Director of the Cape Observatory from 1877 to 1906 and was knighted in 1900 (Gill, Obituary, 1914).

2.2.1 Fearon Fallows (1820–1831)

The Reverend Fearon Fallows (1788–1831) was appointed as the first director of the Observatory by the British Board of Longitude in October 1820. His year of birth had been given as 1789, but this was corrected after research by Warner (1997). He served at the Cape from August 1821 until his premature death in July 1831 but never received timely support from the Board or Admiralty for development of his staff and facilities. Conditions were primitive and undoubtedly contributed to his early death. His life and work have been described by Cameron-Swan (1931).

Soon after his arrival at the Cape "whilst he hunted for a site in conformity with Admiralty requirements, Fallows obtained from the local Government a settler's wooden hut, which he converted into a temporary observatory for his portable instruments." (Cameron-Swan, 1931, 6). The site for the new observatory was approved in July 1822, but he did not receive authority to implement his plans until December 1824. The Observatory buildings had been erected by the end of 1827 but the instruments were not fully operational until 1829.

2.2.2 Thomas Henderson (1831–1833)

Thomas Henderson (1798–1844) was appointed to succeed Fallows in October 1831 and arrived at the Cape in April 1832. He resigned in May 1833, in protest at the conditions under which he had to operate. He returned to Edinburgh, where he later became the first Astronomer Royal for Scotland.

2.2.3 Thomas Maclear (1833–1870)

Thomas Maclear (1794–1879) was appointed to succeed Henderson in July 1833 and arrived at the Cape in January 1834. Like Fallows and Henderson, he found that working conditions were primitive and he too suffered from lack of staff to support his work. During his long tenure, the Observatory was expanded with the erection of new buildings and additional instruments.

From 1834 to 1838 Maclear worked closely with Sir John Herschel at the Cape. Many of their exchanges are recorded in *Herschel at the Cape* (Evans et al., 1969), which contains Herschel's edited diaries and correspondence during an extraordinarily productive period. Robert Wauchope visited Maclear and Herschel on several occasions. Herschel also corresponded with John Augustus Lloyd in Mauritius. A diary entry on 12 March 1835 (Evans et al., 1969, 151) "Packed up & sent in Captn Loyds (*sic*) papers" is probably a reference to the paper about the first observatory in Mauritius that was communicated to the Royal Astronomical Society by Herschel and presented in April 1836 (Herschel, 1836). Charles Piazzi Smyth (1819–1900), later to succeed Henderson as the second Astronomer Royal for Scotland, became Maclear's assistant in 1835 at the age of sixteen. Smyth's descriptions of the Edinburgh time ball arrangement were rooted in his experience at the Cape (Kinns, 2011).

To quote Gill: "In the days of Maclear, a time ball was erected near the Observatory which was dropped in lieu of the pistol signal" (Gill, 1913, 143). This 1836 time ball was supplemented by a repeater ball on Lion's Rump (now called Signal Hill) in October 1853, moved a short distance in May 1860 and replaced by a new apparatus in June 1863. Other signals were distributed around the coast of South Africa during Maclear's tenure, with electrical control from the Cape Observatory.

2.3 *The First Visual Signals at the Cape*

The following statement was written by the editor of the *Nautical Magazine* in 1835 as an introduction to the time ball service at St Helena that started in January 1834 (Editorial, 1835, 658). It provoked a reaction from Robert Wauchope that led to acceptance of his claim to invention of visual time signals (Wauchope, 1836).

> … The plan of communicating time by signal from observations, being coeval with the improvement of chronometers, is of recent date. The advantages of it are great to seamen, and it has been a matter of some surprise to us, that even within the few last years it has not been more generally adopted. We remember it to have been employed successfully by the Rev. Mr. Fallows, when he was astronomer at the Cape of Good Hope, about the year 1820. His plan was to eclipse a light at the moment of eight o'clock (mean time) by means of a shutter. The light was distinctly seen by the shipping in the roads, and the officers being on the look-out, were enabled to obtain a rate for the chronometers on board. …

The article also mentioned rocket signals proposed by Captain Owen before going on to describe the St Helena arrangement. Wauchope's response did not contradict the statement about the shuttered lamp (Wauchope, 1836).

> I had pointed out the advantages to be derived from the plan for communicating time by means of telegraphs so far back as 1818, in my remark-book transmitted to the Admiralty when in command of the Eurydice, at that time on the Cape and St. Helena station. Sir Jahl. Brenton was then naval commissioner at the Cape, and an extract from a letter of his, dated 15th November, 1833, to Mrs. Wauchope, will, I think, establish my prior claim to the invention, before the Rev. Mr. Fallows, (who, it appears, was astronomer at the Cape in 1820), or Captain William Owen, R.N.

This statement suggests that Wauchope had never met Fallows, who arrived at the Cape in 1821 and died in 1831.

2.3.1 Fallows' Shuttered Lamp

The shuttered lamp was an important development. A key feature of the signal would have been the preparatory opening of a shutter to alert ships: closure, which made the lamp disappear, was the exact time signal. Fallows is likely to have used an Argand lamp, which was much brighter than an ordinary oil lamp but was still dim in relation to modern lighting. It was almost another century before electric lights could be used to provide a bright and accurate signal, again using the moment of extinction as the exact time. Its initial date of operation is likely to have been in 1823 (Bartky & Dick, 1981): Fallows did not arrive at the Cape until August 1821. Curiously, this lamp was not mentioned by Gill (1913), suggesting that early records at the Cape Observatory had been lost. The same applies to any discussion at the Cape concerning Robert Wauchope's ideas for visual signals. Instead, Gill referred to a pistol signal and an unsophisticated time ball that is probably a myth.

2.3.2 Mythical Time Ball at the Cape

The following statement suggests that the very first time ball may have been at the Cape (Gill, 1913, 143).

> In the days of Fallows and Henderson, the Astronomer, a few minutes before the appointed hour, ascended to the roof of the Observatory, taking with him a chronometer (of which he had previously determined the error) and a large brass-barrelled pistol. This ungainly weapon is still preserved as an interesting relic. When the second hand of the chronometer reached the appointed instant the pistol was discharged and its flash was observed by a signalman provided with a telescope, and he, by means of a rope attached to his foot, dropped a time ball in the neighbourhood of the Bay.

The same statement was repeated by Cameron-Swan (1931, 13). There is, however, serious doubt about its veracity. The signal would not have been possible until after the Observatory buildings had been erected in 1827 but could have preceded full operation of the Observatory in 1829. If the pistol was fired at night, the flash could

Fig. 16.3 The 1833 Flash Pistol used by Henderson in 1833. (Courtesy Ian Glass)

have been seen by ships while a time ball would have been invisible. Warner (1979) could not find explicit evidence to support Gill's statement and concluded that the time ball was a myth. Bartky and Dick (1981) concurred.

2.3.3 1833 Flash Pistol

There is a January 1833 note that "Flash pistol and powder magazine purchased for visually signalling time" (Harding, 1971). The date is after Fallows had died and not long before Henderson left the Cape. "Early in 1833 Henderson started a new time service … a brass barrel percussion pistol for the making of night signals to vessels in Table Bay, for the regulation of their chronometers … with this gun and a pocket chronometer, Henderson each night climbed onto the roof of the Observatory and fired a charge of black powder at an advertised time. …" (Warner, 1979, 32). The 1833 pistol is shown in Fig. 16.3.

3 Cape Observatory Time Balls

3.1 1836 Time Ball

An Observatory time ball was in operation from October 1836 (Bartky & Dick, 1981), having been erected on 30 September. The apparatus was simpler than the 1833 Greenwich apparatus but used the same basic principles.

An 1852 description of the first time ball is shown below (Maclear, 1852). According to this description, the location had not been changed since its erection in 1836 and high accuracy was achieved. The ball diameter was 5 feet, as at Greenwich. The 1852 notice suggested that it would have to be moved because of new construction that interrupted its line of sight from Table Bay. A "repeater" time ball was constructed on Lion's Rump in July 1853. This repeater ball was initially under manual control and dropped when the Observatory time ball was seen to fall.

> The signal ball is five feet in diameter. It slides upon a rope attached to an arm projecting from the flag-staff at the height of 45 feet from the ground, and commands the outer anchorage in clear weather; but close in, particularly from the decks of small craft, the line of sight is interrupted by windmills and houses.
>
> No alteration has been made in the position of the flag-staff since it was first established; which position is the least objectionable to the sweep of the astronomical instruments, compatible with the command of the harbour. Otherwise, the staff might be placed to the north-west of the Observatory, so as to almost entirely clear the windmills and houses before-mentioned. The removal to the north-west is now under consideration; but it is proper to remark, that no arrangement at the Observatory can obviate the relative interruptions caused by the masts and rigging of a group of vessels at the anchorage. This could only be met by a less distant, and particularly by a more elevated, position for the signal ball than the ground about the Observatory can furnish.
>
> The signal is made daily, Sundays and Good Fridays excepted, (and contingent exceptions that will be noticed presently,) at one o'clock, Cape mean time, corresponding to 11 h. 46 m. 5 s. Greenwich time: and its precision is tested by an observer at the Normal mean time clock. An examination of the registered times of drop, since the beginning of this year, shows that the probable error has been reduced from about two-tenths of a second in time, to one-tenth part of a second in time. There is, therefore, an equal chance that the signal may happen one-tenth of a second too early or too late. This degree of accuracy is more than sufficient for the wants of the seaman, whose ordinary method of noting time signals admits of no such precision.
>
> Furthermore, with regard to the fountain from which the time for signal is derived, it may be well to record, that when the sky is clear, the error of the transit clock (which checks the Normal clock) is determined by astronomical observation as near the time of signal as is practicable.

The notice was extended with further observations concerning time ball reliability. Robert Wauchope pointed out in 1852 correspondence that the original Cape time ball had used his design for the trigger arrangement (Wauchope, 1852).

Figure 16.4 shows a detail from a camera lucida drawing that was made by Sir John Herschel in January 1837, not long after the time ball was first erected. The

Fig. 16.4 Detail from a January 1837 drawing by Sir John Herschel. (Courtesy Willie Koorts)

complete drawing is included in *Herschel at the Cape* (Evans et al., 1969: Plate 12). Herschel's diary entry for 2 January 1837 headed "Observatory" reads "Rose at 7.–Took sketch of the building from across the Salt River. …".

The time ball and mast can be seen to the left of the Observatory. The diameter of the ball appears to be larger than the 5 feet (1.5 m) noted by Maclear (1852), but it is possible that the first time balls at the Observatory had a larger diameter than later. Large balls, probably made of wickerwork and covered in black-painted canvas, would have been difficult to manage in high winds and prone to damage. Smyth (1853b) noted the frequent self-destruction of early time balls at the Cape.

The Observatory time ball was relocated north on 21 May 1860 because it had become hidden by trees (Evans, 1993).

3.2 *1853 Lion's Rump Time Ball*

Details of the time ball at Lion's Rump, now known as Signal Hill, were issued on 13 December 1853 (Maclear, 1853). The same notice was published in the 1854 Nautical Magazine and is transcribed below. It points out that the signal was available from 14 October 1853 and that observers should subtract one second from the time of ball release to give the exact time of 1 pm.

> Mr Thomas Maclear, the Royal Astronomer at the Cape of Good Hope, has given notice, that the Time Ball attached to the Cape Observatory, not being generally visible by the Shipping on the eastern side of Table Bay, owing to the intervention of buildings, another has been established by the liberality of the Colonial Government, at the opposite side of the Bay. It stands on the Lions Rump, and commenced work on the 14th of last October.
>
> The Observer should note the time by his chronometer when this Ball *begins* to fall; and by subtracting one second from that time, he will have the moment of One o'clock p.m. by mean time at the Cape Observatory.
>
> The Cape Observatory stands in latitude 33° 56′ 3″ S., and longitude 18° 28′ 45″ or 1 h 13 m 55 s E. of Greenwich.

In a letter to Airy, the Astronomer Royal at Greenwich, Maclear noted that "an order for a trigger to drop the lever arm ball on the Lion's Rump was given by the Colonial Government" during 1861 (Maclear, 1863b). This suggests that the Lion's Rump ball was then dropped electrically. A "lever arm ball" would be a possible description of the arrangement that was used at Port Elizabeth (see later).

3.2.1 Development of Time Ball Apparatus

In the early years of operation of the 1836 apparatus, the ball had to be replaced on several occasions. This would have been witnessed by Charles Piazzi Smyth, when he was assistant to Thomas Maclear at the Cape. The following comment relates to the selection of the time ball apparatus for Edinburgh (Smyth, 1853a).

> The earliest signal-balls which were made, though provided with ropes passing over pulleys by which they were enabled in their descent to raise a series of weights in order to check in

> a gradual manner the velocity of their fall, were yet invariably found, after a short time, to pull or to smash themselves to pieces. Steel springs were next tried to break the force of the concussion, but were pretty sure to be themselves snapped with a heavy ball, while a light one would not descend quick enough on a windy day. Recourse was finally had to compressed air, a spring of perfect temper never injured by time …

He made similar observations in a presentation to the Royal Society of Edinburgh, in December 1853, which referred specifically to his experience at the Cape (Smyth, 1853b).

> The author had several years' personal experience within 1837 and 1845 with this ball or balls, for several were made, and literally used up, so difficult was it found, with mere simple workmanship, to secure the perfect action, which Mr Field, of the firm of Maudslay and Field, had obtained by the adoption of a cylinder of compressed air to break the fall of the ball's descent.

An arrangement using an air cylinder was more expensive than other alternatives, but its reliability benefits led it being used by many other manufacturers from the 1850s onwards. Many of these systems were imported from England.

Table 16.2 indicates that three time balls were released electrically from 1861. One was at the Observatory and one was at Simon's Bay. The third was probably the repeater ball at Lion's Rump.

3.2.2 1863 Observatory Time Ball

The 1863 time ball apparatus was described in a notice published by Sir Thomas Maclear, which is reproduced in Fig. 16.5. The notice pointed out that the new apparatus was erected about 45 m SW of the original. Its design was generally similar to that used at Edinburgh, Deal and Sydney, with a rack and pinion mechanism for hoisting and a cylinder of air to arrest the fall of the ball. The apparatus was supplied by Sandys & Co. in London, not by Maudslay, Sons & Field. It featured in two letters from Maclear to Airy, which pointed out defects in the delivered system that had to be rectified before the apparatus could be used and suggested that a similar apparatus for New Zealand should be inspected in London (Maclear, 1863a, b). The other apparatus was erected at Wellington, NZ and provided the first time ball service in New Zealand from 1864 (Kinns, 2017). It was designed to give a drop height of 18 feet (5.4 m), which was unusually large for a rack and pinion hoisting arrangement and required strengthening of the mast to allow operation in a windy environment (Maclear, 1863a).

The notice in Fig. 16.5 shows that the time ball was raised by 10 feet at 5 min before the signal, paused there for about 2 min and then raised to its maximum height. The 1863 notice indicated that the ball was painted red with a black central band and had a diameter of 5½ feet (1.65 m).

Figure 16.6 shows the Observatory time ball after 1863. The locations of stays for the mast in Fig. 16.6 suggest that it was decided to reduce the drop height from 18 feet to a more usual 10 feet (3 m) after initial service. The mast cross-section differs from that described by Maclear (1863a), indicating that the hoisting

NEW TIME-BALL AT THE ROYAL OBSERVATORY.
Cape of Good Hope

A new Time-ball, with elaborate machinery, has been erected at a spot distant about fifty yards south-west of the old signal-mast.

The Ball is five and a half feet in diameter, the upper and lower thirds of which are painted red, the middle third black. It slides upon a slotted Shaft, within which works a rack-edged gun-metal Rod, armed with a metallic Piston.

The Piston, when descending, plunges into an iron tube, where the contained air serves by its elasticity to break the shock of the fall.

The weight of the Ball, Rod, and Piston is about three hundred and sixty pounds. One hundred and sixty turns of the Windlass raises the Ball for signal, where it is supported by the catch of an Electro-magnetic Trigger, until the instant of one o'clock, Cape Observatory mean time, at which instant a Contact Lever attached to the train of the Mean Time Clock closes the galvanic circuit, and the Ball falls.

The Ball is raised ten feet above the top of the Tower at five minutes before one o'clock, where it rests about two minutes before raising it to the top of the Shaft. One o'clock, Cape mean time is coincident with $1^h\ 28^m\ 35^s$ Port Elizabeth time, and with $11^h\ 46^m\ 5^s$ a.m. Greenwich time.

The Port Elizabeth time is in terms of the meridian passing through the Elizabeth Monument on the hill, as determined by the triangulation of Captain Bailey, R.E.

THOS. MACLEAR.

Fig. 16.5 Notice of a new time ball at the Cape in 1863 (RGO 6615)

arrangement had been changed in other respects. The original would have had the square cross-section used at Wellington, NZ (see Chap. 23).

Neither the Observatory nor Lion's Rump time balls featured in the first (1880) or subsequent Admiralty lists. Table 16.2 shows that the time ball at the Observatory was still operating in 1878, or at least in 1877 when the report is likely to have been prepared (Royal Observatory, 1878). Time balls were also in operation at Simon's Town and Port Elizabeth, with an inland time ball at Kimberley. These were dropped at 1 pm, when the gun was fired. A ball at the Docks was dropped at noon, accompanied by a privately operated time ball.

> … Time-balls are dropped at one o'clock each day at the Observatory, Simon's Town, Port Elizabeth, and Kimberly, and a gun is fired in Cape Town. At noon a ball is dropped in the Cape Town Docks for the use of the shipping, and a private ball, erected by a Cape Town clockmaker, is dropped by a secondary current. A clock has been established at Port

Fig. 16.6 The Observatory time ball after 1863. (Courtesy Ian Glass)

> Elizabeth, which is regulated daily after a comparison with the mean solar clock of the Observatory. These arrangements have generally worked well, and through them accurate Cape mean time is distributed over a large portion of the Colony. In long circuits, like those in use, failures will occasionally take place through breakage or carelessness, but such failures are not frequent. ...

Port Elizabeth appears to have used a ball at the end of a lever arm (see later).

4 Cape Town (Alfred) Docks Time Ball

4.1 First Docks Time Ball

The time ball at the Docks is variously described as the Cape Town time ball, the Table Bay time ball and the Alfred Docks time ball. The first Admiralty list (List of Time Signals, 1880) gave the place as "Table Bay" and the location as "At Alfred Docks". The early history of this time ball is obscure and there are various inconsistencies in Admiralty descriptions.

The first Docks time ball is believed to have been installed in 1873 (Spencer-Jones, 1993). The entry in the 1880 Admiralty list indicated that the ball was located 47 feet above high water and 36 feet above ground with a small drop height of 6 feet (1.8 m). The location was given as 33° 54′ 27″ S, 18° 25′ 15″ E. No other details of the Docks arrangement before 1894 have been found and its manufacturer is presently unknown. Maudslay, Sons & Field received an order for a mechanical system from Siemens Brothers in 1873, which was thought to be destined for the Cape (Sells, 1883). However, there is strong evidence, reviewed in Chaps. 5 and 23, that this system was sent to Lyttelton, New Zealand rather than to the Cape (Kinns, 2009, 2017).

The Docks time ball was dropped at noon, rather than 1 pm as at other locations in South Africa.

4.2 Second Docks Time Ball

In 1894, a new time ball had been "erected in a conspicuous position near Resident Engineer's Office of the Cape Town Docks" (Evans, 1993). Messrs. Gillett and Johnston in London claimed to have supplied a time ball for The Cape, in an offer of a new time ball for New Zealand (Wellington Harbour Board, 1910. Ledgers still retained by the firm show that an order for a Table Bay time ball was received in November 1893 (Gillett and Johnston, 1893). The change of installation had not been noted in Admiralty lists up to 1919, which gave the same location, ball height and drop height as in 1880, despite addition of an extra story to the tower in 1904. The only listed changes were to the drop time, which was noon GMT (1 h 30 m pm Cape Colony mean time) in 1898. This had changed to 10 am GMT (noon Cape Colony mean time) after the change of time zone in February 1903 (see Table 16.2).

Table 16.2 gives the time ball location in 1900 as 33° 54′ 24″ S, 18° 25′ 10.4″ E, about 100 m further west than in Admiralty lists. This is likely to be a longitude correction that had not been registered by the Admiralty. The ball was 30 feet above ground level and dropped 10 feet. From 18 April 1895, return signals to the Observatory confirmed that the ball had commenced and finished its drop.

Evans (1993) recorded that an extra storey was added to the tower in 1904, so this change should have been reflected in the 1908 list. The ball diameter and drop height were recorded by Evans as 5 feet and 10 feet in 1900. Extracts from drawings of the 1894 installation are reproduced in Fig. 16.9, confirming these dimensions. The height of the signal was stated to be 30 feet above ground and 74 feet above low water in 1900. The 1894 drawing shows that the above-ground height of the centre of the ball in its raised position was indeed 30 feet (9 m). It was then a single-storey building, with part of the time ball apparatus in a deep basement. Its height was increased by adding another storey in 1904, so the height of the ball above water and above ground were then about 92 and 48 feet.

The Admiralty lists after 1922 gave the height above water, height above ground and drop height as 74, 37 and 6 feet. They were never corrected. The 1933 list gave the location as 33° 54′ 21″ S., 18° 25′ 12″ E. The Docks time ball was discontinued on 1 February 1934 and the ball itself was removed on 7 September (Evans, 1993). The tower remained in place.

4.2.1 Restored Docks Time Ball

The tower was declared a National Monument in 1982, but the time ball was then an immovable fibreglass replica. The Cape Town waterfront was redeveloped during the 1990s and it was decided to recreate the time ball that was withdrawn in 1934, using a replica of the original apparatus shown in Fig. 16.7 (Glass et al., 1997). Restoration started in 1997 under the management of the late Gabriël Fagan (1925–2020), a renowned architect and conservation specialist. The original time ball apparatus had been lost, but the Department of Mechanical Engineering at the University of Cape Town built a working replica. The time ball could be dropped again at 1 pm under electrical control from the South African Astronomical Observatory. The GPS coordinates of the time ball are given as 33° 54′ 21.3″ S, 18° 25′ 09″ E, almost exactly those reported in 1900.

Figure 16.8 shows the Docks time ball after its restoration to the 1934 condition. The height of the centre of the raised time ball above ground is now about 14.5 m, 5.5 m higher than its 1894 elevation of 9 m.

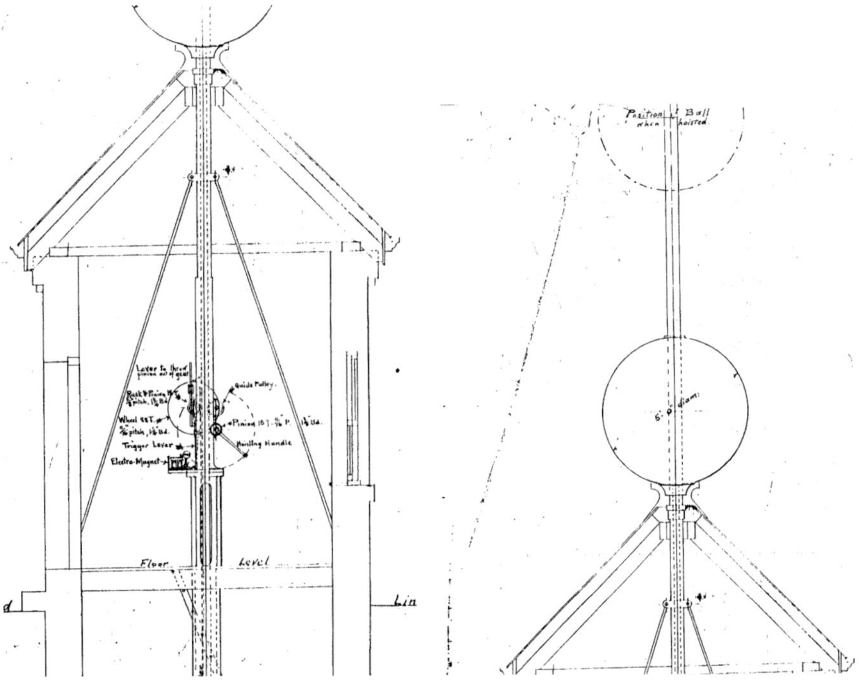

1894 Mechanism 1894 Ball

Fig. 16.7 The 1894 time ball at the Docks. (Courtesy Gabriël Fagan)

Fig. 16.8 Docks time ball tower after restoration. (Photograph by Michael Peel)

5 Simon's Bay Time Balls and Discs

The Admiralty list entries referred to Simons bay and then Simon's Bay, rather than Simon's Town. A common spelling is now Simonstown. It is the home of the South African Navy's largest base.

The first time ball at Simon's Bay was erected in 1857. It was regulated using a portable transit telescope, as indicated in Table 16.2, until telegraph lines had been constructed between the Observatory and Simon's Bay. Electric operation was announced in *The Times* (Cape Town, 1861):

> … The [telegraph] line from Cape Town to Simon's Bay is now made available for dropping the time-ball in the Admiralty-yard, thanks to Sir Thomas Maclean (*sic*), the Astronomer Royal …

The time ball had been replaced by a time disc before 1878, but the disc arrangement had become worn and failed frequently in 1896 and 1897, as shown in Table 16.2. A sketch of the arrangement is shown in Fig. 16.9. The disc was used for the last time on 11 December 1897 and replaced by a time ball which became operational on 14 April 1898.

The 1880 Admiralty list included the circular disc. The list for 1898 was prepared before the change of signal type, so still included the disc signal. There were small corrections to the latitude and longitude between 1880 and 1898, but no further changes afterwards. The list for 1904 indicated that the time ball was initially painted white, but this was changed to a chequered black and white paint scheme

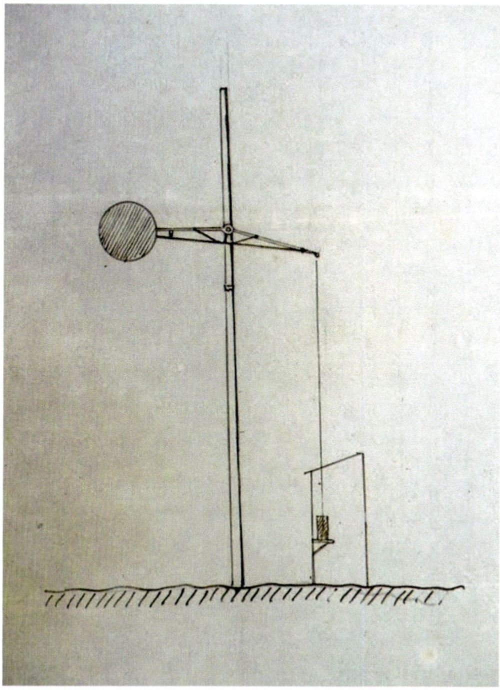

Fig. 16.9 Sketch of the time disc at Simon's Bay. (Courtesy David Erickson, Simonstown Historical Society, via Ian Glass)

before 1908. There was no reported change in the signal location or the drop height of 6 feet with the change from a disc to a ball. Table 16.2 indicates that the time ball was dismantled on 8 February 1929. Inclusion in the 1930 Admiralty list was therefore incorrect.

The design of the time ball apparatus has not been identified. It may have been similar to those erected in Singapore in 1893 and Auckland in 1901, using weights, rather than an air cylinder, to control the descent (Kinns, 2021). The small drop height is also consistent with installation of a lever arm ball, which appears to have been the arrangement at Port Elizabeth.

6 Port Elizabeth Time Balls and Discs

Time signals at Port Elizabeth, Port Alfred and East London were close together along the South coast. The Port Elizabeth signal was established first in 1865 and remained in use longest of the three signals. It was followed by the Port Alfred signal and then the signal at East London.

The origin of the Port Elizabeth time ball is known from an article describing a new time ball for Calcutta (Electric Time-Ball, 1880):

> … The first time-ball apparatus constructed on the polarized system was designed by Mr Varley, at the instance of the late Sir Thomas Maclear … for the Port Elizabeth time-ball which was erected in 1865 and gives time to shipping in Algoa Bay. This ball is discharged

> from the Cape Town Observatory, which is 500 miles distant, and the return signal recording the falling of the ball was stated by Sir Thomas Maclear to have been received 1–15th of a second after the time current was sent. …

The polarized system designed by Samuel Alfred Varley (1832–1921), who was a distinguished electrical engineer, used a preliminary signal to lock the time ball in place, so that it could not drop accidentally due to weak, false signals prior to receipt of the time signal. The return signal was sent when the ball started its fall. The linear separation is about 400 miles, so the cable was not routed directly.

An electric telegraph link from the Cape was in operation by 26 August 1865 (Warner, 1979). The time ball continued to provide a return signal to the Observatory as shown in the following report (Royal Observatory, 1873):

> A return-signal has been arranged for after the drop of the Port Elizabeth time-ball. … The return-signal reaches the Observatory from 3/10 to 6/10 of a second after the current leaves the Observatory.

The 1880 Admiralty list referred to a black disc with a drop of 5 feet (1.5 m). Table 16.2 indicates that a time ball was still in use during 1889, but the description had been changed to a disc in 1893. The 1898 Admiralty list referred to a ball as the signal type, but the detailed description referred to a disc. The 1904. 1908, 1911 and 1916 lists referred only to a black ball. This was also specified by the US Navy in 1916 (Africa Pilot, 1916, 141–142) but Admiralty lists from 1919 referred to a black disc without any change to the elevation and drop height.

Figure 16.10 shows the disc or ball in raised and lowered positions at the lighthouse. it appears that the signal was actually a ball attached to a lever arm, so that it

Disc raised *Disc lowered*

Fig. 16.10 The semaphore disc or ball at Port Elizabeth. (Courtesy Klaus Hülse collection)

could be observed from any direction. This would explain why the same location and drop height were given for both the ball and the disc and why the 1898 list of time signals refers to both: the change was a matter of description. Maclear had described the time ball at Lion's Rump as a lever arm ball in 1863. The essential difference from a conventional time ball was the dropping arrangement. The "disc" was dismantled on 30 September 1930.

7 Port Alfred

The time balls at Port Alfred and East London did not appear in the report for 1878 (Royal Observatory, 1878). The ball at Port Alfred was, however, noted in the 1880 Admiralty list.

There was a small correction to the longitude between 1880 and 1898, corresponding to a change of 0.3 s in astronomical time. The corrected signal location was 1 h 47 m 36.3 s fast on Greenwich. The drop time changed in combination with time zone modifications. There was no description of the time ball arrangement in Admiralty lists, other than the heights of 54 and 26 feet above water and above ground, and the large drop height of 18 feet. The signal was not included in the 1911 or later lists. From 1903, the ball was dropped at 10 am GMT (noon Cape Colony mean time), having been dropped at 1 pm, Cape mean time in 1880 and at noon GMT in 1898.

Table 16.2 indicates that the Port Alfred time ball had been discontinued in 1898, but the following notice shows that this was either incorrect or a temporary withdrawal. The signal was not discontinued until 1911 (Notices, 1911). The Admiralty notices appear to be correct.

> The German Government has given notice that the time ball which has hitherto been dropped from a staff near the West Mole, Port Alfred, southeast coast of Africa, has been discontinued. (N. M. 44 1911).

8 East London

The East London ball was noted in the Table 16.2 entry for 1889 and in later entries up to 1929. It had been withdrawn by 1933. The Port Alfred and East London signals were not far apart.

The time ball at East London was dropped from an iron frame at a high elevation of 160 feet above high water. The drop height was 15 feet. The apparatus would have been much simpler than those used at the Observatory in 1863 and at Alfred Docks after 1894. The time ball at East London was still listed by the Admiralty in 1939 as controlled from the Cape Observatory, when other time balls had been withdrawn by 1934. From 1911 onwards it was pointed out that "The ball falls slowly, and is not to be relied on within one second. Not visible alongside the wharves." Its long period of operation is surprising, given the limited value of the signal. Its extended inclusion may be an error.

9 Durban, Natal

In 1882, David Gill, Director of the Cape Observatory, asked the government of Natal to establish an astronomical observatory at Durban, in anticipation of the transit of Venus on 6 December that year. A site for the observatory was chosen in the southwest corner of the Natal Botanic Gardens. Initial equipment at the Natal Observatory included a 75 mm Troughton & Simms transit instrument and a sidereal clock by Dent (Natal Observatory). Edmund Nevill was appointed Government Astronomer of Natal and Director of the Observatory, with hardly any time to prepare for the Transit. His contribution was extraordinary.

9.1 Edmund Nevill aka Edmund Neison

Edmund Neville Nevill (1849–1940) also used the name Edmund Neison. The following extracts relating to his choice of name and work in South Africa are from his obituary (Spencer Jones, 1941).

> EDMUND NEVILLE NEVILL was born at Beverley, Yorkshire, on 1849 August 27. … and was elected a Fellow of the [*Royal Astronomical*] Society in 1873 under the name of Edmund Neison, having the curious idea that it was derogatory to the holder of an ancient name to make a career in science.
>
> … Nevill had gone to South Africa as Government Astronomer for Natal. There had been under consideration for several years a proposal for the establishment of an observatory in Durban. The observatory was started in 1882, with the aid of sums voted by the Corporation of Durban and by the Legislative Council of Natal, a Grubb 8-inch equatorial being presented by Mr. Harry Escombe. Mr. (later Sir David) Gill, H.M. Astronomer at the Cape, had been consulted and Nevill received an urgent telegram from Gill, offering him the post of Government Astronomer. He sailed almost on twenty-four hours' notice on 1882 October 27, reached Durban on November 27 and, in accordance with instructions furnished by the Colonial Secretary, he took possession of the Observatory on December 1 as Astronomer to the Natal government. On December 6 he obtained successful observations of the transit of Venus, for which the conditions were exceedingly fine.
>
> … But lack of financial support soon made itself felt and work had more and more to be restricted to routine observations–determinations of time for the provision of a time-service, meteorological and tidal observations. In 1888 Nevill was appointed Government Chemist and Official Assayer for Natal, and combined this office with that of Government Astronomer until his retirement.
>
> … The Natal Observatory came to an end in 1911 and Nevill, who had been elected a Fellow of the Royal Society in 1908, returned to England to live in retirement at Eastbourne. … Nevill reverted to his family name in 1888, in accordance with the conditions of a will. … He died on 1940 January 14 in his ninety-first year.

Following the formation of the Union of South Africa in 1910 the post of Government Astronomer of Natal was abolished and the Observatory was closed in 1911.

9.2 Durban Time Balls

It is believed that a time ball was established in 1883, controlled by Natal Observatory. It was positioned initially "At 3 cables N.N.W. ½ W. from Sandy point, North side of Entrance to port" with co-ordinates 29° 52′ 30″ S, 31° 03′ 00″ E. The first time ball at Durban was included in Admiralty lists for 1898 and 1904. The time ball was relocated in 1904, as indicated below (Notices, 1905, 22).

> The Government of Natal has given notice, dated November 3, 1904, that the time ball at port Natal bas been removed from the point to a position on the bluff, from which the Bluff lighthouse bears N. 59 ° E. true … distant 260 yards, and the south leading mark N. 22° W. true … The time ball will be dropped as before, on every day except Sunday, at 1 h. 00 m. 00 s. p.m. Natal standard mean time, corresponding to 23 h. 00 m. 00 s. Greenwich mean time. Approx. position: Lat 29° 52′ 44″ S., Long. 31° 03′ 42″ E.

The stated latitude and longitude were 29° 52′ 44″ S, 31° 03′ 42″ E, identical to those given in later Admiralty lists. This suggests that the same time ball was used from 1904 until the signal was finally withdrawn, with a change in regulation from Durban to The Cape in 1911. The drop height of 8 feet (2.4 m) remained the same in all lists from 1908 onwards. The elevation of the time ball was given as an implausible 95 feet above high water in the 1908 and 1911 lists. This had changed to 283 feet above high water in the 1915 and later lists. The Bluff itself is 195 feet high (McCallum, 2014). The time of the drop was changed on 1 September 1912 from 1 h. 00 m. 00 s. to noon South Africa standard time (Notices, 1912, 839).

Figure 16.11 shows a view of Natal Observatory from the North East. The small veranda was constructed around the Transit Room. Director Nevill appears in the foreground. The photograph was published in the Natal Illustrated Railway Guide of 1903 (Natal Observatory). Figure 16.12 shows the lighthouse in a tinted photograph. The time ball may have been suspended from the signal mast, which was clearly used for a large number of different signals.

Acknowledgements This chapter builds on historical studies by distinguished astronomers in South Africa, using international lists and notices to mariners. Dr. Ian Glass, Willie Koorts and Chris de Coning, of the South African Astronomical Observatory, kindly provided additional illustrations and references. Willie Koorts and Chris de Coning have produced excellent videos concerning time signals at the Cape:

https://youtu.be/BKU32cTe75o
https://youtu.be/nkBOkyl2hJw
The late Gabriël Fagan provided extensive information in 2009 about the 1894 time ball at the Docks.

Fig. 16.11 Natal Observatory in 1903 (Don Africana Library, Durban)

Fig. 16.12 The Durban lighthouse. (Courtesy Klaus Hülse collection)

References

Africa Pilot, Volume I. (1916). H. O. No. 105. Washington Government Printing.

Bartky, I. R., & Dick, S. J. (1981). The first time balls. *Journal for the History of Astronomy, xii.*

Bisset, C. W. M. (1984). Cape Town's Time-Guns, Scientia Militaria. *South African Journal of Military Studies, 14(*4), 67–71.

Cameron-Swan, C. D. (1931). The Rev. Fearon Fallows, M.A., F.R.S., F.R.A.S. (Presidential Address, Session 1930–31). *The Journal of the Astronomical Society of South Africa, 3*(1), 1–14.

Cape Town. (1861). Notice headed "Cape Town, Cape of Good Hope, Nov. 21, 1861.", published in *The Times*, 7 January 1862.

Editorial. (1835). The Time-Ball of St. Helena. *The Nautical Magazine, 4*, 658–660.

Electric Time-Ball for India. (1880, January 2). Article published in *The Times*, London.

Evans, G. P. (1993). *History of time guns and time-balls in South Africa.* Unpublished memorandum, South African Astronomical Observatory.

Evans, D. S., Deeming, T. J., Hall Evans, B., & Goldfarb, S. (1969). *Herschel at the Cape: Dairies and correspondence of Sir John Hershel, 1834–1838.* University of Texas Press.

Gill, D. (1913). *A history and description of the Royal Observatory, Cape of Good Hope.* HM Stationery Office.

Gill, D., Obituary. (1914). Sir David Gill. *The Observatory: A Monthly Review of Astronomy, 37*, 115–117.

Gillett and Johnston. (1893). *Ledger entry* showing an order for Table Bay Harbour on 6 November 1893.

Glass, I. S., Evans, G. P., & Lastovica, E. (1997). Waterfront time ball to drop again, 1997. *Monthly Notes of the Astronomical Society of Southern Africa, 56*, 108–109.

Harding, G. (1971). *The Cape Town Noon Gun.* [This reference has not been sighted by the author. It was included in the 1993 list by Evans, without details other than a publication date of 15 October 1971. The dates have been confirmed using other sources].

Herschel, J. (1836). Account of an Observatory constructed at the Mauritius. Communicated by Sir J. Herschel. *Monthly Notices of the Royal Astronomical Society, 3*, 157–160.

Kinns, R. (2009). Time keeping in the antipodes: A critical comparison of the Sydney and Lyttelton time balls. *Journal of Astronomical History and Heritage, 12*(2), 97–107.

Kinns, R. (2011). The early history of the Edinburgh Time Ball and Time Gun. *International Journal for the History of Engineering & Technology, 81*(2), 264–290.

Kinns, R. (2017). The principal time balls of New Zealand. *Journal of Astronomical History and Heritage, 20*(1), 69–94.

Kinns, R. (2021). Time signals for mariners in South Africa. *Journal of Astronomical History and Heritage, 24*(2), 285–314.

Maclear, T. (1852). The observatory time ball – Cape of Good Hope, dated 7 July 1852. *Notice in the Nautical Magazine and Naval Chronicle, 21*, 611–612.

Maclear, T. (1853). Establishment of an additional time ball at the Cape of Good Hope. Notice to Mariners [No. 153]. Issued by the Hydrographic Office, 13 December 1853. Reproduced in the 1854 *Nautical Magazine and Naval Chronicle, 23*, 54.

Maclear, T. (1863a). *Letter to G. B. Airy* dated 16 June 1863. Papers of George Airy, RGO 6 615, File 7, Leaves 145–146, Cambridge University Library.

Maclear, T. (1863b). *Letter to G. B. Airy* dated 18 June 1863. Papers of George Airy, RGO 6 615, File 7, Leaf 147, Cambridge University Library.

McCallum, G. L. (2014). Bluff Lighthouse, Durban. The article includes several photographs of the lighthouse and signal mast. https://grahamlesliemccallum.wordpress.com/2014/05/27/bluff-lighthouse-durban/

Natal Observatory. http://assa.saao.ac.za/sections/history/observatories/natal_obs/

Nevill, E. N. (1940). Obituary notice. *Nature, 145*, 339–340.

Notices to Mariners of 1905, Nos. 1 to 52, Africa – Southeast coast – Port Natal – Time ball – Position altered. – Washington Government Printing Office, 1906. Issued as Notice to Mariners No. 1209. Admiralty, London. 1905

Notices to Mariners of 1911, Nos. 1 to 52. Africa – Southeast coast – Port Alfred – Time signal discontinued. Washington Government Printing Office, 1911.

Notices to Mariners of 1912, Nos. 1 to 52. Washington Government Printing Office, 1912.

Royal Observatory at the Cape of Good Hope. (1873). Report of the Council to the fifty-third annual general meeting: Proceedings of Observatories, February 1873. *Monthly Notices of the Royal Astronomical Society, 33*, 229.

Royal Observatory at the Cape of Good Hope. (1877). Report of the Council to the fifty-seventh annual general meeting: Proceedings of Observatories, February 1877. *Monthly Notices of the Royal Astronomical Society, 37*, 176–177.

Royal Observatory at the Cape of Good Hope. (1878). Report of the Council to the fifty-eighth annual general meeting: Proceedings of Observatories, February 1878, *Monthly Notices of the Royal Astronomical Society, 38*, 186–187.

Sells, C. (1883). Notebook 12: April 1878 to March 1883. Table inside end cover of notebook showing five time balls supplied by Maudslay, Sons & Field (see Chapter 5), archived in the Science Museum Library, Swindon. Charles Sells was the Chief Draughtsman at Maudslay Sons & Field for 49 years.

Smyth, C. P. (1853a). Remarks on the erection of the time-ball of the Royal Observatory, Edinburgh. *Monthly Notices of the Royal Astronomical Society, 14*(1), 23–25.

Smyth, C. P. (1853b). Notice of the time ball at the Royal Observatory at Edinburgh. 'Given at the request of the Council on 12 December 1853. A working model and diagrams were exhibited. *Later Published in Select Papers of the Society of Arts*, 1856, *4*, 191–196.

Spencer Jones, H. S. (1941). Obituary notices: Fellows: Nevill, Edmund Neville. *Monthly Notices of the Royal Astronomical Society, 101*(3), 137–139.

Spencer Jones, J. (1993). Time and the sailor. *Waterfront Review*, 26–27.

Warner, B. (1979). Astronomers at the Royal Observatory, Cape of Good Hope. Published by Balkema, Cape Town. A review by P. B. Byrne was published in the *Irish Astronomical Society Journal, 14*(5/6), March–June 1980, 105–107.

Warner, B. (1997). The age of Fallows. *Monthly Notes of the Astronomical Society of Southern Africa, 56*, 107.

Wauchope, R. (1836). Letter to the Editor of The Nautical Magazine, 2 April 1836, with supporting testimonials concerning time ball invention. *Published in The Nautical Magazine, 5*, 460–464.

Wauchope, R. (1852). Letter to the Editor of The Naval and Military Gazette 13 September 1852, with supporting testimonials concerning time ball invention, previously published in *The Nautical Magazine, 5*, 460–464. Republished with a supporting editorial in *The Witness*, 20 October 1852.

Wellington Harbour Board. (1910). *Report on proposed new time ball*, dated 30 November 1910, Wellington City Archives, AC020:32094.

Chapter 17
West and East Africa, and Yemen

Roger Kinns

Abstract The sparsely distributed signals in West and East Africa, together with ports on the Red Sea and the Gulf of Aden, are considered in this chapter. Dakar had the most northerly time ball in Sub-Saharan Africa. A short-lived time ball on the Gold Coast was announced in 1839. A combined flag and gun signal was later used at Accra, replaced in turn by maroons, and a combined ball and gun signal at Takoradi. There is also thought to have been a time ball at Cotonou. An 1879 time ball at St Paul de Loanda was supplemented by time lights before 1927.

A time ball at Lorenzo Marques was replaced by electric lights in 1909, thought to be their earliest use as a principal time signal. There was also a light signal at Beira, time guns at Dar es Salaam and Djibuti, and a time ball at Massawa. Aden used flag signals at Ras Marbut and Shamshan.

1 Introduction

Signals in South Africa from Simonstown in the west to Durban in the east have been described in Chap. 16. The signals elsewhere in Africa were more sparsely distributed. Those in West and East Africa are considered in this chapter, together with ports on the coasts of the Red Sea and the Gulf of Aden.

Table 17.1 shows the evolution of signals in Admiralty lists between 1880 and 1947. Figure 17.1 shows the locations of signals in West Africa, with the addition of the 1839 signal at "Cape Coast Castle". Figure 17.2 shows locations in East Africa. Figure 17.3 shows the locations of signals at Massawa, Djibouti and Aden on a larger-scale map. Those at Massawa and Djibouti are in northeast Africa, but nearby Aden (now in Yemen) is in West Asia.

R. Kinns (✉)
The Maudslay Society, Pembroke College, Cambridge, England

R. Kinns (ed.), *The World of Visual Time Signals for Mariners*, Historical & Cultural Astronomy, https://doi.org/10.1007/978-3-031-57334-7_17

Table 17.1 Signals in Yemen, East and West Africa

Area	Location	1880	1898	1904	1911	1915	1919	1922	1928	1934	1939	1947
Middle East (Yemen)	Aden: Ras Marbut				Flag	Flag	Flag	Flag	Flag	Flag	Flag	
					Gun	Gun	Gun	Gun	Gun	Gun	Gun	
	Aden: Shamshan				Flag	Flag	Flag	Flag	Flag	Flag	Flag	
Northeast Africa	Eritrea: Massawa									TB	TB	
	Djibouti							Gun	Gun	Gun	Gun	
East Africa	Mozambique: Laurenzo Marques			TB								
					Light	Light	Light	Light	Light	Light	Light	Light
	Mozambique: Beira									Light		
	GEA, Tanganyika: Dar es Salaam					Gun						
West Africa	Angola: St Paul de Loanda		TB	TB	TB	TB	TB	TB	TB	TB	TB	
									Light	Light	Light	Light
									Flag			
									Gun			
	Ghana (was Gold Coast): Accra		Flag	Flag	Flag	Flag	Flag					
			Gun	Gun	Gun	Gun	Gun	Maroon	Maroon	Maroon	Maroon	
	Ghana: Takoradi									TB		
										Gun		
	Senegal: Dakar					TB (black)	TB (black)	TB (black)	TB (black)	TB (black)	TB (black)	TB (black)
Signal type totals		1880	1898	1904	1911	1915	1919	1922	1928	1934	1939	1947
Time ball, including collapsible		0	1	2	1	2	2	2	2	4	3	1
Gun, including powder flash		0	1	1	2	3	2	3	4	4	3	0
Lights		0	0	0	1	1	1	1	2	3	2	2
Other: disc, drum, flag, etc.		0	1	1	3	3	3	2	3	2	2	0

2 West Africa

2.1 Angola, St. Paul de Loanda

St. Paul de Loanda (now Luanda) is *the capital and largest city of Angola. It is located on Angola's coast with the Atlantic Ocean and serves as Angola's chief seaport and administrative centre. Portuguese explorer Paulo Dias de Novais founded Luanda in 1576 as "São Paulo da Assumpção de Loanda". It is the world's third most populous Portuguese-speaking city, behind only São Paulo and Rio de Janeiro in Brazil. Angola was a Portuguese colony during the period of visual time signals.*

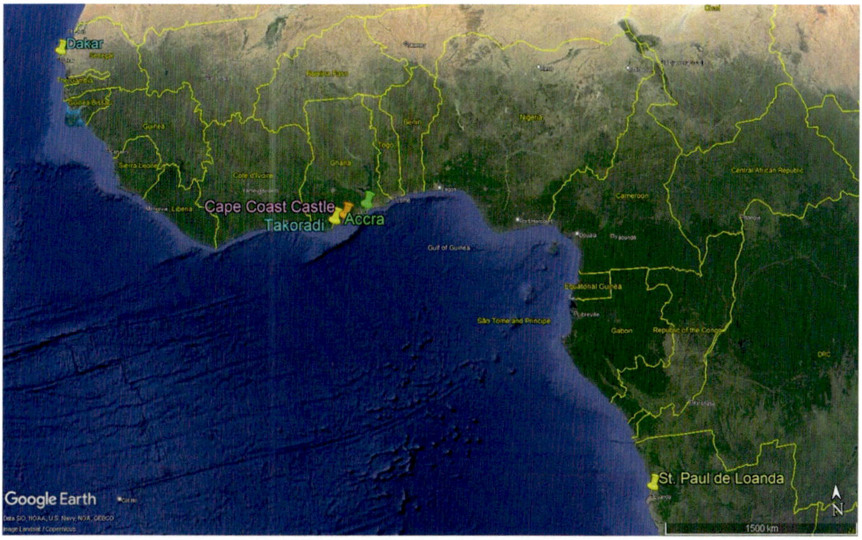

Fig. 17.1 Signals in West Africa (Google Maps)

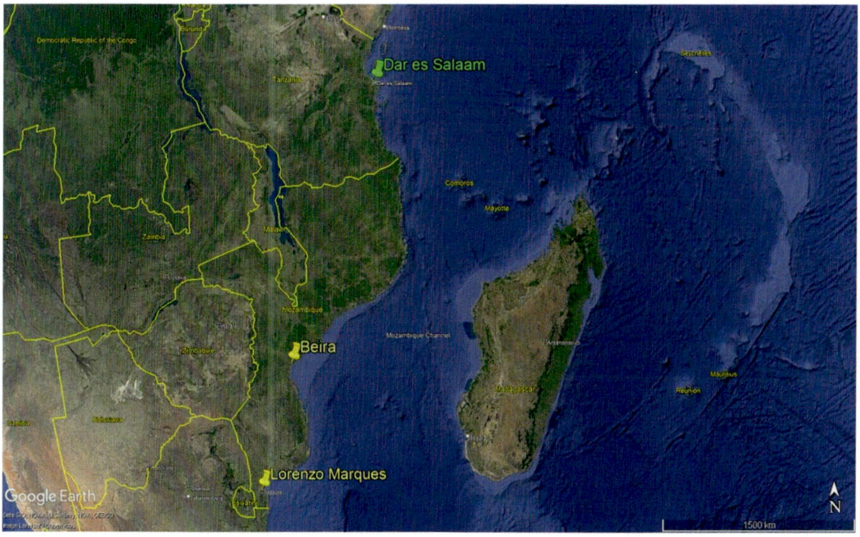

Fig. 17.2 Signals in East Africa (Google Maps)

2.1.1 Time Ball

Although there was no entry in the 1880 Admiralty list, the time ball at Loanda was operational in 1879 (Nautical Notices, 1879).

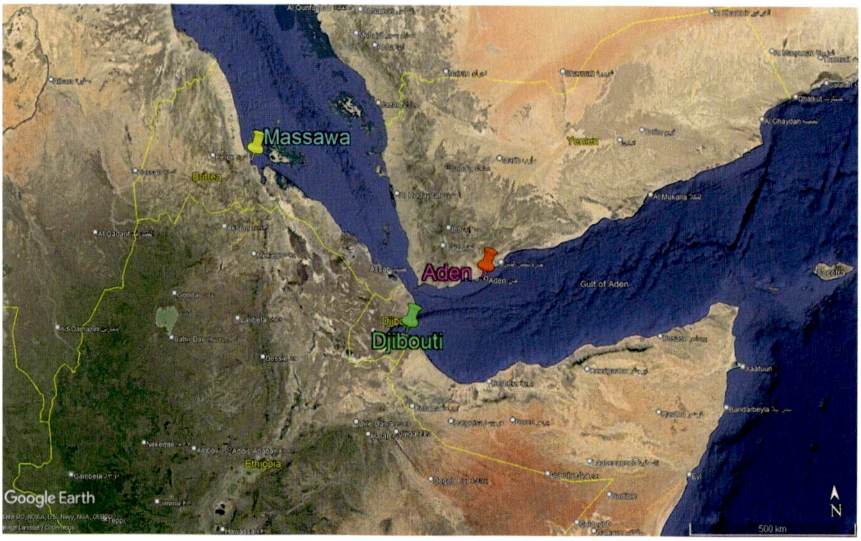

Fig. 17.3 Locations of signals at Massawa, Djibouti and Aden (Google Maps)

> A Time Ball is hoisted half-mast on the staff of the observatory tower at 0 h. 50 m. p.m.; close up at 0 h. 55.m.; and dropped at 1 h. 0 m. p.m. mean time, St. Paul de Loando (*sic*).
> Note. —It is reported that dependence cannot always be placed on the accuracy of this time ball.

The first and second preparatory signals were then at 10 and 5 min before the signal at 1 pm.

The 1898 Admiralty list entry gave the drop time as 1 pm Loanda local time. The indicated times of the preparatory signals at 5 and 2 min before the signal differed from those reported in 1879. However, preliminary signals at 10 and 5 min before the time signal were again noted in the 1908 and later Admiralty lists up to 1939. They were reported as being 5 and 2 min before the signal in 1908 and 1916 notices issued by the United States, so there appears to have been some confusion (West Coast of Africa, 1908; Africa Pilot, 1916). The time of the signal was changed from 1 pm local mean time to noon GMT on 1 January 1912 (Notice, 1912).

Figure 17.4 shows the time ball above the tower in its lowered position.

2.1.2 Time Lights

The following 1932 notice suggested that the time ball had been discontinued and replaced by an evening time light signal (Sailing Directions, 1932). A flag signal, accompanied by the firing of time gun, was made at 1 pm for local use only.

> Time signal. – A time signal is made by means of lights on the observatory tower which are lighted at 8:55 p.m. and extinguished at 9 p.m. There may be an error of 1 second in this signal.

Fig. 17.4 Loanda Observatory. (Courtesy Klaus Hülse collection)

> A time signal, consisting of the lowering of a flag at 1 p.m. accompanied by the firing of a gun, is made for the convenience of the residents.

However, the Admiralty lists up to 1939 included both the time ball and the time lights, which had been introduced by 1927. The lights were still listed in 1947.

2.1.3 Time Gun and Flag

The time gun at Fort Saint Miguel was listed by the Admiralty in 1928, with a stated introduction date of 1926, but not in 1934 or later. It was fired when a flag at the Observatory was dipped at the same time as the ball was dropped. It was reported to be unreliable.

2.2 Ghana

2.2.1 Cape Coast Castle

A time ball on the Gold Coast, now Ghana, was announced on 27 July 1839. It was one of the earliest time ball signals. The notice was published in the *Naval Chronicle* for 1840 (Maclean, 1839). It is reproduced in Fig. 17.5.

George Maclean (1801–1847) was Governor of the Gold Coast from 1830 until 1844 (Wikipedia). The apparatus at the Cape Coast Castle was similar in many

NOTICE TO MARINERS.

Hydrographic-Office, Admiralty, 1st Nov. 1839.

Notice is hereby given, that at *Cape Coast Castle* all Vessels in the Roads may ascertain the errors and rates of their *Chronometers,* without exposing them to the risk of being carried on shore through the surf.

A Flag Staff with a Gaff has been erected on the Southern Turret of the Castle.

At 11h. 30m. Greenwich Mean Time, on each day after 1st of September, 1839, a *Red Flag* with a *White Ball* in the centre, will be hoisted at the Mast Head as a preparatory signal.

At 11h. 55m. Greenwich Time, a *Black Ball,* five feet in diameter will be hoisted to the Gaff end, and the Flag lowered.

At the instant of Noon, Greenwich Mean Time, the Ball will be dropped from the end of the Gaff, and will immediately disappear.

N.B.—The falling of the Ball must occupy upwards of a second of time, but the instant of Noon is to be reckoned from the separation of the Ball from the Gaff end.

The latitude of the Flag Staff of Cape Coast Castle is 6° 5′ 25″ N., the longitude 1° 12′ 5″ W.*

George Maclean,
Cape Coast Castle, 27th July, 1839. *President of Council.*

HEAUX-DE-BREHAT.

(Communicated by the French government.)

Notice is hereby given, that after the 1st of February, 1840, instead

* In Capt. Vidal's Survey, the longitude of the Flag Staff on the West bastion of the Castle appears to be 1 deg. 13 min. 41 sec.

Fig. 17.5 Announcement of the Cape Coast time ball (Naval Chronicle, 1840, 128)

respects to the 1836 arrangement at the Royal Observatory, Cape of Good Hope, with a ball having a diameter of 5 feet that was dropped from a gaff. The footnote in Fig. 17.5 suggests that the stated longitude was too far east by almost 3 km. The time ball may have had a short life.

2.2.2 Accra

Extracts from Admiralty list entries for Accra are shown in Table 17.1. There was no entry in 1880 but a combined flag and gun signal was listed between 1898 and 1919. The gun, but not the flag, was included in Admiralty lists up to 1939. Entries in the 1898, 1911, 1916, 1922 and 1939 lists are shown in Table 17.2. Standard time in Accra was GMT after 1912. Before that, Accra mean time was 56 seconds later. The signal accuracy was reduced because of the need for telegraph re-transmission between London and Accra.

The location of the flag changed from the "Direct Company's Telegraph Office" between 1898 and 1911 to the "African direct company's compound" between 1916 and 1919, with a significant change in latitude and longitude from 5° 31′ 48″ N, 00° 11′ 30″ W to 5° 32′ 24″ N, 00° 12′ 23″ W. The revised location appears to have been 36″ further north and 53″ further west than before, a distance of about 2 km. The

Table 17.2 Time signals in Accra. (From Admiralty lists between 1898 and 1939)

Location	Latitude Longitude	Signal	Additional details
1898	5° 31′ 48″ N	Flag and Gun	Flag dropped at 11 h 00 m 00 s Accra mean time. Greenwich time is received daily, Sundays excepted, at 10 h 00 m 00 s a.m., from London by Post Office Chronograph; but being re-transmitted through several stations, is not exact
Direct Company's Telegraph Office	00° 11′ 30″ W		
1911	5° 31′ 48″ N	Flag and Gun	Flag dropped at 11 h 00 m 00 s Accra mean time. Greenwich mean time is received daily, Sundays excepted, at 10 h 00 m 00 s a.m., from London by Post Office Chronograph; but being re-transmitted through several stations, is not exact
Direct Company's Telegraph Office	00° 11′ 30″ W		
1916	5° 32′ 24″ N	Flag and Gun	Flag dropped and gun fired at 23 h 00 m 00 s Greenwich mean time. Reported to be inaccurate. Greenwich time is received daily, Sundays excepted, at 22 h 00 m 00 s from London by Post Office Chronograph; but, being re-transmitted through several stations, is not exact
African direct company's compound	00° 12′ 23″ W		
1922	5° 32′ 24″ N	Maroon	Maroon fired at Noon Standard time.
Post Office	00° 12′ 23″ W		Greenwich mean time is received daily, Sundays excepted, at 22 h 00 m 00 s from London by post office chronograph, but, being re-transmitted through several stations, is not exact
1939	5° 32′ 24″ N	Maroon	Maroon fired at Noon Standard time.
Post Office	00° 12′ 23″ W		Greenwich mean time is received daily, Sundays excepted, at 10 h 00 m 00 s from London by post office chronograph, but, being re-transmitted through several stations, is not exact

location description changed again to the "Post Office" in 1922 but the latitude and longitude were not altered and remained the same until 1939. From 1922 onwards, the gun was described as a maroon and the flag signal was deleted. A maroon was usually a small gun that fired a rocket, giving a flash and a loud bang.

2.2.3 Takoradi Time Ball and Gun

Takoradi, now joined with Sekondi in Ghana, is about 200 km WSW of Accra. A time ball and time gun were in operation at Takoradi in 1932, as shown in the following notice issued by the United States (Sailing Directions, 1932).

> Time signal. – A time signal is made from the Prince of Wales Clock Tower. The time used is that of Greenwich meridian from January 1 to August 31 and that of 5° E. (0h. 20m. fast

> on Greenwich mean time) during the remainder of the year. Simultaneously with the dropping of the time ball a gun is fired from the harbor signal station.

According to this notice, the local standard time was advanced by 20 min between 1 September and 31 December each year. The drop time was not stated in the notice.

The Admiralty list for 1933 gave the same location and specified a drop time of noon GMT throughout the year, also indicating that a gun was fired. The latitude and longitude were given as 4° 53′ 7″ N, 1° 45′ 11″ W. It was noted that the time signal had been introduced in 1930. The standard time change by 20 min from GMT at Takoradi was, however, stated to be from 1 January to 1 September, contradicting the notice reproduced above. The chosen difference between standard time and GMT for part of the year is hard to understand, given that Takoradi is slightly west, not east, of the Greenwich meridian.

No signal at Takoradi was included in the 1939 list, so the time ball had a short life.

2.3 *Nigeria, Kotonu*

Kotonu (now spelled Cotonou) is a port on the Bight of Benin, about 110 km from Lagos, the Nigerian capital. No time ball at Kotonu was listed by the British Admiralty, but it was noted by the United States in 1893 (West Coast of Africa, 1893).

> Time ball. – A time ball is hoisted on the flagstaff of the telegraph office; the time is given from Loanda. …

The time was regulated from Loanda in Angola, but no further details were specified in the notice. It was not mentioned in the subsequent edition (West Coast of Africa, 1908), so it probably had a short life.

2.4 *Senegal, Dakar*

Dakar is at the western end of the Cape Verde peninsula, with the Cape Verde Islands further west. It was the most northerly of the signals in Sub-Saharan Africa. Senegal was a French colony, becoming independent in 1960.

Time signals at Dakar did not feature in Admiralty lists up to 1911, but there had been a time ball there for a limited period. The following notice was published in 1911, indicating that an earlier time ball had been re-established (Notices, 1911).

> Referring to Notice to Mariners No. 19(966) of 1909, the French Government has given further notice that the time signal at Dakar, west coast of Africa, has been reestablished.
>
> The signal consists of a black ball 3 feet in diameter dropped on a mast located at the southeastern angle of the Arsenal Grounds.

> It is hoisted 5 minutes before and dropped at 11 h. 09 m. 41.5 s., Greenwich mean time, corresponding to 10 h. 00 m. 00 s., Dakar mean time. The ball is immediately rehoisted and dropped again 2 minutes later.
>
> Should the signal fail a red flag is shown from the mast and the signal will be repeated 10 minutes after the first ball was dropped.

In the above notice, times were based on zero hours at midnight, not zero hours at noon as usually expected for time ball data before 1925.

Admiralty lists included similar information in lists from 1915 to 1947, noting that the signal had been introduced in 1911. Its latitude and longitude were given throughout as 14° 40′ 40″ N, 27° 25′ 52″ W. By 1916, the ball was dropped at 10 am and 10.02 am, Standard mean time, corresponding to 11 am and 11.02 am GMT with the introduction of a new time zone. The ball continued to be raised 5 min before the first signal.

The location was described initially as being at the "South-east angle of Arsenal". This had been changed by 1927 to "Tower of Captain of Port's office, south-east angle of Arsenal". Figure 17.6 shows the Port Office at Dakar.

Fig. 17.6 Dakar port office (https://www.senegal-online.com/gallery/cartes-postales-anciennes-du-port-de-dakar/)

3 East Africa

3.1 Mozambique

3.1.1 Lorenzo Marques Time Ball

A time ball had been in operation at the Lorenzo Marques Observatory before 1904 and was still listed in 1908, but not in 1911. It was at the "Port Captain's Office" at 25° 58′ 5″ S, 32° 34′ 35″ E. The ball was hoisted 5 minutes before the signal at 11 am GMT, corresponding to 1 h 10 m 18.3 s local mean time and 1 pm standard time of 30th degree of east longitude.

3.1.2 Lorenzo Marques Time Lights

The time signal changed to electric lights in 1909. This is thought to be the earliest use of electric lights as a principal time signal worldwide and one of the longest lived. It was still listed in 1947.

The unusual light signal remained the same throughout its life. A horizontal line array of 7 lights was illuminated within a black triangle whose corners were highlighted with additional green lights. The array was illuminated 5 min before each time signal at 3 h intervals, giving 8 signals every day without intermediate flashes. The lights were located on "Top of eastern customs shed, Gorjao wharf".

3.1.3 Beira Time Light

Another light signal was introduced at the port of Beira in 1931, further north and east than Lorenzo Marques. The simple time light at Beira was illuminated 3 min before extinction at 10 am and 9 pm local standard time, 2 h ahead of GMT. It had been discontinued before 1947.

3.2 Tanzania

3.2.1 Dar es Salaam Time Gun

A time gun was established in 1913 at Dar es Salaam, then in German East Africa. The colony became British in 1916 and was renamed Tanganyika, now part of Tanzania. The gun was still listed in 1917, but not in 1919 or later. It was fired at noon from "Battery near harbour office". Figure 17.7 shows the time gun in operation.

Fig. 17.7 Time gun at Dar es Salaam. (Courtesy Klaus Hülse collection)

4 Northeast Africa and Yemen

4.1 Eritrea

A time ball at Massawa on the Red Sea was established on the mast of the wireless station at Abd-el-Kadir in 1930 and was still listed in 1939. The ball was hoisted 5 min before the signal and dropped by hand at noon, standard time, 3 h ahead of GMT.

4.2 Djibouti

A time gun at the French colony of Djibouti was established in 1919 (then listed as Sibuti) and was still listed in 1939. It was regulated by Paris Observatory and fired by wireless from the Eiffel Tower at 5 pm GMT, 8 pm standard time.

4.3 Aden

Aden (now in Yemen) used flag signals at two locations: Ras Marbut and Shamshan. Both were listed from 1911 to 1939. The flag pendant at Shamshan was listed as unofficial. It was noted that an accurate telegraph signal was received daily at the

Eastern Telegraph Company's office at Ras Marbut and chronometers could be compared there.

The flags were hoisted 5 min before the signal and dipped at noon, local mean time, 2 h 59 m 54 s ahead of GMT. In that respect, it differed from the signals at nearby Massawa and Djibouti which were at standard time, 3 h ahead of GMT. A gun was also fired when the flag was dipped at Ras Marbut, but was listed as unreliable.

References

Africa Pilot, Volume I. (1916). H. O. No. 105, Washington Government Printing Office, 1916.
East Atlantic Pilot, H. O. No. 134, First Edition, 1916; Second Edition, 1918; Third Edition, 1920, Washington Government Printing Office.
Maclean, G. (1839). Notice to Mariners, dated 27 July 1839. *The Naval Chronicle, 1840*, 128.
Nautical Notices. (1879). Notice 333. *The Nautical Magazine, 48*, 989.
Notices to Mariners of 1911, Nos. 1 to 52, Washington Government Printing Office, 1911.
Notices to Mariners of 1912, Nos. 1 to 52, Washington Government Printing Office, 1912.
Sailing Directions for the SW Coast of Africa, Third Edition, 1932. H. O. No. 105, United States Government Printing Office, Washington, 1932, 75.
West Coast of Africa from Cape Spartel to Cape Agulhas …, Second Edition, 1893. No. 105, United States Government Printing Office, Washington, 1893, 232.
West Coast of Africa, Cape Palmas to the Cape of Good Hope …, Third Edition, 1908. No. 105, Washington Government Printing Office, 1908, 373–374.

Chapter 18
Mauritius

Roger Kinns

Abstract The time ball service at Port Louis, Mauritius started in April 1833, six months before the Greenwich service. It was established by John Augustus Lloyd who had been appointed to Mauritius in 1831. His chosen arrangement used a stationary black ball or disc on a white background behind a shutter. The time service later deteriorated, with only an intermittent flag signal for an extended period, although a time ball had been erected on Signal Mountain in 1866. It was rejuvenated when a new observatory became operational in 1874. It was then operated electrically, but the apparatus was destroyed in the 1892 hurricane. Another ball was introduced as the new principal signal near the Port Office in March 1907 and was the official signal thereafter. It was regulated using wireless signals after the astronomical observatory closed. The earlier time ball was then dropped manually in response until 1945. Streetlights were dimmed as another time signal in the 1920s and 1930s.

1 Introduction

Mauritius is part of the African continent in the Indian Ocean. The main island has an area of 1871 km^2, with islets that take the total to 2011 km^2. Despite its small size, Mauritius played an important role in time signal innovation. All the recognised time signals in Mauritius were at Port Louis, the capital. The existence of an 1833 time signal at Mauritius was noted by Bartky and Dick (1981).

Figure 18.1 shows the location of Mauritius in relation to East Africa and the islands of Reunion and Madagascar. Figure 18.2 shows the locations of the three time balls used in Mauritius. Table 18.1 show entries for Port Louis, Mauritius in Admiralty lists from 1880 to 1947.

R. Kinns (✉)
The Maudslay Society, Pembroke College, Cambridge, England

R. Kinns (ed.), *The World of Visual Time Signals for Mariners*, Historical & Cultural Astronomy, https://doi.org/10.1007/978-3-031-57334-7_18

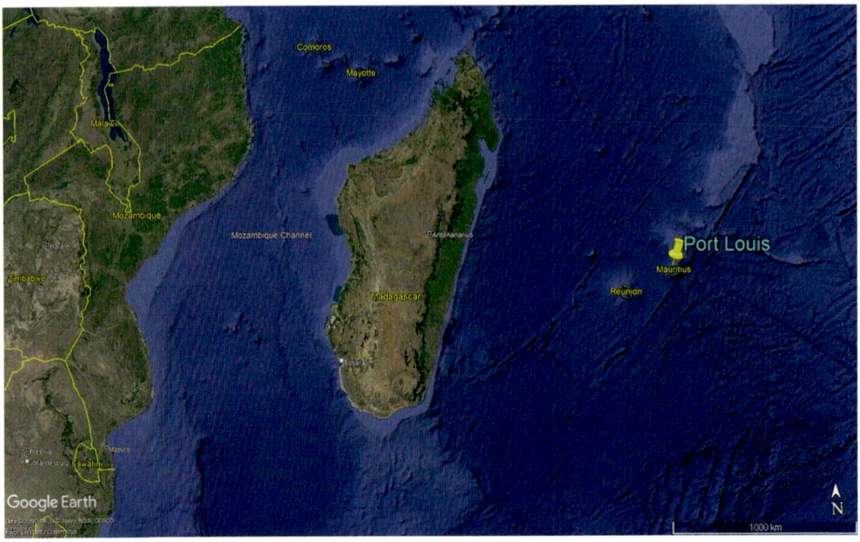

Fig. 18.1 Location of Mauritius (Google Maps)

Fig. 18.2 Locations of time balls and observatories in Mauritius (Google Maps)

1.1 A Brief History of Mauritius

The then uninhabited island(s) of Mauritius are believed to have been discovered by the Moors before the sixteenth century. Three islands (now named Réunion, Mauritius and Rodrigues) were discovered by the Portuguese between 1507 and

Table 18.1 Signals in Mauritius

Area	Location	1880	1898	1904	1911	1915	1919	1922	1928	1934	1939	1947
Mauritius	Port Louis: Signal Mountain	TB (black)	TB (black)	TB (black)	TB (black)	TB (black)	TB	TB	TB	TB	TB	
	Port Louis: Port Office				TB (white)	TB (white)	TB (white)	TB (white)	TB (white)	TB (white)	TB (white)	TB (white)
Signal type totals		1880	1898	1904	1911	1915	1919	1922	1928	1934	1939	1947
Time ball, including collapsible		1	1	1	2	2	2	2	2	2	2	1
Gun, including powder flash		0	0	0	0	0	0	0	0	0	0	0
Lights		0	0	0	0	0	0	0	0	0	0	0
Other: disc, drum, flag, etc.		0	0	0	0	0	0	0	0	0	0	0

1513. The group was named Mascarene but there was no permanent Portuguese colony. It was rediscovered by the Dutch in 1598 and used as an occasional refuge during long sea voyages. It was first colonised in 1638 when a small garrison was established. This was withdrawn temporarily before a second attempt was made to establish a thriving colony. The main island was named Mauritius. The islands were abandoned by the Dutch in 1710.

Mauritius became a French Colony in 1715 when it was named "Isle de France". The period of British administration, starting in 1810 near the end of the Napoleonic wars, covered rapid social and economic changes. The island was renamed Mauritius. The French language continued to be used widely and this is reflected in the notice of the first time ball in 1833. The development of time signals in Mauritius was during the period of British rule.

British rule ended in 1968, when Mauritius achieved independence with Elizabeth II as Queen of Mauritius. Mauritius became a republic within the British Commonwealth in 1992. English is now the official language, but French continues to be used extensively and Creole is widely spoken. The total population is now about 1¼ million.

2 The First Signal in Mauritius

The first service for rating chronometers at Port Louis, Mauritius started in April 1833, 6 months before the Greenwich time ball was operational. The time signal was established by John Augustus Lloyd (1800–1854) who had been appointed Surveyor-General and Civil Engineer in 1831.

Fig. 18.3 Lt. Col Lloyd in 1851. (illustrated London News, 28 June 1851)

2.1 *John Augustus Lloyd*

Lloyd was an extraordinarily talented army officer, then serving as a Captain in the Royal Engineers, who soon developed a well-equipped observatory and a time service (Kinns, 2020a, b). He was widely respected in the scientific community for his earlier survey work in Panama and had presented papers to the Royal Society and the Royal Geographic Society. His reputation was noted by Charles Darwin when he visited Mauritius and stayed with Lloyd during the voyage of HMS *Beagle* (Darwin, 1836). Lloyd departed Mauritius in 1849, leaving a legacy of engineering and astronomical achievements. Lloyd was also an adventurer, showing extraordinary bravery in hostile environments worldwide (see Dash, 2012, for example). As a Lieutenant-Colonel, he was later a Special Commissioner for the Great Exhibition (Supplement, 1851). The engraving of him in Fig. 18.3 is taken from that supplement in the *Illustrated London News*. He died of cholera in Constantinople, now Istanbul, on military service in 1854 (Obituary, 1855).

2.2 *Lloyd's Time Signal*

Lloyd's notice of the signal is reproduced in Fig. 18.4. The precise nature of the signal is unclear from the notice, but a paper read to the *Royal Astronomical Society* on behalf of Sir John Herschel clarified how it worked (Herschel, 1836). At that time, Herschel was working at the Cape of Good Hope and there is a diary entry confirming that he had received a package from Lloyd in March 1835 (Evans et al., 1969, 151). This is likely to have included material Herschel used in preparing his 1836 paper. Part of that paper (Herschel, 1836, 189) is transcribed below.

> … A contrivance has been made for indicating the correct time to the ships in the harbour. A lever has been constructed which supports a heavy slide, the working of which may be

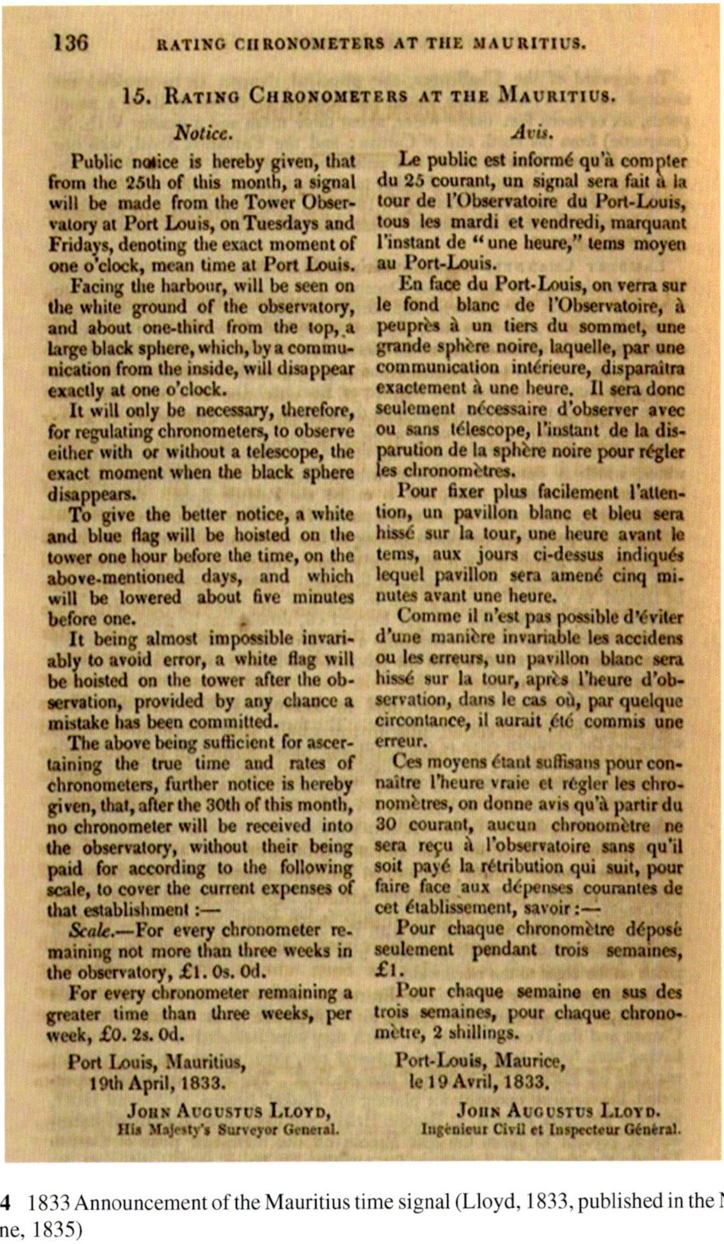

136 RATING CHRONOMETERS AT THE MAURITIUS.

15. Rating Chronometers at the Mauritius.

Notice.

Public notice is hereby given, that from the 25th of this month, a signal will be made from the Tower Observatory at Port Louis, on Tuesdays and Fridays, denoting the exact moment of one o'clock, mean time at Port Louis.

Facing the harbour, will be seen on the white ground of the observatory, and about one-third from the top, a large black sphere, which, by a communication from the inside, will disappear exactly at one o'clock.

It will only be necessary, therefore, for regulating chronometers, to observe either with or without a telescope, the exact moment when the black sphere disappears.

To give the better notice, a white and blue flag will be hoisted on the tower one hour before the time, on the above-mentioned days, and which will be lowered about five minutes before one.

It being almost impossible invariably to avoid error, a white flag will be hoisted on the tower after the observation, provided by any chance a mistake has been committed.

The above being sufficient for ascertaining the true time and rates of chronometers, further notice is hereby given, that, after the 30th of this month, no chronometer will be received into the observatory, without their being paid for according to the following scale, to cover the current expenses of that establishment :—

Scale.—For every chronometer remaining not more than three weeks in the observatory, £1. 0s. 0d.

For every chronometer remaining a greater time than three weeks, per week, £0. 2s. 0d.

Port Louis, Mauritius,
19th April, 1833.

John Augustus Lloyd,
His Majesty's Surveyor General.

Avis.

Le public est informé qu'à compter du 25 courant, un signal sera fait à la tour de l'Observatoire du Port-Louis, tous les mardi et vendredi, marquant l'instant de "une heure," tems moyen au Port-Louis.

En face du Port-Louis, on verra sur le fond blanc de l'Observatoire, à peuprès à un tiers du sommet, une grande sphère noire, laquelle, par une communication intérieure, disparaîtra exactement à une heure. Il sera donc seulement nécessaire d'observer avec ou sans télescope, l'instant de la disparution de la sphère noire pour régler les chronomètres.

Pour fixer plus facilement l'attention, un pavillon blanc et bleu sera hissé sur la tour, une heure avant le tems, aux jours ci-dessus indiqués lequel pavillon sera amené cinq minutes avant une heure.

Comme il n'est pas possible d'éviter d'une manière invariable les accidens ou les erreurs, un pavillon blanc sera hissé sur la tour, après l'heure d'observation, dans le cas où, par quelque circontance, il aurait été commis une erreur.

Ces moyens étant suffisans pour connaître l'heure vraie et régler les chronomètres, on donne avis qu'à partir du 30 courant, aucun chronomètre ne sera reçu à l'observatoire sans qu'il soit payé la rétribution qui suit, pour faire face aux dépenses courantes de cet établissement, savoir :—

Pour chaque chronomètre déposé seulement pendant trois semaines, £1.

Pour chaque semaine en sus des trois semaines, pour chaque chronomètre, 2 shillings.

Port-Louis, Maurice,
le 19 Avril, 1833.

John Augustus Lloyd.
Ingénieur Civil et Inspecteur Général.

Fig. 18.4 1833 Announcement of the Mauritius time signal (Lloyd, 1833, published in the Nautical Magazine, 1835)

seen in the drawings. This slide, when down, is supported by a **T** formed of a very strong spar resting on the solid coral beneath, in order to prevent concussion when falling, and covers a large black sphere. Twice a-week, at about twelve o'clock, a flag is hoisted, and the slide hauled up from the inside, and the lever pushed out so as to just support the edge of the slide; at 1¼ beats of a pocket chronometer, equal to 5–10ths of a second before one o'clock, mean time, the lever is smartly struck on a round plate for the purpose and the slide

> immediately hides the ball. This period of time is the exact quantity the slide takes to entirely cover the black disc from the moment it is struck inside, and which was determined by a mean of several trials.
>
> The tower being all white, and even the windows whitened, at a distance of four miles the black disc is easily discerned as a beautifully clear spot on a large white ground; and the disappearance of the spot is so apparently instantaneous at that distance, that it has almost a magical effect. …

Lloyd's chosen arrangement was clearly unlike the 1833 design for Greenwich (The Greenwich Time Ball, 1835) or the arrangement proposed by Wauchope (1830). It used a stationary black "sphere" on a white background behind a shutter whose complete closure, not the moment of release, signalled the exact time. Both the paper and the original notice (Fig. 18.4) refer to a sphere or ball, but these may have been euphemisms for a painted disc on a white background. Lloyd was a talented watercolourist and his set of paintings of the proposed observatory have survived in the archives of the *UK Hydrographic Office* (Lloyd, 1831–32). Two of these paintings are shown in Figs. 18.5 and 18.6. The black-painted disc behind a shutter would have been a simple and effective solution. A shutter in free-fall for 0.5 s would drop 4 feet, so the disc is likely to have had a diameter of 4 feet (1.2 m), or slightly less.

A notice concerning withdrawal of unpopular charges for chronometer calibration was published in 1836 (Lloyd, 1836). This is shown in Fig. 18.7. No further evidence of continued operation has been found.

One of the earliest daguerreotype cameras was used in Mauritius to create images of the observatory buildings. The known images are from the opposite side to the time ball or disc, so do not provide any clarification of the time signal design. They

Fig. 18.5 Lloyd's observatory proposal in 1831–32. (Courtesy, UK Hydrographic Office)

Fig. 18.6 Lloyd's time signal proposal in 1831–32. (Courtesy, UK Hydrographic Office)

CHRONOMETERS AT THE MAURITIUS.

IN the 4th vol. of the *Nautical Magazine*, p. 136, will be found the particulars of a notice respecting the rating of chronometers at the Mauritius. It will be seen by the following that the expense of sending them to the observatory, has been done away. For the particulars of the time-signal, which it appears will be continued, we must refer our readers to the notice above-mentioned.

Notice.—In order to afford the greatest facility to masters of vessels arriving at this port, to have their chronometers properly rated and examined, notice is hereby given to all merchants and ship-owners, that from and after the 1st of October next, chronometers belonging to shipping trading to this port, will be received and rated at the observatory during the stay of the vessel, free of any charge or expense whatever.

Masters of vessels are also at liberty to send their sextants, circles, or barometers, to be verified.

Time-keeper will be received every day from 12 to 1, and ½ past 3 to 4 o'clock; and one day's notice will be required before being taken away.

The exact mean time 1 o'clock signal will as usual be given on the stated days, viz., Tuesdays and Fridays.

J. A. LLOYD, Surveyor-General and Civil Engineer.

Observatory, Port Louis, 22d Sept., 1836.

Fig. 18.7 1836 Announcement of fee changes (Nautical Magazine, 1836)

show a large square building next to the observatory tower which was not included in the watercolours. Figure 18.8 shows an engraving that was made from an 1840 lithograph by Dureau. Figure 18.9 shows another lithograph, dating from 1853. The

Fig. 18.8 1840 Lithograph by Dureau. (Courtesy Jacques Pougnet)

Fig. 18.9 1853 Lithograph. (Courtesy Jacques Pougnet)

time signal is likely to have ceased operation by that time and the flagstaff had been removed. The harbour was developed extensively after the Old Colonial Observatory was demolished, so the approximate location shown in Fig. 18.2 is no longer on the waterfront.

2.3 Changes to the First Mauritius Observatory

Changes are known to have been made to the Old Colonial Observatory by the Army in Mauritius during the 1850s. Another observatory was established about 200 m from the first one in 1852, primarily for meteorological observations. This was demolished in 1858, to make way for railway development (Mülnier & Padya, 1974). The army observations were continued at the old observatory). This may explain an approach to Airy about a new time ball for Mauritius (Airy, 1861). By 1860, there were increasing environmental problems at the old observatory, including dust, noise and vibration from dock traffic that disturbed the instruments. It was subsequently decided to build a new observatory at Pamplemousses, next to the Botanic Garden, about 10 km north of Port Louis. The first observatory building was sold to the New Mauritius Docks Company in October 1866. The instruments were moved first to a temporary location at a house in Port Louis and then in 1871 to a house in the village of Pamplemousses. They were used there until the new observatory could be completed (Mülnier & Padya, 1974).

2.3.1 Possible Erection of a Conventional Time Ball at the Old Observatory

There is a photograph of the old observatory in *Le Musée de la Photographie* in Port Louis which shows a mast and what appears to be a conventional time ball. The photograph is believed to date from the late 1860s. This may have been an experimental installation to give a more elevated signal, possibly for experimental purposes before the tower was demolished to facilitate harbour development. It was not recorded in Mauritius almanacs during the 1860s.

3 Signal Mountain Time Ball

The *Mauritius Almanac* was first published in Mauritius in 1851 and re-issued annually. Editions were also published in London. They usually included extensive summaries of meteorological observations, but the time ball service was not mentioned before 1866 when a new time ball service was announced (*The Mauritius Almanac*):

> Mean time is given by means of a ball on the Signal Mountain every day at 1 o'clock p.m.; Sundays and holidays excepted.
>
> The ball will be hoisted half mast at 5 minutes before 1 o'clock, and will be mastheaded at 2 minutes before 1 o'clock.
>
> The disconnection of the ball from the cross head will mark the precise instant of its fall.
>
> In the event of accident or error, the ball will again be hoisted half mast and remain so till 2 minutes to 2 p.m. when it will be mastheaded and let fall at 2 p.m. precisely, and so on from hour to hour till the correct time is given.
>
> Chronometers are received to be rated every day from noon to 2 p.m., at the Transit Chamber at the base of the Signal Mountain.

This appears to be the first mention of a time ball on Signal Mountain and of a Transit Chamber below. A similar notice was published in 1868 (The Mauritius Almanac). Charles Meldrum had been appointed "Meteorological Observer" and as "Officer in charge of Time Ball". He was on leave in 1868, so staff were appointed to deputise for him.

The 1870 entry shows that Meldrum had returned to Mauritius (The Mauritius Almanac). There was no description of the time ball in 1870. A published letter showed that the time ball often failed to operate between about 1866 and 1870. Reliable operation of a time ball that was remote from a temporary home for transit instruments may not have been achievable within available budgets: laying of permanent telegraph lines, for example, may have been deferred until after construction of the new observatory. Meldrum's period of leave in England enabled him to procure new instruments and he became the first Director of the Royal Alfred Observatory (Mülnier & Padya, 1974).

3.1 Problems with the Time Ball

A letter published in 1870 was highly critical of the then time service at Mauritius ("Caleb Quotum", 1870). Caleb Quotum is a pseudonym for "Jack of all trades". The relevant part of the letter is reproduced below. The author failed to appreciate that logbooks were retained for copying of meteorological observations (Mülnier & Padya, 1974).

> Now I know they have an observatory there and a time-ball on the signal mountain. But what is the use of either to the ships that go there. The observatory is of no use to the shipping further than depriving every English ship of her log book during her stay in port, and you have to go and seek it yourself when getting ready for sea! and the timeball, to my certain knowledge, has not been worked for two years, and if I were to believe the report of others, it is twice two since it was at work. The time is given now by dropping a flag, – on some days you get it, and on others you don't. I managed to get time on seven occasions in three weeks, which was as often as the flag was dropped during my stay in port. I know that they have had a transit instrument there this last three or four years, but it is not in its place yet, and I don't think it ever will be placed, unless there is some one that understands what it is for, and will take it in hand. I can't believe that it will take such a time to get it into position if they thoroughly understand it at the observatory.

The Editor responded "Our correspondent has done a good turn to his brother ship-masters in reporting the discontinuance of the time signal at the Mauritius. We trust that his representation will lead to the subject being looked into. ...".

3.2 The Royal Alfred Observatory

The foundation stone for the new observatory was laid in person by Prince Alfred, Duke of Edinburgh on 30 May 1870, accompanied by the Governor, Sir Henry Barkly (New Observatory, 1870). It was to be known as the Royal Alfred Observatory and was fully operational by 1875. Figure 18.10 is a photograph of the observatory, taken in 1920.

Brief entries concerning the time ball were included in most Mauritius almanacs from 1876 onwards. The initial entry is transcribed below (The Mauritius Almanac). It was noted under the heading "Position of the Colony in 1874".

> ... A time ball on the Signal Mountain at Port Louis is dropped by electricity from the Observatory, three times a week. This is of great importance to the ships in harbour.

Automatic electric operation ceased on 29 April 1892, the date of one of the most severe cyclones ever experienced on the island, and was never re-established on Signal Mountain (Annual Report, 1903). After the cyclone, the time ball operator would have released the signal by hand on receipt of a signal from the Observatory.

Fig. 18.10 Royal Alfred Observatory in 1920 (https://www.lexpress.mu/article/280955)

The Time-Ball on Signal Mountain was dropped by hand three times a week until 30 September 1900. After that, it was dropped every day except Saturdays, Sundays and public holidays (Royal Alfred Observatory, 1901). Although the time ball was dropped by hand, the accuracy of the signal was impressive. Statistics were quoted in annual reports. Those for 1903 are transcribed below (Annual Report, 1903).

> **Time ball.** The time-ball on the signal mountain in charge of the military signallers, was dropped by hand daily at 1 p.m. every week day (Saturdays and Public Holidays excepted). The fall of the ball is observed from the Observatory, and the exact time of passage across the yard arm is telephoned to the Harbour-Master and also communicated to the local press.
>
> In the year 1903 the error was less than 0.3 secs on 119 occasions; from 0.3 secs to 0.5 secs on 49 occasions, from 0.6 secs to 0.9 secs on 27 occasions, and above 1 sec on 19 occasions. The current was defective on 19 occasions, and the signal mountain not visible from the Observatory on 13 occasions.
>
> I understand that a sum of Rs. 3000 is to be born on this year's estimates for the erection of an Electrical Time-ball to replace the apparatus destroyed in the cyclone of 1892, April 29.

The new electric time ball was to be erected near the Port Office.

The time ball on Signal Mountain was no longer mentioned in *Mauritius Almanacs* after 1907, but it continued to appear in Admiralty lists from 1911 as a repeater ball "Dropped, by hand, when the ball on the port tower is dropped". It appears that it was no longer the responsibility of the Observatory and was operated independently by military signallers at the Station on top of Signal Mountain. Its latitude and longitude, which appear to have been stated correctly in 1880 and then incorrectly in some later editions, were not stated at all in 1911 or later. It was finally withdrawn altogether in 1945, but the Port Office ball was still in use in 1947. It had been controlled using wireless signals since 1929, following cessation of astronomical observations at the Royal Alfred Observatory.

The 1908 Admiralty list was prepared before introduction of the new service, but it indicated that the Signal Mountain time ball had been changed to a pair of cones, whose moment of separation was the time signal (see Kinns, 2020b). It was described as "2 cones, bases together; upper cone white, lower cone black (diamtr. 6 feet)." Additional information still referred to a ball. No confirmation of this arrangement has been found in Mauritius. Other announcements refer only to a ball.

4 Time Ball at the Port Office

A white ball with a reduced diameter of 1.2 m and a larger drop height of 6 m was introduced as the new principal signal near the Port Office in March 1907 after its erection in the previous year. It is shown in Fig. 18.11.

The following extract from the 1913 *Almanac* described the official signals.

> The new time ball erected, in 1906, on the Port Tower at an elevation of 80 feet is dropped electrically by the Standard clock of the Royal Alfred Observatory at 1 p.m. 60th meridian time 4 hours East of Greenwich …

Fig. 18.11 Port Office time ball. (Courtesy Jacques Pougnet)

> Hourly signals are also sent to the Port Office and to Messrs. Guillemin & Co., Jewellers, Port Louis for the purpose of regulating Marine Chronometers.

4.1 Light Signals in Mauritius

Another signal was added in the 1915 *Almanac*:

> … and the electric lights are lowered at 8 p.m. all over Port Louis, Beau Bassin, Rose Hill, 4 Bornes, Moka and St. Pierre.

The *Mauritius Almanac* for 1920 included the following statement about the time service.

> **THE ROYAL ALFRED OBSERVATORY**
> The time service has been greatly extended of recent years. A complete equipment of electric clocks has been set up in a specially constructed clock chamber which forms the basement of the transit observatory. There are three services for the distribution of time. The first by the dropping of a 1 p.m. time ball at the Port Tower in Port Louis, the mechanism of which is actuated electrically by the Standard Clock of the Observatory; the second by flashing signals along the telegraph wires to all stations … and the third by an electric light signal. The latter is arranged to lower the electric lights throughout the townships of Port Louis, Beau Bassin, Moka, St. Pierre and Quatre Bornes exactly at 8 p.m. on a signal from the Observatory clock.

The practice of dimming lights, apparently unique to Mauritius, continued until at least 1931. It was never mentioned in Admiralty lists.

4.2 *Time Signals After 1920*

4.2.1 Decline of Signal Accuracy

The accuracy of time signals deteriorated during the 1920s, but new equipment was installed in 1925 and there was optimism that the quality of the time service would be restored to previous standards (Annual Report, 1925). The new station was at Vacoas.

> **Time equipment and clocks.** The new transit instrument was finally set up and its constants determined in November. The erection of the two new clocks was also completed in November and ratings commenced early in December. The Wireless receiving set was successfully brought into use and the time signals from Bordeaux received daily. A preliminary determination of the Longitude of the new station was determined in December as 3 h. 49 m. 55 s. 8 fast on Greenwich …
>
> The preliminary rates of the new clocks are exceptionally good and it may be confidently excepted [*sic*] that the period of inefficiency in the Time Service of the Observatory has now been permanently closed. …

Unfortunately, the optimism in 1925 seems to have been misplaced. Later notices indicated that transit observations were no longer made and that the time ball was no longer sufficiently accurate for rating of chronometers (Annual Report, 1927).

> **Time equipment and clocks.** Transit Observations have not been made and two new clocks erected in 1925 have not been used. A time signal has been received daily on the Wireless receiving set at Vacoas Aerological Station. The time ball is dropped by hand from the Observatory. No measurements of the accuracy of the time ball have been attempted but errors of two seconds must be frequent in spite of all reasonable care, and although such errors are unimportant for most purposes in the life of the Colony, the rating of accurate chronometers for navigational or other purposes should not be attempted from the Time Ball.

A similar notice was included in the report for 1928.

4.2.2 Withdrawal of the Observatory Time Service

It was announced that the Observatory had ceased to be responsible for the time service in November 1929 (Annual Report, 1929).

> **Time services.** Transit Observations have not been made. Until November a time signal was received daily on the Wireless Receiving set at Vacoas and the time ball at the Port Tower was dropped by hand from the Observatory. In November the Wireless set together with an oil motor and dynamo were handed over to the Port Office and the Observatory ceased to be responsible for timekeeping in the Colony.
>
> Although some astronomical equipment remains at the Observatory no use has been made of it for many years and the handing over of the time service formally closes the astronomical side of the Observatory's activities.

4.2.3 Operation by Wireless Signals

By 1931, the time ball was operated electrically from the Port Office using a wireless signal (Mauritius Almanac, 1931–32). The time light signals continued.

> **Time Ball and Time Signals.** – The time ball erected in 1906 on the Port Tower at an elevation of 60 [*sic*] feet is dropped electrically by the Wireless set now installed at the Pilots' Office, at 1 p.m., 60th meridian time; (4 hours East of Greenwich) in accordance with Ordnance No.6 of 1906.
>
> Time signals are sent at 1 p.m. through the telegraph department and flashed to all stations on the Railway.
>
> Hourly signals are sent to the Port Office relay from which they are distributed to the Post Office, electric light companies &c. The electric lights are lowered at 8 p.m. all over Port Louis, Beau Bassin, Quatre Bornes, Moka and St. Pierre.

Acknowledgements I am grateful to Rob Allan at the Meteorological Office in London for his help in locating illustrations and articles relating to the history of observatories in Mauritius. He leads the international ACRE (Atmospheric Circulation Reconstructions over the Earth: http://www.met-acre.net) initiative. Several images were kindly provided by Jacques Pougnet, Secretary of the Mauritius Meteorological Society. He has also found notices published in Mauritius that are inconsistent with international lists of signals.

References

Airy, G. B. (1861). *List of Places at which Time Balls have been projected, the erection of none of which is known to have been proceeded with, Papers of George Airy, RGO 6 615, Item 3.* Cambridge University Library.

Annual Report of the Director of the Royal Alfred Observatory. Reports are referenced as *Annual Report*, with the year to which they refer. They were published in Mauritius.

Bartky, I. R., & Dick, S. J. (1981). The first time balls. *Journal for the History of Astronomy, xii*, 155–164.

Caleb Quotum. (1870). About Mauritius Hurricanes, time-balls, and transit instruments, letter to the editor of the *Nautical Magazine and Naval Chronicle for 1870*, 155–157.

Darwin, C. (1836). *Diary written during the voyage of HMS Beagle*, entries between 29 April and 9 May 1836. http://darwinbeagle.blogspot.com/2011/05/5th-may-1836.html

Dash, M. (2012). Run out of Town on an Ass. *Smithsonian Magazine*, dated 4 June 2012. Republished online in 2018. https://brewminate.com/run-out-of-town-on-an-ass/

Evans, D. S., Deeming, T. J., Hall Evans, B., & Goldfarb, S. (1969). *Herschel at the Cape: Diaries and correspondence of Sir John Herschel, 1834–1838.* University of Texas Press.

Herschel, S. J. (1836). Account of an Observatory constructed at the Mauritius by Capt. J. A. Lloyd. Communicated by Sir J. Herschel and read on his behalf. *Monthly Notices of the Royal Astronomical Society, 3*(22), 157–160.

I-DARE Mauritius Project. (2015). *National Archives of Mauritius*, dated 13 May 2015. https://www.idare-portal.org/data/mauritius-project

Kinns, R., (2020a). The 1833 time ball at Port Louis, Mauritius: the forgotten service for chronometer calibration. *International Journal for the History of Engineering and Technology, 89* (1–2), 264–275.

Kinns, R., (2020b). The time balls of Mauritius. *Journal of Astronomical History and Heritage, 23*(2), 281–296.

Lloyd, J. A. (1833). Rating chronometers in the Mauritius. Notice published in the *Nautical Magazine*, 1835, v. 136, dated 19 April 1833.

Lloyd, J. A. (1836). Chronometers in the Mauritius. Notice published in the *Nautical Magazine and Naval Chronicle for 1837*, 365, dated 23 September 1836.

Mülnier, K., & Padya, B. M. (1974). Mauritius: Two hundred years of meteorology. *WMO Bulletin, 23*(4), 228–232. https://library.wmo.int/doc_num.php?explnum_id=6621

New Observatory at the Mauritius. (1870). *Nature*, 7 July 1870, 201–202.

Obituary (1855). John Augustus Lloyd 1800–1854. *Minutes of the Proceedings of the Institution of Civil Engineers, 14*, 161–165.

Royal Alfred Observatory Results of Magnetical and Meteorological Observations. Published in Mauritius (1901).

Supplement concerning Special Commissioners for the Great Exhibition. (1851). *The Illustrated London News,* 28 June 1851.

The Greenwich Time Ball. (1835). *The Nautical Magazine, iv*, 384–586.

The Mauritius Almanac was published in Mauritius from 1851 onwards. The complete title of the first edition was *Bolton's Mauritius Almanac and Official Register for 1851*. The complete title was later changed to *The Mauritius Almanac and Civil Service Register* and then to *The Mauritius Almanac and Colonial Register*. Editions were also printed in London.

Wauchope, R. (1830). Plan for ascertaining the rates of chronometers by signal. *Edinburgh New Philosophical Journal, 8*, 160–162 and 289–291 [The second part, which included the diagram of a time ball arrangement, was submitted on 1 February 1830].

Chapter 19
Indian Subcontinent

Roger Kinns

Abstract The geographical area now includes modern India, Pakistan, Bangladesh and Sri Lanka. Many cities have changed names since independence in 1948. A first time ball service in India was established at Calcutta in 1835. Later, there were three Calcutta time balls in parallel operation. Another early time balls were erected at Madras but it had a chequered history and Madras preferred a time gun as the principal signal for extended periods. Another early time ball was operational at Colaba Observatory in Bombay during 1840, later moved to Bombay Castle. Another time ball was installed at Bombay docks in 1891 and became the principal signal. A time ball at Karachi was announced in 1874 and time balls later existed there at three different locations. Another was established in 1919 in the Andaman Islands. Colombo in Ceylon initially had a time ball, replaced by a semaphore signal which was in turn replaced by another time ball.

1 Introduction

The Indian Subcontinent includes modern India, Nepal, Bangladesh, Pakistan, Bhutan, Sri Lanka, and the Maldive Islands. The Andaman Islands are part of India. Time signals in Admiralty lists were located in what is now India, Sri Lanka and Pakistan.

The partition of India after the end of British rule led to the creation of Pakistan in 1947. Many cities in India have since been re-named. Bangladesh, formerly East Pakistan, became a separate country in 1971. Ceylon was re-named Sri-Lanka in 1972.

Burma had been administered as part of British India, although it was located in Southeast Asia. It is now an independent country in Southesst Asia and was renamed Myanmar in 1989.

R. Kinns (✉)
The Maudslay Society, Pembroke College, Cambridge, England

R. Kinns (ed.), *The World of Visual Time Signals for Mariners*, Historical & Cultural Astronomy, https://doi.org/10.1007/978-3-031-57334-7_19

Table 19.1 Signals in India and Ceylon

Area	Location	1880	1898	1904	1911	1915	1919	1922	1928	1934	1939	1947
India: Bombay (now Mumbai)	Castle: NE Bastion	TB	TB	TB	TB	TB	TB	TB	TB			
		Clock	Clock	Clock	Clock	Clock	Clock	Clock	Clock			
	P&V Docks		TB	TB	TB	TB	TB	TB	TB	TB	TB	
India: Calcutta (now Kolkata)	Fort William	TB	TB	TB	TB	TB	TB	TB	TB	TB	TB	TB
	Port Office		TB	TB	TB	TB	TB	TB	TB	TB	TB	TB
	Port Office		Flag	Flag	Flag	Flag	Flag	Flag	Flag	Flag	Flag	Flag
	Kidderpore Docks			TB	TB	TB	TB	TB	TB	TB	TB	TB
India: Madras (now Chennai)	Port Office	Sema.	Sema.	Sema.	Sema.	Sema.						
	Harbour Office						TB (red)	TB (red)	TB (red)			
	Fort St George					Gun	Gun	Gun	Gun	Gun	Gun	Gun
Andaman Islands:	Port Blair							TB	TB	TB	TB	TB
India (now Pakistan): Karachi	Merewether Pier		TB	TB								
	Manora Point			TB								
	Baba Channel				TB	TB	TB	TB	TB	TB		
Ceylon (now Sri Lanka)	Colombo	TB										
			Sema. (r, w)	Sema. (r, w)	Sema. (r, w)	Sema. (r, w)						
							TB	TB	TB	TB	TB	
Signal type totals		1880	1898	1904	1911	1915	1919	1922	1928	1934	1939	1947
Time ball, including collapsible		3	5	7	6	6	8	9	9	7	6	4
Gun, including powder flash		0	0	0	0	1	1	1	1	1	1	1
Lights		0	0	0	0	0	0	0	0	0	0	0
Other: disc, drum, flag, etc.		2	4	4	4	4	2	2	2	1	1	1

Signals in the Indian Subcontinent from 1880 onwards are shown in Table 19.1. Figure 19.1 shows their locations in India, Pakistan and Sri Lanka. Many cities had more than one time signal. These are shown in separate larger-scale maps. The history of time signals was reviewed in a recent paper which includes transcriptions and quotations from original sources (Kinns, 2020).

2 India

The earliest time balls in India were erected at Calcutta, Madras and Bombay, but many cities had a strong preference for time guns.

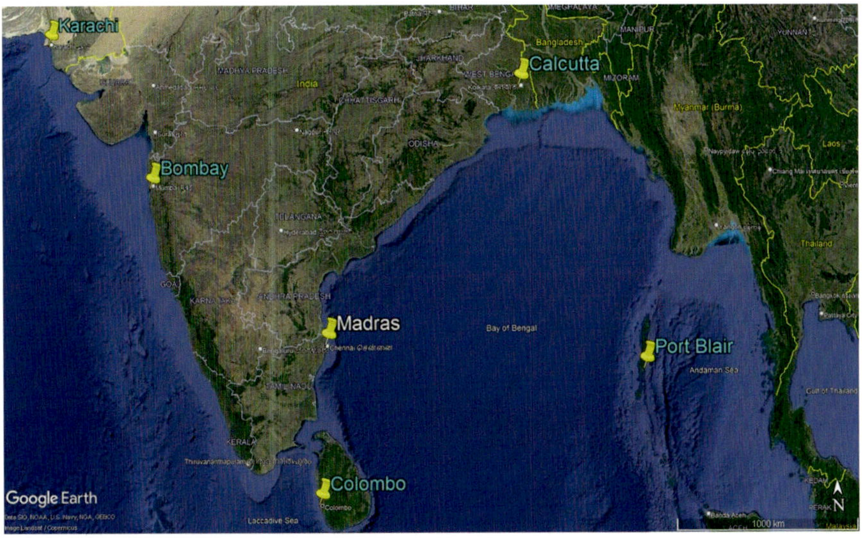

Fig. 19.1 Signals in the Indian Sub-Continent (Google Maps)

2.1 Calcutta

2.1.1 Calcutta: Early Signals

The first time ball in Calcutta, now Kolkata was described in *Historical Records of the Survey of India.* (Phillimore, 1958, 114–115). It started operation in January 1835 and is one of the earliest known time balls. The archived records relate to what is now known as the Great Trigonometrical Survey of India, led by Sir George Everest (1790–1866). Everest, after whom the mountain is named, was Surveyor General of India from 1830 to 1843. Everest saw time signals for mariners as a distraction from the main work of the survey. The following extract concerning Calcutta shows his reluctance to sanction the diversion of staff to time ball provision. It also shows that the primary time ball was at the Surveyor General's Office and that a repeater time ball was operated by the Army. "The Semaphore" was at Fort William.

> … the signal for showing mean time commenced on the 28th January last [1835], and has been continued every day since (Sundays excepted). The signals are made by hoisting a black ball to the top of the staff on the office 5 minutes before noon, and as a preparatory signal to be followed up immediately at the Semaphore. Then at the exact mean noon, as indicated by the clock and chronometer, the ball is dropped down instantaneously, while the same is done at the Semaphore under the management of a non-commissioned officer. … The service was interrupted from 16th July, owing to the move of the office to new quarters and the signals were not resumed until 10th May 1837 … The signals now proceeded regularly, even though Government refused any special allowance for the staff employed, or for the engagement of an extra assistant on Hindu holidays. In April 1838 De Penning's reference to the extra work thrown upon the computing office called down an indignant order

> from Everest to the effect that nothing whatever should interfere with the computations, and the signals were only allowed to continue on that understanding.

Another extract refers to occasional failures to drop the repeater time ball promptly in 1839. "From 1855 the signal was changed to 1 p.m. instead of noon, and in 1858 a new time ball was installed at Fort William operated from the Surveyor General's office by electricity." (Phillimore, 1958, 115). It appears from this extract that "The Semaphore" and "The Fort" are both references to Fort William and that a repeater time ball there started operation in 1835.

A time ball at Calcutta had been noted in 1861 (Airy, 1861). The installation had suffered significant deterioration by 1862 (East India, 1863, 74). The braking arrangement appears to have used a piston in a wooden cylinder to provide an air spring. The use of wood suggests that it was designed and built in India and may have been part of the 1835 arrangement.

2.1.2 Calcutta: Signals After 1880

The Admiralty lists do not mention a time ball at the Surveyor General's Office, which would have been invisible to mariners and was unnecessary when electrical connections were established. Another time ball at the Port Commissioner's Office was listed in 1898. It was accompanied by a flag that was used both as a first preparatory signal and as a listed signal. Another time ball had been installed at Kidderpore (Kidderpur) Docks by 1904.

Figure 19.2 shows the locations of time balls at Calcutta that appeared in Admiralty lists. The signal at Fort William was included in the 1880 list. A

Fig. 19.2 Locations of Calcutta time balls (Google Maps)

combined flag and time ball signal at the Port Office was included in the 1898 list and the time ball at the Prince's and Victoria Docks was included in the 1904 list. All were still listed in 1947 and continued for many years after that.

2.1.3 Calcutta: Fort William

There was a significant exchange of letters concerning proposed time signals at Calcutta between the India Office in London and the Astronomer Royal at Greenwich in April 1878. The request from the India Office is shown below (Shaw, 1878).

> With reference to a requisition which has been received from the Meteorological Office Calcutta I have the honour to request that you will inform me what is the most appropriate description of Time Ball and Dropping Apparatus and also from whom it can be procured. The description sent from India of what is required is as follows; "Time Ball and Dropping Apparatus with a Clock to start the signal and an apparatus for firing a gun by an electrical current and sending a return signal."

The request from the India Office reflects the arrangement of a time ball and time gun at Edinburgh, which were noted for the high accuracy of the time signals (Kinns, 2011). The detailed reply from Airy, sent on the following day, contains useful detail from his own experience (Airy, 1878).

> Acknowledging your letter of April 12, relating to the proposed erection at Calcutta of a Time Signal Ball and Dropping Apparatus with apparatus for firing a gun, and with clock communication for effecting these objects: –
>
> 2. I have to state that the Time Signals with which I have been personally connected (at Greenwich, Deal, Sydney) were constructed by Messrs. Maudslays & Field, with the usual skill of those eminent engineers, and have been perfectly successful.
> 3. The Time Signal at Edinburgh (which, in combination with gun-firing approaches nearest to the proposed apparatus at Calcutta) was constructed, I think, by the same engineers, but on this and other points relating to Edinburgh, it would probably be best to address directly to Professor C. Piazzi Smyth, The Observatory, Calton Hill, Edinburgh.
> 4. For the time signal at Bidston (for the harbour of Liverpool), it would be best to inquire of John Hartnup Esq., The Observatory, Bidston, Birkenhead.
> 5. In regard to the works at Calcutta, I would submit that it may be very desirable to ascertain whether it will be necessary to build a tower from its foundation, or whether a tower applicable to such purpose exists. At Greenwich, advantage was taken of one of Sir Christopher Wren's turrets, and a proper modification of apparatus was made by Messrs. Maudslays. At Deal, I remarked a disused Semaphore Tower, and immediately adapted it for the Time-Signal.

A new time ball at Fort William was announced in 1880, using an electric apparatus that would also fire a gun (Electric time-ball for India, 1880). The apparatus was designed by Alfred Varley and supplied by Messrs. Siebe and Gorman. It was similar to that designed by Maudslays in the 1850s, using a rack and pinion hoisting arrangement and an air spring to control its descent. Part of the article is reproduced below.

> The Indian Government having decided upon erecting an electric time-ball and gun-firing apparatus at Fort William, Calcutta, the same has been designed by Mr. S. Alfred Varley,

Fig. 19.3 Calcutta time ball at Fort William. (Courtesy Klaus Hülse collection)

> who has had special experience in this class of electrical mechanism. … The time-ball is 5 ft. 6 in. in diameter and weighs, with its attachment, about 5 cwt. It is wound up a mast by means of a rack and pinion, and when released falls through a space of 10 ft. Attached to the ball, and beneath it, is a rod having a piston at its lower end. After the ball has fallen 4 ft. this piston enters a pneumatic cylinder, in which it meets with resistance from the contained air. The air forms a cushion, and allows the ball to finish its descent without any shock or concussion being experienced, this result being assisted by a pressure relieving valve in connexion with the pneumatic cylinder. The ball, as regards its size and the space through which it falls, is similar to those which were erected at Charing-cross and Liverpool in 1853 (*sic*) and 1854 at the suggestion of the Astronomer-Royal, to whom we are indebted for the idea of utilizing electricity for discharging time-balls … We may add that a set of time-ball apparatus similar to that just described is in course of construction for another Indian station.

The gun was not mentioned in Admiralty lists. The tower is shown in Fig. 19.3.

2.1.4 Calcutta: Port Commissioner's Office

The time ball at the Port Commissioner's Office had an elevation of 35 m above high water. The drop height and diameter were not stated.

A flag was hoisted twenty minutes before the time ball was dropped at the Port Commissioner's Office. This was an additional preparatory signal, heralding the raising of the time ball close-up fifteen minutes later.

2.1.5 Calcutta: Kidderpore Docks

The Kidderpore (Kidderpur) Docks time ball was on a tower, located between two lock entrances. It was first listed in 1904. It had an elevation of about 25 m above high water, with a drop height of 1.8 m.

All three Calcutta time balls were hoisted close-up at 5 min before the signal at 1 pm Calcutta (Fort William) mean time.

2.2 *Bombay*

2.2.1 Bombay: Colaba Observatory and the Castle

Samant and Samant (2016) noted that the first time ball at Bombay was erected at Colaba Observatory in 1840. It was moved subsequently to the "Bombay Castle Flag Tower" and continued until 1928 (Samant & Samant, 2016, 5). Figure 19.4 shows the relative locations of the Observatory, the Castle and the Prince's and Victoria Docks.

The following notice was published in 1853 (The Time Ball, 1853).

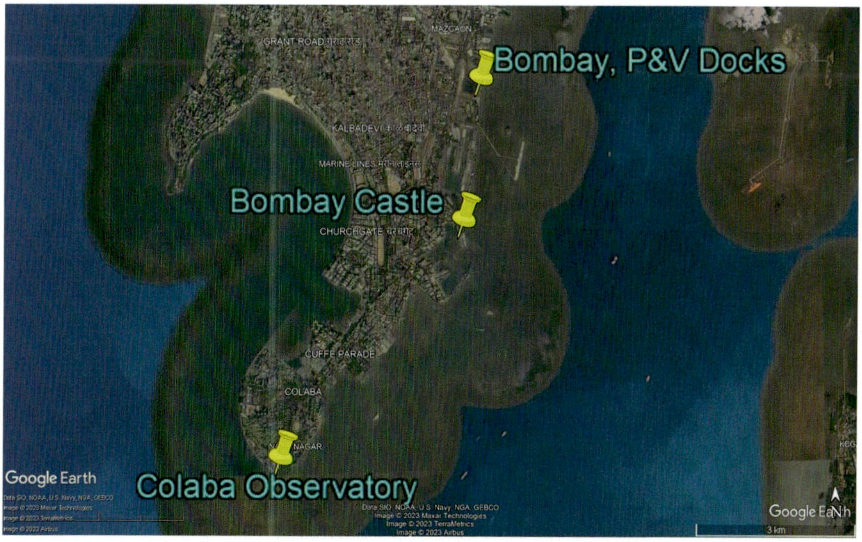

Fig. 19.4 Locations of Bombay time balls (Google Maps)

> THE TIME-BALL at the Observatory having been taken down for the purpose of making a larger one, and refitting the gear, apparatus, &c. there will be no Ball dropped, or Time given to the shipping in harbour (except the flash of the evening Gunfire) till the alterations &c. have been completed.

The evening gun was not an accurate signal, often being in error by over one minute, but the exact time of the flash was usually published a day or more later.

A significant article was published in July 1936 (The Time Ball, 1936). Its author has not been identified, but it contains an apparently authoritative history of the Bombay time balls. The content relating to the Castle is transcribed below:

> … The ball was superseded by a flag, but, following complaints that this was not easily visible, the ball was restored and remained in operation until 1865 when it was removed to the Castle Flag Staff Tower. Here it continued to work until 1928, together with a large clock fitted with a second hand and electrically connected to the master clock at the Observatory.

A report on the performance of the time ball from 1 May 1861 to 30 April 1862 was favourable "… At Bombay, the time ball has been dropped daily, Sundays and holidays excepted, without a single failure." (East India, 1863, 74).

Tenders for construction of a time ball tower at the Castle were invited in March 1866 (Morland, 1866a). A flag was used as the substitute signal until the transfer had been completed (Morland, 1866b):

> Notice is hereby given that until the completion of the removal of the Time Ball to the new Tower on the North East Bastion of the Castle, time will be given daily, from the Colaba Observatory, by dropping a Flag at 1 p.m.

The controlled clock near the Castle time ball was one of very few noted in Admiralty lists. It cannot be seen in Fig. 19.5 so it may have been introduced at some time after the time ball. It was listed from 1880 until 1928, with the following description in the 1908 list:

> Clock (having a seconds hand) beats by electricity in unison with the clock at Kolaba Observatory, and indicates practically exact India standard mean time.

The clock time had been Bombay mean time in lists from 1880 to 1904.

2.2.2 Bombay: Docks

An additional time ball was erected at the Prince's and Victoria Docks and was operational in 1891 (The Time Ball, 1936). Figure 19.6 shows the Dock Offices at Bombay, with the time ball on top of the tower. It was a large ball with a diameter of 1.8 m. The colour of the ball was not stated in Admiralty lists, but it was painted red in 1942 (Sailing Directions, 1942, 266). A prominent feature of the tower is the clock associated with the time ball. This clock was not electrically controlled from the Observatory (Time Ball, 1936) and was not recognised by the Admiralty as a time signal for mariners.

Fig. 19.5 Bombay Castle time ball. (Courtesy Klaus Hülse collection)

Fig. 19.6 Bombay docks time ball. (Courtesy Klaus Hülse collection)

An accidental explosion in 1944 led to cessation of the Docks time ball service but remains of the apparatus still exist (Samant & Samant, 2016). The remains of the ball have red paint.

2.3 *Madras*

2.3.1 Madras: Early Signals

No design details of a pre-1850s time ball at Madras, now Chennai, have been found. It was, however, under consideration and may have operated for a short period, even before the 1835 time ball at Calcutta. An intent to provide a time ball service appears to have been undermined by the desire of George Everest (1790–1866) to prioritise resource allocation to the Madras-based Great Trigonometrical Survey of India (Kinns, 2020). A Madras time gun was certainly operating in 1841, whereby the precise time of firing was available to ships on the following day (Madras, 1841). A time ball at Madras, but not the time ball at Calcutta, was mentioned when a time ball was being considered for Edinburgh in 1846 (Kinns, 2011, 267). Bartky (1987) suggested "after 1839" and Howse (1997, 83) suggested "1840s" as dates of introduction of a time ball at Madras.

A Madras time ball and a telegraph connection from Madras Observatory to the Beach Office was established after August 1853, according to a notice in *The Journal of the Society of Arts* (Kenrick, 1853). John Henry Kenrick (1816–1895) was Secretary to the Madras Polytechnic Institution in 1854 (Dutt, 1854). He appears to have funded development of a time ball and telegraph link from his own resources, perhaps hoping for later recovery of his costs (see Kinns, 2020). The following notice indicated that a Madras time ball was working in January 1854 (Elliott, 1855), but the service may have been suspended soon afterwards:

> At Madras, a time ball has been established over the Master-Attendant's office, in electrical connection with the observatory. The ball is hoisted up five minutes before one o'clock, and is dropped at one p.m., observatory mean time. The observatory is two miles and a quarter west of the Master-Attendant's flag-staff.

Kenrick proposed in July 1854 that the apparatus should be transferred to the Government of Madras. This proposal was considered in London (India Office Records, 1855) but the outcome is uncertain and any transfer is likely to have been delayed by turmoil in India that led to creation of the India Office. A time gun was still the principal signal in 1858 (Toynbee, 1859). The following extracts from a report covering the period from 1 May 1859 to 30 April 1860 show that a time ball was in use by 1859 but that it was initially unreliable and had to be rebuilt (East India, 1861, 71, 102, 110).

> The time ball, which was put on the centre of the sea customs house in 1858-59, has not worked satisfactorily, owing to the apparatus being defective. It is now under repair, and it is expected will soon mark the time more correctly than heretofore. …

> … The time ball has sustained severe damage from heavy gales to such an extent, that it was deemed advisable to construct a new one on a much more substantial principle. …
>
> … The correct time of the firing of the evening gun has been regularly published, and the time ball has also been in use; but till February its mechanism was very defective, and failures from this cause were frequent. In April the ball itself threatened to give way, and a new one has been constructed. No future difficulty in the matter is now anticipated.

2.3.2 Madras, Signals After 1880

The Madras time ball had been replaced by a semaphore signal by 1880. It was replaced in turn by another time ball in 1917, but the service was not listed after 1932. Madras tended to favour the time gun at Fort St. George which was listed in 1915 and was still operating in 1947.

2.4 Andaman Islands: Port Blair

A time ball was established at Port Blair in the Andaman Islands in 1919. It was still listed in 1947. The time ball was only dropped on Thursdays and Fridays and was considered to be unreliable.

The Andaman and Nicobar Islands were governed by India. The government headquarters were at Port Blair (Bay of Bengal Pilot, 1916, 293). There was no time signal at Port Blair in 1916, but there was a reference to time signals at Madras, Calcutta and Rangoon on the Bay of Bengal (Bay of Bengal Pilot, 1916, 105).

> **Port Blair** is an excellent harbor and the headquarters of the representative of the Indian Government at the Andaman and Nicobar Islands …
>
> Vessels are not allowed to visit Port Blair for commercial purposes except under Government contracts, and in order to visit and trade at the Nicobars a permit must be obtained from Port Blair.

The entry was repeated in 1923 (Bay of Bengal Pilot, 1923, 314). An additional paragraph in 1923 confirmed a time ball at Port Blair:

> **Time signal.** – A ball is hoisted on the radio tower at Port Blair 10 minutes before the signal and dropped at lh. 40m. 00s. local mean time, corresponding to 19h. 30m. 00s. Greenwich mean time. The signal is made on Thursdays and Fridays only and is unreliable.

The Admiralty lists for 1919 onwards included the Port Blair time ball and indicated that it was established in 1919. Standard time was then 6 hours 10 min fast on Greenwich. It was still listed in 1954 and still dropped at 1h 40m pm with the qualification "The signal is made on Thursdays and Fridays only. Unreliable". Standard time had changed to be 6h 30m fast on Greenwich by 1947.

3 Pakistan

Pakistan was part of India before Partition in 1947. The only visual signals in Admiralty lists were time balls at Karachi.

3.1 Karachi

The first notice of a time ball at Karachi, then spelled Kurrachee and now in Pakistan, was dated 3 August 1874 and published in the *Nautical Magazine* (Parker, 1875, 80). It was signed by G. C. Parker, Acting Master Attendant, Kurrachee, Master Attendant's Office, Manora,

> *Time Ball.* – Masters of ships are informed that the time will be given from the Master Attendant's Office, Manora, by means of a time ball, on Tuesdays and Fridays, at one o'clock p.m., precisely (Kurrachee mean time). At ten minutes before one, the ball will be raised half-mast high; at five minutes to one, it will be hoisted to the yard-arm, and at one o'clock mean time the ball will drop. In the event of any accident or error, No. 1 Marryatt's Code will be hoisted at the masthead, and will remain flying until the ball is again hoisted as before for giving out the time in a similar manner at two o'clock p.m., mean time. Longitude of Manora Point 66° 58′ 15″ E., or 4 h 27 m 53 s fast of Greenwich mean time.

This time ball did not, however, feature in the 1880 Admiralty list.

In 1898, a time ball was located on Merewether Pier, but there appear to have been concerns about its reliability. Figure 19.7 shows the Pier time ball in 1902. Another time ball at Manora point had been established, or re-established, by 1904 as a daily service, excluding Sundays. They were both dropped at 1 pm local mean time (8 h 32 m 6.7 s am GMT) at Merewether Pier. Manora Point was only 5″ further East, so the astronomical time difference was 0.3 s.

The Merewether Pier and Manora Point time balls had both been withdrawn by 1908 and replaced by a time ball at the south end of the Baba channel. The signal position was described in the 1916 and later Admiralty lists as "Beacon on low sandy spit, south side of Baba channel". It had been described incorrectly as being on Baba island in the 1908 and 1911 lists, with the same latitude and longitude. It was 34″ further West than Merewether pier, so local astronomical time was 2.3 seconds later. The relative locations of the three time balls are shown in Fig. 19.8. The Baba channel ball was always dropped at 8.30 am GMT, corresponding to 2 pm India standard time (local time of 0 h 57 m 51 s pm). The time signal was still extant in 1935 but had been withdrawn by 1939.

Fig. 19.7 Karachi time ball at Merewether Pier. (Courtesy Klaus Hülse collection)

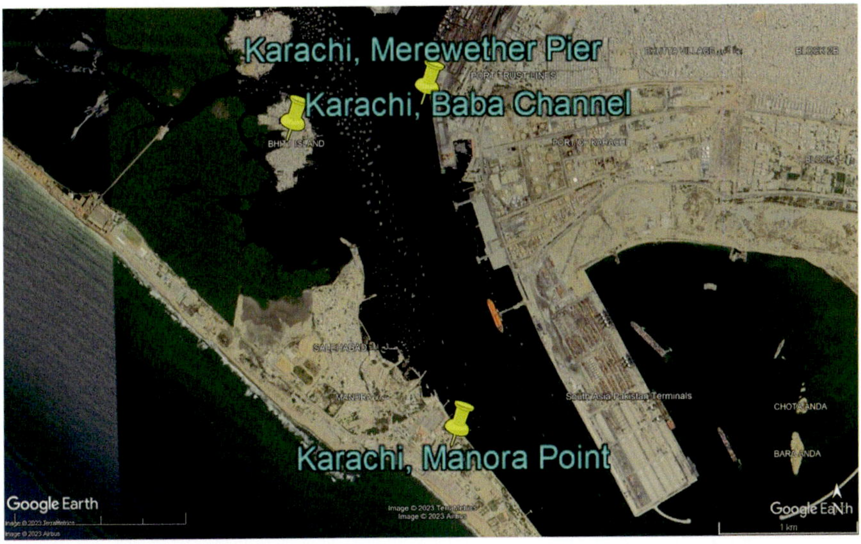

Fig. 19.8 Karachi time ball locations (Google Maps)

4 Ceylon

Colombo was the only port in Ceylon, now Sri Lanka, with a time signal in Admiralty lists. The first time ball at the Master Attendant's office had a small diameter of 3 feet (0.9 m). The drop height was stated to be 40 feet (12 m) in the 1880 Admiralty list. The first time ball was dropped at 1 pm local mean time on weekdays, 5 h 19 m 22 s ahead of GMT.

By 1898, the ball had been replaced by a semaphore signal coloured red and white, apparently at the same location and same height as the time ball. It was operated using a clock that was regulated by telegraph from Madras Observatory. It was raised to about 45°, 5 minutes before the signal, then horizontal 2 min before the signal. Poor visibility and irregularity of the semaphore signal had been noted in the 1904 list with an additional note in 1908 that a local observatory in Colombo was being erected. In 1904, it was dropped every day at 4.15 pm Madras Standard time, which became India Standard time, 5½ h ahead of GMT. It was also dropped at 8.15 am on weekdays. The same timings were given in the 1916 Admiralty list. With a continuing note that "it was difficult to make out from the outer parts of the harbour".

The semaphore signal was in turn replaced by another time ball at the Master Attendant's office, which was controlled from Colombo Observatory. It had a diameter of 1.8 m and a drop height of 4 m at a higher elevation than the semaphore signal. The time ball is believed to have been erected in 1914 but it was not listed until 1917. It may have been subject to trials over an extended period to establish its reliability. In 1917, the time ball was dropped at 9 am, 1 pm and 4 pm Standard Time on weekdays, with a drop at 9 am only on Sundays and public holidays. This was changed to 1 pm daily in 1918 but reverted to the earlier timings after 1919, which were retained until 1931. After that, the signal was made at 9 am only until at least 1939. The signal had been withdrawn by 1942.

Acknowledgements I have appreciated the support of libraries throughout the UK, including the British Library in London, The National Library of Scotland in Edinburgh, The National Maritime Museum at Greenwich and Cambridge University Library. I am particularly grateful to Dr. Margaret Makepeace at the British Library, who located records of an 1850s time ball in Madras. I am grateful too to the authors of papers published in India who provided local insights and details.

References

Airy, G. B. (1861). *List of time balls in actual operation* (Papers of George Airy, RGO 6 615, item 2). Cambridge University Library.

Airy, G. B. (1878). *Letter to The Director General of Stores, India Office, Westminster, dated 13 April* (Papers of George Airy, RGO 6 618, Section 8, Sheets 209 and 210). Cambridge University Library.

Bartky, I. R. (1987, January). The bygone era of time balls. *Sky and Telescope*, 32–35.

Bay of Bengal Pilot: *Bay of Bengal and the Coasts of India and Siam, including the Nicobar and Andaman Islands*. (H. O. No.160.) (1916). Government Printing Office.

Bay of Bengal Pilot: Comprising Bay of Bengal and the Coasts of India and Siam, including the Nicobar and Andaman Islands (H. O. No.160, 2nd Ed.) (1923). Government Printing Office.

Dutt, M. S. (1854). Letter to J.H. Kenrick from Vepery Castle, expressing a desire to dedicate a lecture to Kenrick as Secretary to The Madras Polytechnic Institution. Dated 12 April. https://en.wikisource.org/wiki/Page:The_Anglo-Saxon_and_the_Hindu.djvu/2

East India *(Progress and Condition): Statement Exhibiting the Moral and Material Progress and Condition of India during the Year 1859–60*. (1861, May 14). India Office.

East India *(Progress and Condition): Statement Exhibiting the Moral and Material Progress and Condition of India during the year 1861–62* (1863, May 14). India Office.

Electric Time-Ball for India. (1880, January 2). Article in *The Times*, London.

Elliott, C. H. J. B. (1855). Nautical remarks on the Hooghly: Madras to the Sandheads, River Hooghly, Jan 17th to 30th, 1854. *The Nautical Magazine and Naval Chronicle for 1855*, 401.

Howse, D. (1997). *Greenwich time and the longitude*. Philip Wilson Publishers.

India Office Records and Private Papers. (1855). *Madras, electric telegraph, proposed transfer of time ball, established by Mr. Kenrick, to Government* (IOR/Z/E/4/26/M131). See also IOR/Z/E/4/26/T84 and IOR/E/4/830, 296.

Kenrick, J. H. (1853). Electric telegraph in Madras. *Journal of the Society of Arts and of the Institutions in Union, 1*, 554.

Kinns, R. (2011). The early history of the Edinburgh time ball and time gun. *International Journal for the History of Engineering & Technology, 81*, 264–290.

Kinns, R. (2020). Time signals for mariners in India, Burma and Ceylon. *Journal of Astronomical History and Heritage, 23*(3), 523–552.

Madras. (1841). Entry in the *Nautical Magazine and Naval Chronicle for* 1841, 363.

Morland, H. (1866a, April 3). Harbour and Pilotage Board Notice. *The Bombay Gazette*, 1.

Morland, H. (1866b, July 26). Harbour and Pilotage Board Notice. *The Bombay Gazette*, 1.

Parker, G. C. (1875). Time Ball, *Nautical Magazine, 80*. [Parker signed the notice dated 3 August 1874 as Acting Master Attendant, Kurrachee].

Phillimore, R. H. (1958). *Historical records of the Survey of India* (Vol. IV). Survey of India.

Sailing Directions for the West Coast of India (H. O. No. 159, 3rd Ed.). (1942). United States Government Printing Office.

Samant, H., & Samant, A. (2016). The Ghadiyal Godi Timeball – A maritime legacy of Bombay Port. *The Bombay Explorer, 55*, 2–11.

Shaw (India Office). (1878). Letter to G.B. Airy, dated 12 April. Papers of George Airy, Cambridge University Library (RGO 6 618, Section 8, Sheet 208).

The Time Ball. (1853). Notice to the commanders of vessels in Harbour, Government Observatory, 28 April 1853. *The Bombay Gazette*, 29 April 1853, 2.

The Time Ball. (1936, July). *Port of Bombay*, 302.

Toynbee, C. (1859). The Westerly Route for crossing the line. *Letter to the Editor of the Nautical Magazine*, 170–177.

Chapter 20
Southeast Asia

Roger Kinns

Abstract The first time ball outside the British Empire was erected by the Dutch in Batavia in 1839. The only time signals listed by the Admiralty in 1880 were disc signals in at Fourth Point, Batavia and Surabaya, all in Java. A time ball at Makassa was established in 1915. A time ball at Rangoon in Burma was announced in 1893 and changed to a light signal in 1931. Time guns were also used at Rangoon and Moulmein. Singapore used an extraordinary range of time signals, starting in 1893 with two time balls, followed by a gun and time bells. A multi-armed signal was added in 1930 plus time lights in 1931. Time signals in Indo-China (now Vietnam) were first at Haiphong in 1883 and then Saigon from 1908. A signal in the Philippines was established in 1885 and continued after the Spanish-American war of 1898 with addition of other signals.

1 Introduction

Southeast Asia includes Brunei, Myanmar, Cambodia, Timor-Leste (East Timor), Indonesia, Laos, Malaysia, the Philippines, Singapore, Thailand and Vietnam. Of these, Burma (now Myanmar), Indonesia, the Philippines, Singapore and Vietnam had visual time signals in Admiralty lists.

The time signals in SE Asia have been described previously, including images and signal details from various sources (Kinns, 2021). This chapter has evolved from that previous study. The development of astronomy in Southeast Asia and the main influences on it have been described by Ôhashi and Orchiston (2021).

Entries in Admiralty lists are shown in Table 20.1. Their locations are shown in Fig. 20.1. The multiple signals in Singapore, Jakarta and Manila were too close together to be separated on such a small-scale map. Their relative locations are shown on larger-scale maps. Important early signals were established by the Dutch

R. Kinns (✉)
The Maudslay Society, Pembroke College, Cambridge, England

R. Kinns (ed.), *The World of Visual Time Signals for Mariners*, Historical & Cultural Astronomy, https://doi.org/10.1007/978-3-031-57334-7_20

Table 20.1 Signals in SE Asia

Area	Location	1880	1898	1904	1911	1915	1919	1922	1928	1934	1939	1947
Burma (now Myanmar)	Moulmein					Gun	Gun	Gun	Gun			
	Rangoon		TB	TB	TB	TB	TB	TB	TB			
	Rangoon				Gun	Gun	Gun	Gun	Gun	Gun	Gun	
	Rangoon									Light	Light	
Indonesia	Java: Fourth Point	Discs										
	Java: Batavia: Tanjong Priok	Disc										
	Java: Batavia: Tanjong Priok		Boards	Boards	Boards	Boards	Boards	Boards	Boards	Boards	Boards	
	Java: Surabaya	Discs	Boards	Discs	Discs	Discs	Discs	Discs	Discs	Discs	Discs	
	Makassar						TB (black)	TB (black)	TB (black)	TB (black)	TB (black)	
Philippines	Kavite				TB	TB	TB	TB	TB	TB		
	Manila: Observatory		TB (black)	TB (black)	TB (black)	TB (black)	TB (black)	TB (black)	TB (black)	TB (black)		
	Manila		Gun									
	Manila: Engineer Island				TB	TB	TB	TB	TB	TB		
Singapore	Fort Canning		TB	TB	TB	TB	TB	TB	TB	TB	TB	
	Fort Canning								Gun			
	Pulo Brani		TB	TB (red)								
	Mount Faber Observatory				TB (red)	TB (red)	TB (red)	TB (red)	TB (red)	TB	TB	
	Mount Faber Observatory									Light		
	Fullerton Building									Lever discs	Lever discs	
Vietnam (was Cochin China)	Haiphong		TB									
	Saigon				TB	TB	TB	TB	TB	TB		
	Saigon				Gun	Gun				Gun		
Signal type totals		1880	1898	1904	1911	1915	1919	1922	1928	1934	1939	1947
Time ball, including collapsible		0	5	4	7	7	8	8	8	7	3	0
Gun, including powder flash		0	1	0	2	3	2	2	3	2	1	0
Lights		0	0	0	0	0	0	0	0	2	1	0
Other: disc, drum, flag, etc.		3	2	2	2	2	2	2	2	3	3	0

authorities in Java, preceding other signals in SE Asia by many years. Their evolution has been studied by Willems (2023) and many of his recent findings are included in this chapter.

Sections about specific signals are in the order: Myanmar (Burma), Singapore, Indonesia, Vietnam (Indo-China) and the Philippines.

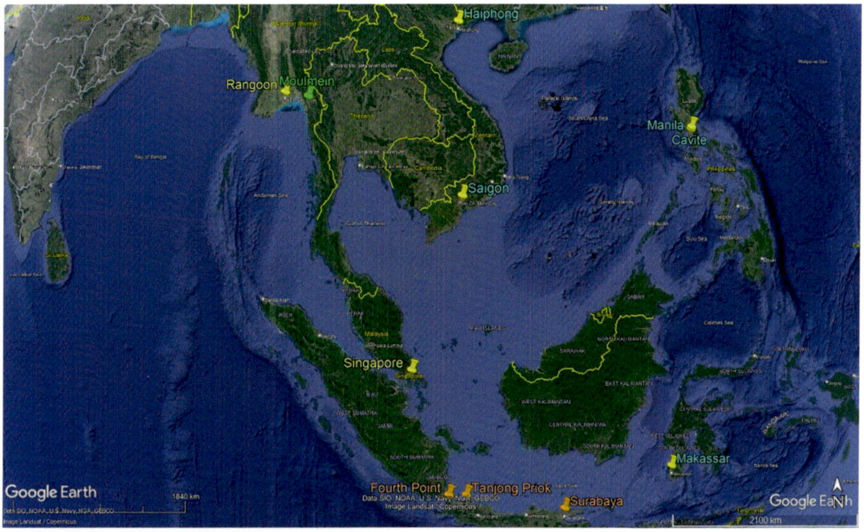

Fig. 20.1 Locations of Signals in Southeast Asia (Google Maps)

2 Burma

Burma, now Myanmar, was administered as part of India during the period of British rule up to 1947.

2.1 Rangoon

There is a mysterious photograph, taken in 1855, which shows a ball suspended from a yard arm on top of an ancient pagoda. It is shown in Fig. 20.2. It has been stated to be a time ball, but no supporting evidence has been found and it may have served some other purpose. The origin of the photograph is given by Kinns (2021).

2.1.1 Rangoon Time Ball

The first official time ball was gazetted in 1893 (Rangoon, 1893) as operational from 1 October 1893 and listed from 1898 to 1908 as being on the tower of the Mayo Sailors' Home. This description had changed by 1911 to the tower of the Mayo Marine Institute when a time gun was also fired nearby. There was no change in the stated longitude or in the elevation of the time ball. It had a drop height of 3 m and was 24.3 m above water. The stated latitude in Admiralty lists was reduced by 1 min of arc in 1911 but this may have been correction of an error, rather than a change of location.

Fig. 20.2 1855 Photograph showing the Signal Pagoda of Rangoon. (By Linnaeus Tripe, British Library on-line Gallery No. 102)

2.1.2 Rangoon Time Gun

Until at least 1922, the Rangoon gun was fired at noon Burma Standard time (5.30 am GMT) while the time ball was dropped at noon, Rangoon (local) mean time, 5 min 20 s later. This had changed by 1927 when the time ball was dropped and the gun fired at 1 pm Standard time, an hour later than before.

2.1.3 Rangoon Time Light

The Rangoon time ball was replaced by a simple time light in 1931, switched on 2 min before extinction at 1 pm, Burma standard time, when the gun was fired. The time gun at Rangoon was discontinued in 1939, but the light remained in use (Sailing Directions for the Bay of Bengal, 1941).

2.2 *Moulmein*

A time gun had been introduced at Moulmein by 1915 but was not recommended for rating chronometers. It was fired initially at the same time as the Rangoon gun, but this had changed to 10 am by 1930. It was no longer listed in 1931.

3 Singapore

An extraordinarily wide range of time signals existed in Singapore (Kinns, 2021). Two time balls were established in 1893 and appeared in the 1898 list. These were later complemented by electric bells at various shore locations, a time gun and an unusual lever-arm arrangement. The time gun was not listed by the Admiralty.

3.1 *Time Balls*

There was a visit to Devonport in 1890 by Saxby & Farmer, who had been chosen as suppliers of the time ball apparatus for Singapore (Homes, 2009). The time balls at Singapore were to use the "Devonport principle" (Lewis, 1910). The two installations in Singapore had still to be completed in March 1892 (The proposed time balls, 1892).

> For the past three or four years there have been rumours of apparatus for signalling correct time to the shipping, but no decision was come to erect time balls till a couple of years ago. Now matters have so far advanced that tenders are called for by the P. W. D. for the erection of two time balls, one at Fort Canning for ships in the roads, and a second on Pulau Brani facing Tanjong Pragar, so as to be visible to all ships at the Wharves and in the New Harbour.
>
> The Observatory is the small building on the spit of land behind the Drill Hall. The chief feature in it is a massive pedestal of granite, about 4 feet high. … The observatory is to be the home of the clocks, when they are completed, which will probably not be until the end of the year, and of the transit instruments.
>
> The time ball apparatus consists of a steel mast, about 50 feet in height, duly supported by guys and an iron base. Projecting from the mast are two arms one above the other, about twenty two or three feet apart. Two strong wire ropes are stretched as guides to a time ball six feet in diameter and weighing 77 lbs.
>
> The ball is raised by a wire rope and then held in position by another single strand wire rope connected below with a magnet. When the standard clock makes connections, the ball is released by the single strand wire and falls 13 ft. 6 in. At that point in its fall, the main hauling rope brings into play a leaden check weight weighing 86 lbs. and the last six feet of the fall is accomplished at a diminishing rate of speed, until the ball is brought up by spiral springs with a compressibility of 80 lbs. and an indiarubber cushion.

Each ball had a diameter of 1.8 m and a mass of 35 kg with a large drop height of about 6 m. It was anticipated that the balls would not be operational until at least the end of 1892. They were certainly operational during July 1893, as shown in the following announcement (Time-balls, 1893).

> TIME-BALLS on Fort Canning and Pulau Brani are dropped daily (Sundays excepted) at 1 p. m. Observatory mean time. They are hoisted five minutes before 1 o'clock. When the ball drops at 1 p. m. local mean time, the corresponding Greenwich mean time is 6 h. 4 m. 34.95 sec. a.m. …

Announcements about the time ball drop times were published regularly in the local press, occasionally noting that there had been a temporary withdrawal of the service or a signal fault. The red time ball at Pulo Brani was moved to Mount Faber in 1905 and was listed there from 1908 onwards (see Kinns, 2021 for additional details). The locations of the three time balls are shown In Fig. 20.3. The photograph in Fig. 20.4, taken in 1902, shows the Fort Canning time ball to the left (looking at the image) of the tall signal mast at top centre, with the lighthouse to the right of the signal mast. Time balls at Fort Canning and Mount Faber were still listed in 1939.

It had been established that the time balls actually fell 0.7 s after the electrical time signal had been received from the standard clock, so it was decided to adjust the clock time to match. The principal time ball signals then gave a more exact time, but the bells rang slightly early.

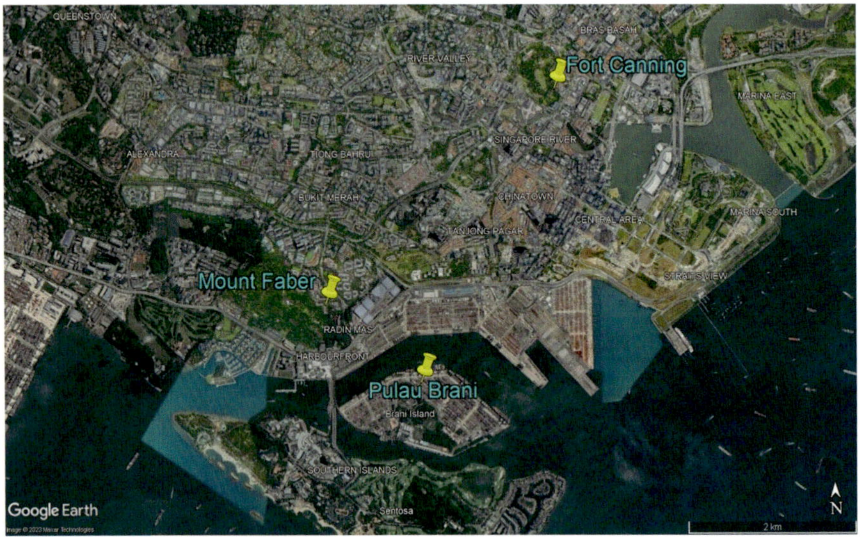

Fig. 20.3 Locations of time balls in Singapore (Google Maps)

Fig. 20.4 Time ball at Fort Canning, 1902

3.2 Lever Arm

In 1930 another signal, designed locally, was installed on the Fullerton building (New Time Signal, 1930). It used lever arms with discs at their ends on a mast above the roof. The arms were extended horizontally before dropping towards the mast as the time signal. The building and time signal arrangement is shown in Fig. 20.5.

3.3 Time Light

A complicated time light signal was added in 1931, using a defined sequence of flashes from three white lights arranged vertically (Night time signals, 1931). It was operated to give signals at 8 pm and 9 pm, local Standard time, when the lights were extinguished. It had been discontinued before 1939.

Fig. 20.5 Fullerton Building, Singapore

3.4 *Time Bells*

An electric bell was installed in an office on Tanjong Pragar wharf in 1905. It rang automatically at every hour of GMT and was included in the 1908 and 1911 lists. It is not included in Table 20.1 as it was not visible or audible to ships in harbour. Time bells had been installed at other locations by 1927 and three were mentioned in the 1930 list as being connected to the standard clock.

3.5 *Time Gun*

A time gun at Fort Canning was in use by January 1907 (Time balls, 1907). The notice is transcribed below:

> TIME BALLS on Fort Canning and Mount Faber drop daily at 1 p.m., Singapore standard time, corresponding to 6 a.m. Greenwich meantime.
>
> The time-gun is fired at 12 o'clock noon, indicating Singapore standard time, on every day excepting Sunday, when it is fired at one o'clock.

The time gun was mentioned in pilot guides issued in the United States until at least 1932 but it was never included in Admiralty lists.

4 Indonesia

4.1 Time Balls in Java

The first known time ball in SE Asia was installed in Java at Batavia (now Jakarta) in 1839. It was noted by Bartky (1987).

4.1.1 Batavia Time Ball

The time ball in Batavia was reported by Horsburgh (1852), who quotes from a Seaman's Guide by Lieutenant Melvill. In the quotation below, Horsburgh was referring to an 1850 publication by Dutch naval officers Lieutenant Baron Melvill of Carnbee and Lieutenant Smits (Melvill and Smits, 1850). Melvill (1816–1856) had been attached to the hydrographical bureau at Batavia in 1839.

> BATAVIA OBSERVATORY is in lat. 6° 9′ S., lon. 106° 51¾′ E., by astronomical observations made by Johan Mauritz Mohr, and this longitude is considered to be very correct ...
>
> Lieutenant Melvill's *Seamans Guide* gives the following information respecting the Time Ball, established at the Observatory in 1839:– "The geographic position of the Time Ball is in 6° 8′ S. and the assumed longitude 106° 52′ E., 7 h. 7 m. 28 sec. in time ... The Time Ball is hoisted every day at five minutes before noon, Batavia mean time, half-way up the pole; at two minutes to mean noon it is hoisted to the top, and precisely at Batavia mean noon it falls. For those ships that wish to rate their chronometers according to Greenwich mean time, the moment of six o'clock A.M. Greenwich mean time is indicated in the same way; the ball is hoisted half-way up at 1 h. 2 m. 28 secs. P.M. Batavia mean time; at 1 h. 5 m. 28 secs. to the top; and exactly at 1 h. 7 m. 28 secs. P.M. Batavia time, which corresponds to six o'clock A.M. Greenwich mean time, it falls."

The time ball was dropped twice every day: the first drop was at noon Batavia mean time, with the second drop 1 h, 7 min 28 s later at 6 a.m. Greenwich mean time.

There was subsequently a correction of 15½ s to the Greenwich mean time (Annual Report, 1876), corresponding to a revised longitude of 106° 48′ 7.5″ E.

> BATAVIA. – There is a time-ball dropped at Batavia mean noon, and also one at 1 h 7 m 12 s.5 Batavia civil time = 18 h Greenwich mean time.

A later guide (Findlay, 1878, 684) gave the corrected longitude, but contained errors of 15½ min (*sic*) in the Greenwich time of the signal. The following details about the signal location correspond to those given by Melvill (1850), apart from the longitude.

> Batavia Observatory, where a time ball has been exhibited since 1839, stands East, 3133 yards from the boathouse near the river, or a little more than a mile and a half from the lighthouse on the pier-head. The geographic position of the time ball is in 6° 8′ S. and the assumed longitude 106° 48′ 7½″ E ...

The reference meridian for signals in the Dutch East Indies was located at the Batavia time ball (later time disc), so its estimated longitude had particular significance (Willems, 2023). The assumed longitude, which also affected the assumed longitudes of other signals in Java, was adjusted on various occasions.

4.1.2 Surabaya

The first time ball at Surabaya (also spelled Sourabaya and Soerabaija) was operated from a guardship. The signal was moved to the naval shore establishment at Surabaya on 16 March 1864 (Soerabaija, 1864; Willems, 2023). Part of the published 1864 notice is shown below, translated from Dutch.

> (*For the country*.) The rear admiral … informs interested parties that on the 16th of March the time ball will be put into effect in Soerabaija instead of on the guard ship, at the naval establishment at the said location.

The time ball was later specified as a signal using 4 discs. It is likely that the 1864 signal used rotating discs from the outset, rather than a dropping ball.

4.1.3 Fourth Point (Sunda Strait)

The Sunda Strait stretches in a roughly northeast/southwest orientation, with a minimum width of 24 km at its northeastern end between Cape Tua on Sumatra and Cape Pujat on Java. It is very deep at its western end, but as it narrows to the east it becomes much shallower, with a depth of only 20 m in parts of the eastern end. It is notoriously difficult to navigate because of this shallowness, very strong tidal currents and sandbanks. It had been an important shipping route for centuries, especially during the period (1602–1799) when the Dutch East India Company used it as the gateway to the Spice Islands of Indonesia. Large ships tended to use the more easily navigable Straits of Malacca, which is now one of the most important shipping lanes in the World.

A time signal at Anyer, the port of the Fourth Point, was erected as a disc signal in 1876 (Time-signal station on Fourth point, 1879).

4.2 *Time Discs in Java*

4.2.1 Entries in Admiralty Lists

The 1880 Admiralty list included three entries for Indonesia and none for other countries in SE Asia. These were all in Java: at the Fourth Point (Sunda Strait), Batavia and Surabaya. The Sunda Strait signal was destroyed in 1883, while the Batavia location was changed in 1886. There have been adjustments to the latitude and longitude of signal locations at Batavia and Surabaya that may be corrections to previously measured coordinates rather than changes of signal locations.

The principal signals in Java from 1880 onwards used one or more discs rotating about horizontal axes, instead of time balls. This followed the practice at the principal ports in the Netherlands, described in Chap. 13. Curiously, the Admiralty changed the description of signals in Java from discs to boards between the 1880 and 1898 lists (Table 20.1). The lists of time signals between 1898 and 1939 all refer to the use of four boards at Tanjong Priok. The 1898 description of four boards at Surabaya had been changed to four discs by 1908. Available photographs suggest that discs were used throughout, as did sailing instructions published in the United States.

Local mean time noon was the time of Java signals up to at least 1908, with an additional Batavia signal at 1 h 7 m 32.6 s pm local time, corresponding to 6 am Greenwich mean time. By 1911 this had been changed to an additional signal at 1 pm local time. In 1922, Java standard time was 7 h. 19 m. 14.5 s. fast on Greenwich. By 1927, it had been changed to 7 h. 20 m. 00 s. fast on Greenwich. By 1935, it had been changed again to 7 h. 30 m. 00 s. fast on Greenwich, corresponding to longitude 112½° E.

4.2.2 Fourth Point (Sunda Strait) Time Discs

The port, and presumably the time signal, were completely destroyed by the Krakatoa (Krakatau) eruption in 1883. The volcano was only about 50 km west of Anyer in the Sunda Strait. No evidence has been found that the Fourth Point signal was ever reestablished. The need for a signal is likely to have diminished when an increasing majority of large ships used the Straits of Malacca.

4.2.3 Batavia (Tanjung Priok) Time Discs

The time ball at Batavia was changed to a single time disc in 1856 (Batavia, 1856; Willems, 2023), but the change was not recognised internationally for a long time. Part of the1856 notice, published initially in a Java newspaper and then in Leiden, is translated below:

> … Lieutenant Commander 1st class. J. Groll, acting port and equipment master, director of the time ball, has made a drawing, which gives a good idea of how the time ball could be organized … . This device consists of a simple but sturdy disk of sheet iron, with a diameter of approximately 7 feet, mounted at the top end of the mast, which is now present at the time ball building. The rotating movement is performed in a very simple manner, by means of an iron rod and a crank, which is moved by hand.
>
> The Government has approved this new institution, in accordance with the above-mentioned drawing, and has made the required funds available for this purpose. The device will be … manufactured in Batavia.

The use of a disc with a diameter of 7 feet was reported in an 1879 notice that emanated from Germany (Time-ball at Batavia, 1879). Its location was that of the earlier time ball. The small difference in location between the original time ball and the first disc signal shown in Fig. 20.6 is therefore an error in latitude.

Work on the Tanjung (also spelled Tandjong) Priok Port began in May 1877 and was completed in 1886. After the inauguration of the Port in 1886, the activities of the main port of Batavia, which was originally located in Ciliwung River around the Batavia castle, were transferred to the Port of Tanjung Priok (Wikipedia). The entry in the 1880 Admiralty list dates from the period of port development, while the 1898 entry is after that development.

The Batavia signal was changed from a single disc to four discs at a new location in 1886 (Willems, 2023). Figure 20.6 shows the change in location according to the 1880 and 1898 Admiralty lists, as well as the stated location of the original time ball. The photographs in Figs. 20.7 and 20.8 show the change of signal type and location.

The following extract confirms that four "network" discs were used (Sailing Directions for Sunda Strait, 1935).

> Standard time in Java is that of the meridian 112° 30″ E.
>
> Time signals are made at the signal station, or "look-out", on the east side of the entrance to the first inner harbor. The installation consists of an iron structure above which are four round, black, network discs. Normally, these discs lie horizontally. At 5 minutes before noon (Java time), the discs are placed at an angle of 45°; at 2 minutes before noon, they are placed vertically. At noon (Java time) the discs are turned horizontally. ...

Fig. 20.6 Batavia time ball and time disc locations (Google Maps)

Fig. 20.7 Batavia time disc, shown arrowed, at 1880 location. (Courtesy Klaus Hülse Collection)

Fig. 20.8 Time discs at Tanjung Priok after 1886. (Courtesy Klaus Hülse Collection)

4.2.4 Surabaya Time Discs

A signal using four discs was listed for Surabaya from 1880 to 1939 (Table 20.1). It was notified in 1879 (Time-ball at Sourabaya, 1879; Willems, 2023) but perpetuated confusion between a dropping time ball and rotating disc signals (see Chap. 13).

The term "time ball" was sometimes used loosely to describe the rotating disc signals favoured in the Netherlands (see Chap. 13). It was used, for example, in the 1879 notification of time discs, transcribed below.

> The time-signals at Sourabaya are made near the new landing-bridge. The building containing the signal apparatus stands on the eastern bank of the harbor, at the east corner of the mouth of the small river Kali-Mas. …
>
> The signal apparatus consists of four iron grating disks, 39 feet above the ground, and attached to a cross-piece fixed to the supporting pole.
>
> At 5 minutes before the signal the disks assume an angle of 45°; at two minutes before they are raised perpendicular and are entirely visible; and at the proper moment fall horizontal.
>
> The chronometer and dropping apparatus are in a round building with a cone-shaped roof, to the westward of the time-balls (*sic*).
>
> The chronometer is regulated by observations of the sun with an artificial horizon, and opportunity is given for regulating and comparing chronometers.
>
> Geographical position of the time-ball (*sic*): Latitude 7° 12′ 10″.1S., longitude 112° 43′ 40.4 E.

As often in the late 1870s, the notice originated in Germany and was circulated widely. The change of signal may have been implemented much earlier.

There was a distinct change in location between 1880 and 1898, with changes to the elevation of the signal above ground from 39 feet (12 m) to 59 feet (18 m). The location description changed from "Post on Rampart, East Point of mouth of Kalimas River" to "Steel mast, 59 feet high and painted white, on the western Mole head of Kalimas River. In 1927 it was moved from a mast to the top of the harbour office building, but it still used four discs. The same elevation was listed from 1927 onwards. The four descriptions in Table 20.2 are taken from 1880, 1898, 1908 and 1927 Admiralty lists.

The photograph in Fig. 20.9 shows four time discs, but its date is uncertain. It is probably the first arrangement.

The description in sailing instructions dated 1916 (East Indies Pilot, Volume 1, 1916, 73) corresponds to the 1898 Admiralty list. It was also noted in Admiralty lists from 1911 onwards that "The apparatus is released by hand in preference to electricity". Its location had been changed in the 1927 list to the "Roof of harbour office, Rotterdam Quay" The location in 1935 remained the same (Sailing Directions for Celebes, 1935).

> *Time Signal*. – The visual time signal located on the roof of the harbor office at the Northeast corner of Rotterdam Quay consists of four black discs which are inclined at an angle of 45° at 5 minutes before the signal; placed in a vertical position 2 minutes before the signal; and let fall to a horizontal position at noon standard time corresponding to 4 h. 30 m. 00 s. Greenwich civil time. …

Table 20.2 Time discs at Surabaya

Location	Latitude Longitude	Signal	Additional details
1880	7° 12′ 10″ S	Four black discs	39 feet above ground
Post on Rampart, East Point of mouth of Kalimas River.	112° 43′ 40″ E		Discs inclined at an angle of 45° at 5 min before signal
			Discs placed vertical at 2 min before signal
			Discs fall into a horizontal position at noon Surabaya mean time
1898	7° 11′ 55″ S	Four boards	65 feet above sea
Mast on the western Mole head of Kalimas River.	112° 44′ 21″ E		59 feet above ground
			Boards inclined at an angle of 45° at 5 min before signal
			Boards placed vertical at 2 min before signal
			Boards fall into a horizontal position at noon Surabaya mean time
1908	7° 11′ 55″ S	Four black discs	65 feet above high water
Steel mast, 59 feet high and painted white, on the western Mole head of Kalimas River.	112° 44′ 21″ E		59 feet above ground
			Discs inclined at an angle of 45° at 5 min before signal
			Discs placed vertical at 2 min before signal
			Discs fall into a horizontal position at noon Surabaya mean time
1927	7° 12′ S	Four black discs	65 feet above high water
Roof of harbour office, Rotterdam quay	112° 44′ E		59 feet above ground
			Discs inclined at an angle of 45° at 5 min before signal
			Discs placed vertical at 2 min before signal
			Discs fall into a horizontal position at Noon standard time

From Admiralty lists: 1880, 1898, 1908 and 1927

4.3 Time Ball at Makassar

Makassar is at the SW end of Celebes Island, now South Sulawesi. There was a time ball at Makassar Lighthouse in 1916, as defined in the following notice. (East Indies Pilot, 1916)

> Time ball. – A black ball, hoisted on the yard of a mast, 66 feet high, at Makassar Lighthouse, is dropped at 23 h. 57 m. 37.2 s. local mean time of the meridian of Makassar Lighthouse, corresponding to 16 h. 00 m. 00 s. Greenwich mean time. …

The time ball at Makassar was noted in Admiralty lists as being established in 1915 with the signal details given above. The lighthouse was not mentioned in Admiralty lists. The latitude and longitude were stated to be 5° 08′ 10″ S., 119° 24′ 18″ E. It was reported unreliable in 1925. The Makassar time ball was included in Admiralty lists up to 1939, but the stated latitude and longitude were 5° 07′ S., 119° 24′ E in lists between 1930 and 1939. The drop time was given as noon, civil standard time,

Fig. 20.9 Time discs in Surabaya. (Courtesy Klaus Hülse Collection)

Fig. 20.10 Makassar lighthouse in about 1905. (Author's collection)

corresponding to 4 h. 00 m. 00 s. Greenwich civil time. The stated signal height remained at 66 feet, but the location was noted as "Roof of Hanger V".

Details given in 1935 sailing directions also indicated a change in latitude and longitude, but these did not correspond exactly to details in Admiralty lists and there was no mention of the lighthouse or hanger (Sailing Directions for Celebes, 1935).

The photograph in Fig. 20.10 shows Makassar Lighthouse in about 1905. This lighthouse is believed to have been demolished in the 1920s.

5 Indo-China, Now Vietnam

There have been two time balls in Vietnam. The first was established in Haiphong in about 1886 and the second was established in Saigon in about 1908. Their locations are shown in Fig. 20.11. A time gun at Saigon but was not listed consistently and was not regarded as an accurate signal.

5.1 *Haiphong*

The Haiphong time ball is believed to have been established in 1886 (Hite, 2014). William Doberck (1852–1941), then Director of Hongkong Observatory, made precise measurements of the longitude of Haiphong in 1887, using undersea telegraphy. These measurements were presented to the Royal Astronomical Society in London in March 1888 and are likely to have been made in support of a time ball at Haiphong (Doberck, 1888). The paper refers to a transit instrument in Haiphong that had been installed in 1886. Doberck's wider work as an astronomer has been described by Mackeown (2007). The time ball location was at the Observatory shown in Fig. 20.12. The time ball was on top of the tower.

The time ball appears to have been discontinued by 1904. It was not included in the 1904 or later Admiralty lists.

Fig. 20.11 Vietnam time balls (Google Maps)

Fig. 20.12 The Observatory at Phu-Lien (Haiphong). (Courtesy Klaus Hülse Collection)

5.2 Saigon

The time ball at Saigon was positioned on the signal mast by the commercial harbour. It was installed in 1908, with modified drop times in 1917. It was listed by the Admiralty between 1911 and 1934. The following notice in 1915 shows that the ball was dropped twice daily, first at noon, Saigon mean time and then 6 min, 48.3 s later at 5 am Greenwich mean time (Asiatic Pilot, 1915).

> **Time signal** (Lat. 10° 46′ 40″ N., Long. 106° 42′ 2″ E.). – A ball is dropped from the signal mast in the commercial harbour at noon, Saigon mean time, corresponding to 16 h. 53 m. 11.7 s. Greenwich mean time. The ball is dropped a second time at 17 h. 0 m. 0 s. Greenwich mean time, or standard time of the 105th meridian east of Greenwich.

The drop times were changed in 1917 to 15 h. 00 m. 00 s. and 15 h. 05 m. 00 s. Greenwich mean time. These drop times were the same in 1930, although they were then reported as 3 h. 00 m. 00 s. and 3 h. 05 m. 00 s. "Greenwich mean time (civil)", corresponding to 10 h. 00 m. 00 s. and 10 h. 05 m. 00 s. "Standard time (civil)". By 1925, radio signals were also available (Asiatic Pilot, 1925).

The time ball was still operating in 1937 but was deemed to be for local use only. The principal time signals were then by radio (Sailing Directions, 1938).

> **Time signal**. – Standard time is that of the meridian of 105° east longitude. – A time ball for local use is dropped at 10 a. m., from the signal mast of the Bureau du Port on the northern side of the mouth of the Arroyo Chinois. The ball is dropped again 5 minutes later. …

The photographs in Fig. 20.13 show the Saigon time ball from river and shore sides. They indicate that the time ball had an open frame construction.

Fig. 20.13 Saigon time ball, from river and shore sides. (Author's collection)

5.2.1 Saigon Time Gun

A time gun at Saigon was listed in 1911 as being fired at noon when the time ball was dropped but both the time ball and gun signals were described as unreliable. The gun was not listed between 1916 and 1930. It was listed again from 1933 to 1935 when it was fired at 10 am Standard time (3 am GMT), two hours earlier than before. Neither the time ball nor the time gun was listed in 1939.

6 The Philippines

Manila Observatory had been founded in 1865 by the Society of Jesus and introduced its time service in 1885, when the Philippines were administered as a Spanish colony (Manila Observatory). This period ended with the Spanish-American war in 1898, but the observatory was undamaged. The American colonial government recognized the importance of the observatory's work and established it as the Philippine Weather Bureau in 1901. The original observatory and its records were destroyed during World War II.

6.1 Manila Time Signals

The 1885 time ball at the Observatory was supplemented by additional time balls at the Cavite (Kavite) naval station in 1906 and at Engineer Island after 1908. All three time balls were listed by the Admiralty up to 1934, as shown in Table 20.1. A time gun of doubtful accuracy was also listed in 1898. The Observatory and Engineer Island time balls were dropped at noon, local standard time, while the Cavite ball was dropped at 11 am. The co-ordinates of the balls are given in Table 20.3 and shown on the large-scale map in Fig. 20.14.

Table 20.3 Time balls in the Philippines

Location	Latitude Longitude	Signal	Additional details
Manila	14° 35′ 12″ S	Black ball	54 feet above high water, 80 feet above ground, drop 25 feet
At the observatory	120° 58′ 35″ E		Ball hoisted at 5 min before the signal, and dropped at Noon, Standard time
Manila	14° 35′ 43″ S	Ball	Ball hoisted at 5 min before the signal, and dropped at Noon, Standard time
On semaphore tower, Engineer island	120° 57′ 21″ E		
Kavite	14° 29′ 00″ S	Ball	Ball hoisted at 5 min before the signal, and dropped at 11h 00m 00s, Standard time
On water-tower, north-west of Fort St. Philip	120° 54′ 45″ E		

From 1927 Admiralty list

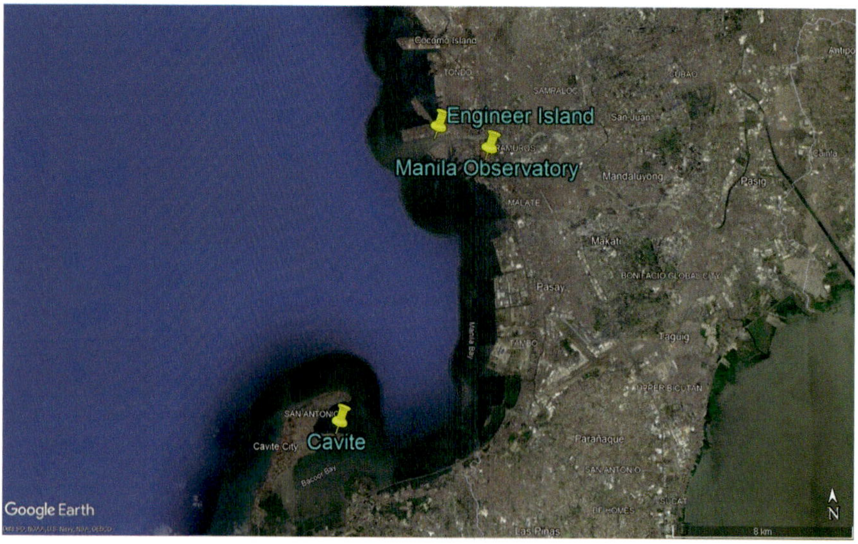

Fig. 20.14 Locations of time balls in the Philippines (Google Maps)

6.1.1 Manila Time Ball and Time Gun

The first time ball was included in the 1898 Admiralty list, together with a time gun. The ball was dropped and the gun was fired at noon local mean time. The gun was stated to be insufficiently accurate for rating chronometers and was not listed by the Admiralty in 1904 or later.

A report showed that the time gun was still in operation in 1902 (Gazetteer, 1902):

> … A time ball, black, which is hoisted on the roof of Manila observatory at Ermita, is dropped daily at noon, one hundred and twentieth meridian (E. lon.), standard time. It is hoisted 5 minutes before noon and in case of failure is slowly lowered 5 minutes after the signal time. A gun is also fired from the Battery of San Diego …

The following extract is taken from a 1905 report (Report, 1905). The report emphasised the accuracy of the time ball but there was no mention of a time gun.

> … As in former years, the correct time has been determined from star transits and daily telegraphed at 11 a.m. mean time of the one hundred and twentieth meridian east of Greenwich to all telegraph stations in the islands, while to shipping in port the time is signaled at 12 noon by means of the time ball. An arrangement has been made with the Daily Bulletin of Manila, according to which this paper is to be immediately informed whenever the error in dropping the ball exceeds half a second. The correction is then printed in the next edition of the paper. Thus far it happened only once.

6.1.2 Cavite Time Ball

In February 1906, a second time ball was established at the Cavite Naval Station (Cavite, 1906). This time ball was located on top of a water tower, shown in Fig. 20.15.

Fig. 20.15 Cavite time ball on the water tower. (Courtesy Klaus Hülse Collection)

6.1.3 Engineer Island Time Ball

A third time ball was erected on the Semaphore Tower at Engineer Island after 1908.

Acknowledgements Generously, Joris Willems has shared the results of his wider research into signals in the Netherlands, Belgium and Indonesia that is presently unpublished, clarifying when signals were erected in Indonesia. I am also grateful to Prof. Patrick Seitzer at the University of Michigan for helping me to locate references to the time balls at Manila.

References

Annual Report of the Secretary of the Navy on the Operations of the Department from the year 1876. (1876). Published by the United States Government Printing Office.

Asiatic Pilot. (1915). *Volume IV: The shores of the China Sea from Singapore Strait to and including Hongkong, H. O. No. 125*. Published by the United States Government Printing Office.

Asiatic Pilot. (1925). *Volume IV: The shores of the China Sea from Singapore Strait to and including Hongkong, H. O. No. 125* (2). Published by the United States Government Printing Office.

Bartky, I. R. (1987). The bygone era of time balls. *Sky and Telescope*, 32–35.

Batavia. (1856). Notice entitled "Oost-Indie" published in the *Javasche Courant* on 1 March 1856, reported in the *Leidsche Courant* on 10 May 1856 (In Dutch).

Bay of Bengal Pilot: Bay of Bengal and the Coasts of India and Siam, including the Nicobar and Andaman Islands, H. O. No. 160. (1916). Published by the Washington Government Printing Office.

Bay of Bengal Pilot: Bay of Bengal and the Coasts of India and Siam, including the Nicobar and Andaman Islands, H. O. No. 160 (2nd Ed.). (1923). Published by the Washington Government Printing Office.

Cavite – Time Ball Established. (1906, February). *Notice to Mariners*. Coasts of the United States, Adjacent Territories and Islands under the Jurisdiction of the United States. Published by Washington Government Printing Office.

Doberck, W. (1888). Telegraphic determination of the longitude of Haiphong (communicated by the Astronomer Royal). *Monthly Notices of the Royal Astronomical Society, 48*, 282–284.

East Indies Pilot, Volume 1, Island of Java (with exception of the Western Part from Java Head to Batavia), Islands East of Java, South and East Coasts of Borneo and Celebes island, H. O. No. 163. (1916). Published by the United States Government Printing Office.

Findlay, A. G. (1878). *A Directory for the Navigation of the Indian Archipelago, China and Japan and the Straits of Malacca and Sunda, and the Passages East of Java* (2nd ed.). Published by Richard Holmes Laurie.

Gazetteer. (1902). *Gazetteer, the Philippine Islands, a Pronouncing Gazetteer and Geographical Dictionary of the Philippine Islands, United States of America*. Published by the Washington Printing Office.

Hite, L. (2014, July 1). How time balls worked; Featuring the Cincinatti Observatory, Birthplace for American Astronomy. https://www.cincinnatiobservatory.org/media/documents/How_Time_Balls_Worked.pdf

Homes, C. (2009). The Astronomer Royal, the Hydrographer and the time ball: Collaborations in time signalling 1850–1910. *The British Journal for the History of Science, 42*(3), 381–406.

Horsburgh, J. (1852). *The India directory on sailing to and from the East Indies, &c* (Vol. 2, 6th ed.). Published by Wm. H. Allen &, Booksellers to the Honourable East-India Company.

Kinns, R. (2021). Chapter 12: Time signals for mariners in SE Asia: Time balls, discs, bells, guns and lights. In W. Orchiston & M. N. Vahia (Eds.), *Exploring the history of Southeast Asian*

Astronomy: A review of current projects and future prospects and possibilities (pp. 411–458). Springer.

Lewis, T. (1910, April 21–25). *Handwritten notes for the Astronomer Royal* (RGO 7 257, Papers of William Christie). Cambridge University Library.

Mackeown, P. K. (2007). William Doberck – Double Star Astronomer. *Journal of Astronomical History and Heritage, 10*(1), 49–64.

Manila Observatory. http://www.observatory.ph/introduction/

Melvill, P. (Baron Melvill van Carnbee and Lieutenant, DRN) and Smits, H. D. A. (Lieutenant DRN) (1850). *The Seaman's Guide*. Published by the Hydrographic Office, Admiralty. Republished as a digitised facsimile edition by the British Library. Printed in Great Britain by Amazon.

Mount Faber and Fort Canning Time Signals. (1927, 23 December). *The Singapore Free Press and Mercantile Advertiser*, 12.

New Time Signal on Post Office: Electric Contact with Mount Faber. (1930, February 4). The Singapore Free Press and Mercantile Advertiser, 17.

Night Time Signals. (1931, October 16). To be Established from Monday at Mount Faber. *The Straits Times*, 8.

Ôhashi and Orchiston. (2021). Chapter 24: The evolution of local Southeast Asian Astronomy and the Influence of China, India, the Islamic World and the West. In W. Orchiston & M. N. Vahia (Eds.), *Exploring the history of southeast asian astronomy: A review of current projects and future prospects and possibilities* (pp. 673–767). Springer.

Rangoon Signal Established. (1893). *Notices to Mariners*, 1893. No. 549—East Indies Station, Gulf of Mataban—Rangoon River.

Report. (1905). *Report of the Philippine Commission to the Secretary of War, Part 2*. Published by the United States Government Printing Office.

Sailing Directions for Celebes, Southeast Borneo, Java (except from Java Head to Batavia), and Islands East of Java (H. O. No. 163, 3rd Ed.). (1935). Published by the United States Government Printing Office.

Sailing Directions for Sunda Strait and Northwest Coast of Borneo, and off-lying Dangers (H. O. No. 126, 3rd Ed., 1934). (1935). Published by the United States Government Printing Office.

Sailing Directions for the Bay of Bengal (H. O. No. 160, 3rd Ed.). (1941). Published by the United States Government Printing Office.

Sailing Directions for the Western Shores of the China Sea (H. O. No. 125, 3rd Ed., 1937). (1938). Published by the United States Government Printing Office.

Soerabaija. (1864, February 29). *Notice published by the Rear-admiral in Batavia* (in Dutch).

The Nautical Magazine and Naval Chronicle for (1850). Published by Simpkin, Marshall.

The Proposed Time Balls. (1892, March 21). *The Singapore Free Press and Mercantile Advertiser*, 3.

Time-ball at Batavia. (1879, February 8). Java-North Coast, *Nachricten für Seefahrer*, No. 6, Berlin. Also Notice (135), in the *Nautical Magazine 34-'79*. [Athough listed as a time ball, the description is of a disc signal.]

Time-ball at Sourabaya. (1879, February 8). *Nachricten für Seefahrer*, No. 6, Berlin. Also Notice (136), in the *Nautical Magazine 34-'79*. [Athough listed as a time ball, the description is of a four-disc signal.]

Time-balls. (1893, July 15). *The Straits Times*, 2.

Time-balls. (1907, January 15). *The Straits Times*, 5.

Time-signal station on Fourth point. (1879, June 1). Published in *Berigt aan zeevarenden, No. 36*, 'sGravenhage, September 6, 1879. Also published as Notice 12 in the *Nautical Magazine, 5-'80*.

Willems, J. (2023). *Presently unpublished research into the history of time signals in The Netherlands, Belgium and Indonesia*.

Chapter 21
China, Japan and Korea

Roger Kinns

Abstract There was an earlier, inaccurate signal at Shanghai in 1877, but there were no listed visual signals in China in 1880. Accurate time balls were established at Hong Kong and Shanghai in 1885. Later, time guns, a flag signal and lights were added at various ports along the coast of China. The time light at Shanghai was operating by 1911. The time lights in China varied substantially in style and timing between locations. Only the Shanghai time ball and the lights at Hong Kong and Macau were still listed in 1947.

There were no listed time signals for Japan in 1880. A time ball and gun at Kobe were the only signals listed in 1898. Later Admiralty lists included a range of different time signals, including a flag, time lights and a whistle or siren at Niigata. Time balls after 1930 were red with a white band. The only Korean signal was a time gun at Chemulpo.

1 Introduction

The range of different time signals in China in Admiralty lists between 1880 and 1947 is shown in Table 21.1. Those in Japan and Korea are shown in Table 21.2. These tables show how time signals evolved at the different locations. Figure 21.1 shows their locations. Figure 21.1 also shows the location of the time signal at the Russian port of Vladivostok on the Sea of Japan. Its evolution is described in Chap. 14.

R. Kinns (✉)
The Maudslay Society, Pembroke College, Cambridge, England

R. Kinns (ed.), *The World of Visual Time Signals for Mariners*, Historical & Cultural Astronomy, https://doi.org/10.1007/978-3-031-57334-7_21

Table 21.1 Signals in China

Area	Location	1880	1898	1904	1911	1915	1919	1922	1928	1934	1939	1947
China	Amoy		Gun	Gun	Gun	Gun	Gun	Gun	Gun	Gun	Gun	
					Flag	Flag	Flag	Flag	Flag	Flag	Flag	
	Chifu			TB (black)	TB (black)	TB (black)	TB (black)	TB (black)	TB (black)	TB (black)	TB (black)	
				Gun	Gun	Gun	Gun	Gun	Gun	Gun	Gun	
	Dairen								Light	Light	Light	
	Hong Kong: Kaulung		TB	TB								
	Hong Kong: Blackhead Hill				TB (red)	TB (red)	TB (red)	TB (red)	TB (red)			
	Hong Kong							Light	Light	Light	Light	Light
	Macau										Light	Light
	Newchwang				TB							
	Shanghai		TB	TB	TB	TB	TB	TB	TB	TB	TB	TB
					Light	Light	Light	Light	Light	Light	Light	
	Swatau		TB	TB	TB	TB	TB	TB	TB	TB	TB	
			Gun	Gun	Gun	Gun	Gun	Gun	Gun	Gun	Gun	
	Tsingtau			TB (black)	TB (black)	TB (black)						
						Gun	Gun	Gun	Gun	Gun	Gun	
Signal type totals		1880	1898	1904	1911	1915	1919	1922	1928	1934	1939	1947
Time ball, including collapsible		0	3	5	6	5	4	4	4	3	3	1
Gun, including powder flash		0	2	3	3	4	4	4	4	4	4	0
Lights		0	0	0	1	1	1	2	3	3	4	2
Other: disc, drum, flag, etc.		0	0	0	1	1	1	1	1	1	1	0

2 China

There were no listed visual signals in China in 1880, but accurate time balls were established at Hong Kong (then spelled Hongkong) and Shanghai in 1885. Later, time guns, a flag signal and lights were added. Table 21.1 shows that most visual signals in China were still operating in 1939, but only the Shanghai time ball and the lights at Hong Kong and Macau remained in 1947. The time lights in China varied substantially in style and timing between locations.

Table 21.2 Signals in Japan and Korea

Area	Location	1880	1898	1904	1911	1915	1919	1922	1928	1934	1939	1947
Japan	Hakodate			Flag	Flag	Flag	Flag	Flag	Flag	Flag	Flag	
	Kagosima					Gun	Gun	Gun	Gun	Gun	Gun	Gun
	Karafuto									Light	Light	
	Kobe		TB	TB (red)	TB (red)	TB (red)	TB (red)	TB (red)	TB (red)	TB (r, w)	TB (r, w)	
			Gun	Gun	Gun	Gun	Gun	Gun	Gun	Gun	Gun	
												Light
	Kure			TB (red)								
	Moji				TB	TB (r, w)	TB (r, w)	TB (r, w)	TB (r, w)	TB (r, w)	TB (r, w)	
	Nagasaki				Flag							
					Gun	Gun	Gun	Gun		Gun		
					TB	TB (red)	TB (red)	TB (red)	TB (red)	TB (r, w)	TB (r, w)	
									Light	Light	Light	
	Niigatako								Whistle	Whistle	Whistle	Siren
	Osaka								TB (r, w)	TB (r, w)	TB (r, w)	
	Sasebo			TB (red)								
	Yokohama Eastern Hatoba			TB (black)	TB (black)	TB (black)	TB (black)	TB (black)				
	Yokohama N. breakwater							TB				
	Yokohama Kanagawa H.B.								TB	TB (r, w)	TB (r, w)	TB (r, w)
Korea	Chemulpo				Gun	Gun	Gun	Gun	Gun	Gun	Gun	
Signal type totals		1880	1898	1904	1911	1915	1919	1922	1928	1934	1939	1947
Time ball, including collapsible		0	1	4	4	4	4	5	5	5	5	1
Gun, including powder flash		0	1	1	3	4	4	4	3	4	3	1
Lights		0	0	0	0	0	0	0	1	2	2	1
Other: disc, drum, flag, etc.		0	0	1	2	1	1	1	2	2	2	1

2.1 *Hong Kong*

The need for a time ball and observatory at Hong Kong had been emphasised in November 1877 (Official reports, 1877):

> The Surveyor General's Report on the proposed Time-Ball for Hongkong, which appears in the *Gazette*, together with Admiral Ryder's letter, furnished much detailed information which tends to support the adoption of the proposal. …
>
> … it would be necessary to build and organise a small Observatory, and to engage a competent professional person from England to take charge of it. A Transit, a Sidereal Clock, electric apparatus and wires, and the mechanism for dropping the ball would be the chief requisites.

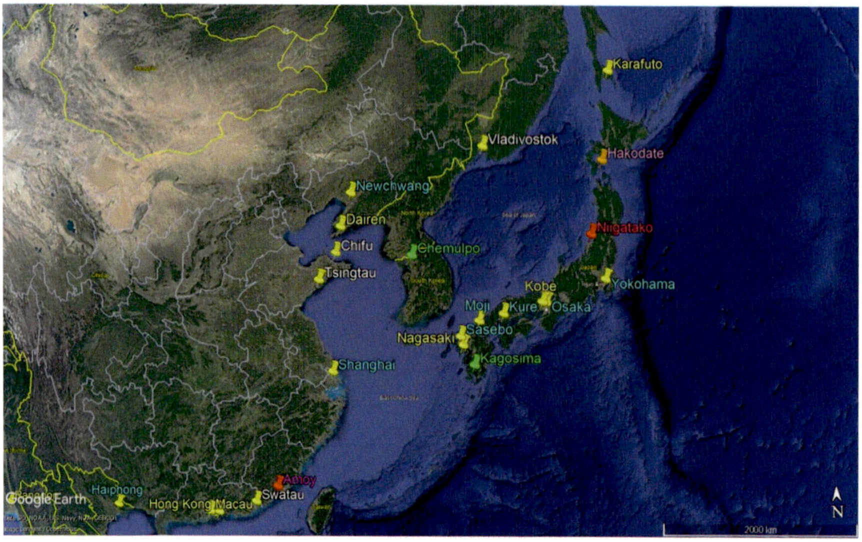

Fig. 21.1 Locations of signals in China, Japan, Korea and Vladivostok (Google Maps)

William Doberck (1852–1941) was appointed as Government Astronomer. His wider work as an astronomer has been described by Mackeown (2007). A description of the new observatory and its equipment was published in 1884 (The Hongkong Observatory, 1884). It included a brief description of the time ball and its operation:

> The mean-time clock, which is to discharge the time-ball automatically, is furnished with a magnetic apparatus for setting it to correct time without touching any part of the clock. The time-ball will be dropped at Tsim-sh'at-sui (*sic*) point, opposite the harbour, about a mile from the Observatory. It is 6 feet in diameter. Opposite the mean-time clock is the sidereal standard clock ….

The first time ball at Kaulung in Hong Kong was described as being near the police station and is shown in Fig. 21.2. It began operation on 1 January 1885 and was moved to a more prominent location at the signal station in January 1908, as shown in Fig. 21.3. The time ball was withdrawn in June 1933 (History of Hong Kong time service).

Time lights were added before 1920 and continued operation after the time ball had been withdrawn. Three lights were arranged vertically on the Observatory wireless mast at Kowloon. Their mode of operation is described below (1922 list):

> The lights are extinguished momentarily at even seconds excepting the 2nd, 28th, 50th 52nd and 54th seconds, of each minute, between 8 h 56 m 00 s and 9 h 00 m 00 s, Standard time.

The lights remained at the same location throughout. The sequence of flashes changed in detail, as shown in the Admiralty list entry for 1933.

> The lights are extinguished momentarily, every second from 20 h 55 m 00 s to 21 h 00 m 00 s Standard time except at the 28th, 54th, 55th, 56th, 57th, 58th and 59th seconds, of each minute (1933 list).

Fig. 21.2 Hong Kong time ball before 1908. (Courtesy Klaus Hülse collection)

Fig. 21.3 Hong Kong (Kowloon) time ball after 1908. (Courtesy Klaus Hülse collection)

These lights are not mentioned in the History of the Hong Kong time service.

The time ball apparatus has been reconstructed and returned to its original site. It is a working arrangement. A display notice records that:

> The time ball served its function admirably from 1885 to 1907. On 7th December 1907 the time ball was moved to Signal Hill in Tsim Sha Tsui, ending the historic mission of the Time Ball Tower. The time ball on top of this Time Ball Tower has been reconstructed based on detailed studies of time ball towers in Greenwich, Lyttelton, New Zealand, and Sydney, Australia and on records of the Hong Kong Observatory.

Although the references are to time balls built by Maudslay, Sons & Field in 1833, 1855 and 1873 which used similar design principles, the supplier of the original Hong Kong apparatus has not been identified.

2.2 *Shanghai*

There was an inaccurate time ball at Shanghai by 1877: "Shanghai has a time-ball once a week, which, though better than nothing, is liable to serious error." (Official reports, 1877). A new time ball was erected in 1884 at the French Concession. It was operated using electrical signals from the observatory at Zikawei (Siccawei) in SW Shanghai which was opened in September 1884. It has been described by Combridge (1989), who included a contemporary sketch that had been reproduced previously in *Nature* (Shanghai, 1885) and is shown in Fig. 21.4. The drawing is not to scale, as the overall mast height was expected to be 90 feet (27 m) with an unusually large ball diameter of 9 feet (2.7 m) (Shanghai, 1884).

The time ball at Shanghai outlived others in China but underwent various changes after its introduction. Its location continued to be given as the Semaphore at the French Concession in Admiralty lists. It was listed as having a diameter of 6 feet (1.8 m) from 1908 onwards, with a drop of 21 feet (6.3 m) from a height of 102 feet (35 m) above ground. Figure 21.5 shows the support arrangement that existed before 1907. It was changed then from a mast to the concrete column shown in Fig. 21.6. The tower has been restored, but the time ball no longer exists.

Shanghai had one of the first known time light signals, operational before 1911 and continuing until after 1939. Four white lights were arranged in the form of a diamond, also at the Semaphore location (1911 list). These were extinguished simultaneously every minute from 8.55 pm to 9.00 pm. Later lists included the statement "A short eclipse of 1 second is given 10 seconds before the signals to attract attention."

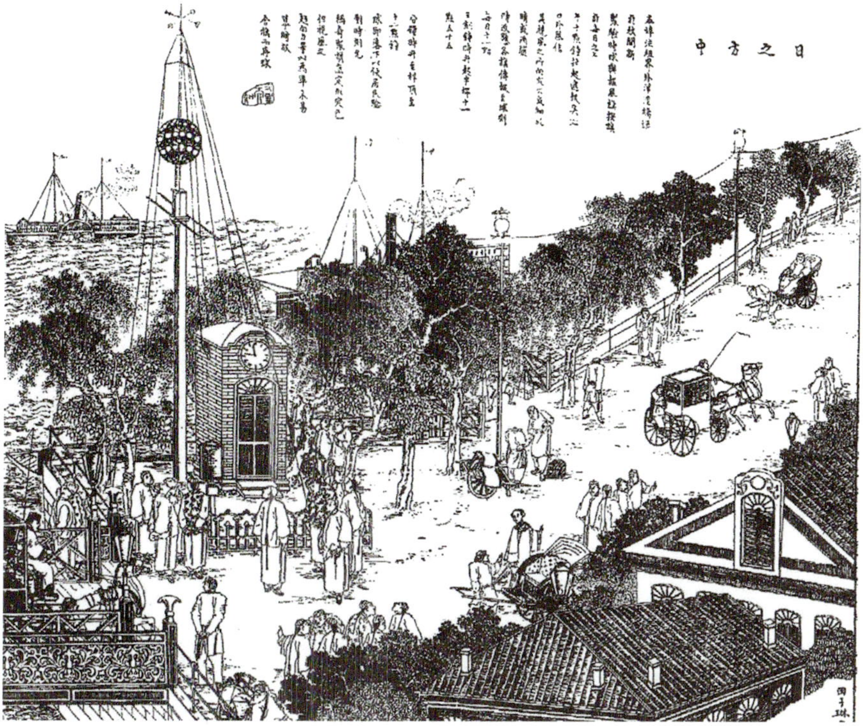

Fig. 21.4 Shanghai time ball sketch (Nature, 1885)

Fig. 21.5 Shanghai time ball before 1907. (Courtesy Klaus Hülse collection)

Fig. 21.6 Shanghai time ball after 1907. (Courtesy Klaus Hülse collection)

2.3 Amoy

A gun had been introduced at Amoy before 1898. Admiralty lists from 1911 onwards indicated that a flag showing the international T code was hoisted to alert observers that the gun would be fired in 5 min. It was lowered when the gun was fired. The gun was, however, reported as unreliable. It was still operating in 1939.

2.4 Tsingtau

A time ball at Tsingtau was established by 1904 at the German Concession. The Tsingtau ball had a large diameter of 2.1 m and a drop of 3 m (1911 list). Its construction can be seen in Fig. 21.7. A gun was added at Tsingtau and continued to operate after the time ball was withdrawn in 1915. It was reported to be inaccurate (1915 list).

Fig. 21.7 Tsingtau time ball. (Courtesy Klaus Hülse collection)

2.5 Dairen

Time lights were introduced at Dairen in 1922 and replaced an earlier time ball, according to a published report from a visiting British ship which included the statement "This signal is reported to be very satisfactory" (Notice, 1922). The arrangement of lights was not stated in Admiralty lists before at least 1927. The following description of an elaborate arrangement was given in lists after 1930.

> The lights are shown in a spherical frame in 8 perpendicular rows of 10 lamps each. The lights are switched on at 20 h 59 m 00 s and extinguished at 21 h 00 m 00 s Standard time. They are switched on again one second later and extinguished at 21 h 01 m 00 s and repeated at 21 h 02 m 00 s.
>
> If the signal fails or is in error, three red horizontal lights are shown from Kwantung observatory ….

The lights were located above the roof of a house on No. 2 pier.

2.6 Macau

The time lights at Macau were not listed by the Admiralty before 1934, but there is an indication in the 1939 list that they had been introduced as early as 1923. This may be a typographical error for 1933. They were located at the Meteorological Observatory (1939 list):

> The lights are in the form of a triangle, and are switched on 4 m 30 s before the hour and extinguished at Noon and every 3 hours Standard time.

2.7 *Other Locations in China*

A time ball and gun at Swatau were operational before 1898 but the service was only provided on Saturdays "and is not to be depended on" (1922 list). By 1904 an additional time ball and gun had been established at Chifu, but "Stated not to be sufficiently accurate for comparing chronometers" (1922 list). A short-lived time ball at Newchwang was listed in 1911 but not in 1915 or later.

3 Japan

There were no listed time signals for Japan in 1880, although representatives of the Japanese Navy had visited Bidston Observatory, Liverpool to study the methods used by John Hartnup to rate chronometers (The Japanese Navy, 1878). A time ball and gun at Kobe were the only signals listed in 1898. A range of different time signals appeared in later Admiralty lists, including time lights and a whistle, later replaced by a siren. Not all existed together. Apart from the time lights, signals were made at noon.

All the time balls in Japan were raised close-up at 5 min before the signal without the initial preparatory signal of raising to half-height. Flags were, however, hoisted in two stages.

3.1 *Hakodate*

A red and white flag signal was listed for Hakodate from 1904 to 1939. It was located on a flagstaff on the east side of the harbour. It was hoisted "on the dip" at 5 min before the signal, raised fully at 2 min before the signal and then dipped as the signal at noon Japanese standard time, 9 h ahead of GMT.

3.2 *Karafuto*

Karafuto is at a remote northern location in Japan, at a latitude 24° N of Nagasaki. A simple T-shaped time light was introduced at the meteorological observatory in 1926. It was included in Admiralty lists from 1929 to 1934. It was illuminated for 5 min before extinction at 8 pm. No reference to a previous time signal at Karafuto has been found.

3.3 *Kobe*

A time ball and time gun had been listed by 1898. In 1908, the ball was on a flagstaff near the Harbour Office. The gun was "On the West Camber". The ball colour was not stated in 1898, but it was listed as red from 1904 to 1928 and then red with a white band between 1930 and 1939. These signals continued until after 1939. All listed time balls in Japan used the same colour combination from 1930 onwards.

The Kobe time light was not introduced until about 1941. It used a T-shaped array of lights that was illuminated for a short period before extinction as the signal at 9 pm. It was still listed in 1947 as the only remaining time light in Japan.

3.4 *Moji*

Moji had a time ball in 1911 with an unstated colour. It was then located "In Shimonoscki road, on the south side of the town". It was repositioned to the summit of the Customs building in the late 1930s. It was red with a white band by 1915 and retained that colour throughout its listing until 1939.

3.5 *Nagasaki*

Flag and gun signals were introduced at Nagasaki before 1911. The red flag signal followed the procedure used at Hakodate. The gun was fired when the flag was dipped. The signals were stated to be inaccurate and a time ball was introduced during 1911, when the flag signal was discontinued. The red framework ball was located on the "northern slope of Nabekamuri yama" (1919 list). It was moved to a different location close to the Observatory in 1923 and was painted red with a white band from 1930.

Electric lights were installed in 1923 near the new time ball. They differed from other installations worldwide in using three green lights at the vertices of a triangle. These were illuminated together 5 min before the signal, twice switched off briefly before being illuminated again and extinguished as the time signal at 9 pm.

3.6 *Niigatako*

A whistle was listed as the signal at Niigatako (Niigata) between 1924 and 1939, replaced by a siren in 1947. The whistle was on a white tower with a yellow dome near the harbour. The signal was made at noon and 8 pm. The following description shows how it operated (1933 list).

> The whistle commences a low note 20 seconds before the hour and ceases exactly at the hour. The blast is worked by compressed air. The time is obtained from a Standard clock corrected every other day by wireless by Niigata Observatory. Height of whistle 110 feet.

This, and the earlier signal at Honolulu, were the only whistles listed by the Admiralty.

3.7 Osaka

A red and white time ball at Osaka was introduced in 1926 and listed until 1939. It was dropped electrically by Tokyo University from a yard arm at the administrative buildings.

3.8 Yokohama

Yokohama had three time balls, but not all existed together. The first, painted black, was on a white mast on the Eastern Hatoba. It was listed for the first time in 1904. A supplementary time ball was positioned on the "Watch house on north breakwater" in 1920, but both were replaced by a new time ball near the Kanagawa Harbour Board building in 1924. It was positioned on a wireless mast, shown in Fig. 21.8. Initially painted black, it was painted red with a white band after 1930 and was still listed in 1947.

The Eastern Hatoba location was given as 35° 26′ 41″ N, 139° 39′ 00″ E. The location of the short-lived time ball on the watch tower was not specified. The ball on the wireless mast was at 35° 26′ 42″ N, 139° 38′ 55″ E, about 150 m W of the first location. All three were operated electrically from Tokyo Observatory.

3.9 Other Time Signals in Japan

There were red time balls at Kure and Sasebo in 1904 and 1908 but neither was listed in 1898 or 1911. Both appear to have been short-lived.

Kagosima introduced a time gun in 1911 which was still listed in 1939 and 1947.

4 Korea

The only Korean signal was a time gun at Chemulpo on the western coast of the Korean peninsula. It was listed between 1911 and 1939 as being located at the "Observatory on summit above town".

Fig. 21.8 Yokohama time ball. (Courtesy Klaus Hülse collection)

Acknowledgements The Hong Kong time ball apparatus and its description were photographed for me by my late son-in-law Andrew Brown (1978–2019).

References

Combridge, J. H. (1989). The Shanghai time ball of 1884. *Antiquarian Horology, 18*(4), 381–385.

History of Hong Kong Time Service. (2022). https://www.hko.gov.hk/en/gts/time/History_of_HK_Time_Service.htm.

Mackeown, P. K. (2007). William Doberck – Double star astronomer. *Journal of Astronomical History and Heritage, 10*(1), 49–64.

Official reports on the proposed time-ball, harbour of refuge, and tree planting. (1877, November 22). Editorial in the *Overland China Mail*, Hongkong, 2.

The Hongkong Observatory. (1884, March 18). Editorial in the *Overland China Mail*, 4.

The Japanese Navy. (1878, September 4). Article in the *Hampshire Telegraph and Sussex Chronicle*, 2.

Chapter 22
Australia

Roger Kinns

Abstract The first time balls in Australia was introduced in 1853. Time balls at Sydney and Newcastle in New South Wales, Semaphore in South Australia, Williamstown and Geelong in Victoria, and Hobart in Tasmania were listed in 1880, together with a flag signal at Queenscliff in Victoria and a time gun at Hobart. A new time ball at Brisbane in Queensland was added in 1894, earlier time ball and gun signals having proved unreliable. Four more, at Fremantle in Western Australia, Devonport in Tasmania, and Port Pirie and Port Adelaide in South Australia, were added in 1900, 1907, 1917 and 1920. A flag signal at Geelong was listed and then discontinued before a new time ball was installed there in 1922. The time gun at Hobart was the only one in Australia listed by the Admiralty as a primary signal. Unlisted time guns had existed at various locations throughout Australia.

1 Introduction

The first time balls in Australia had been introduced in the 1850s and many others had been introduced by the 1870s. Time balls at Sydney and Newcastle in New South Wales (NSW); Semaphore, near Adelaide in South Australia; Williamstown and Geelong in Victoria; Hobart in Tasmania were included in the 1880 Admiralty list. A flag signal at Queenscliff in Victoria and a time gun at Hobart were also listed. Time balls at Brisbane in Queensland and Fremantle in Western Australia were added in 1894 and 1900. Three more: at Devonport in Tasmania, Port Pirie and Port Adelaide in South Australia were added in 1907, 1917 and 1920. A flag signal at Geelong was listed and then discontinued before a new time ball was installed there in 1922.

The time gun at Hobart was the only one in Australia listed by the Admiralty as a primary signal. A time gun at Fremantle was noted as a secondary signal. Many others had existed at various locations throughout Australia but never appeared in Admiralty lists.

R. Kinns (✉)
The Maudslay Society, Pembroke College, Cambridge, England

R. Kinns (ed.), *The World of Visual Time Signals for Mariners*, Historical & Cultural Astronomy, https://doi.org/10.1007/978-3-031-57334-7_22

Table 22.1 Signals in Australia

Country	Location	1880	1898	1904	1911	1915	1919	1922	1928	1934	1939	1947
Australia	Newcastle NSW	TB	TB	TB	TB	TB	TB	TB	TB	TB	TB	
	Sydney NSW	TB	TB	TB	TB	TB	TB	TB	TB	TB	TB	TB
	Brisbane Queensland		TB (black)	TB (black)	TB (black)	TB (black)	TB (black)	TB (black)	TB (black)			
	Semaphore SA	TB	TB	TB	TB	TB	TB	TB	TB			
	Port Adelaide SA							TB	TB			
	Port Pirie SA						TB	TB	TB	TB		
	Hobart Tasmania	TB	TB	TB	TB	TB	TB	TB				
		Gun	Gun	Gun	Gun	Gun	Gun	Gun				
	Devonport Tasmania				TB	TB	TB	TB	TB			
	Williamstown Victoria	TB (black)	TB (black)	TB (black)	TB (b&r)	TB (b&r)	TB (b&r)	TB (b&r)				
	Geelong Victoria	TB	TB	TB					TB	TB		
					Flag	Flag						
	Queenscliff Victoria	Flag	Flag	Flag	Flag	Flag	Flag	Flag	Flag			
	Fremantle WA			TB (black)	TB (black)	TB (black)	TB (black)	TB (black)	TB (black)	TB (black)	TB (black)	
									Gun	Gun	Gun	
Signal type totals		1880	1898	1904	1911	1915	1919	1922	1928	1934	1939	1947
Time ball, including collapsible		6	7	8	8	8	9	10	9	5	3	1
Gun, including powder flash		1	1	1	1	1	1	1	1	1	1	0
Lights		0	0	0	0	0	0	0	0	0	0	0
Other: disc, drum, flag, etc.		1	1	1	2	2	1	1	1	0	0	0

Table 22.1 shows entries for mainland Australia and Tasmania in Admiralty lists after 1880. Figure 22.1 shows their locations. There were two time balls for Port Adelaide after 1920 that were too close together to appear separately on a small scale map.

2 New South Wales

2.1 Sydney

The time balls at Sydney and Newcastle have been described in detail by Kinns and Abell (2009). The time ball apparatus for Sydney was manufactured by Maudslay, Sons & Field in 1855 and was operational in June 1858 after observatory completion (Pickett & Lomb, 2001). It was a development of the 1853 apparatus for

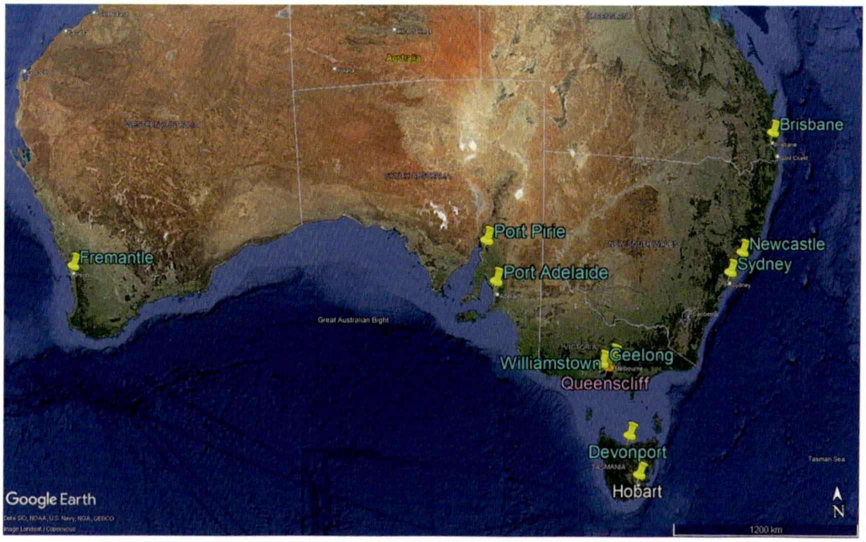

Fig. 22.1 Locations of Signals in Australia. (Google Maps)

Edinburgh and Deal which became operational in 1854 and 1855 respectively. The mechanical features of the apparatus made by Maudslays for these locations have been described in Chap. 4. Charles Todd (1826–1910) saw the Sydney apparatus under construction before he departed for Adelaide in 1855 and noted its similarity to the apparatus for Deal (Kinns & Abell, 2009; Abell & Kinns, 2010). Details of the apparatus, including photographs of the mechanical components which show design development for Sydney, have been presented by Kinns (2009). Todd's subsequent responsibility for the time ball at Semaphore in Australia is described later.

The Sydney ball and mechanism were modified in the 1870s by Henry Russell (1836–1907), then the NSW government astronomer, but most of the original features were retained. Russell listed some of the modifications he had made in a letter to Sir Charles Todd in Adelaide, during an exchange about possible designs for Western Australia (Russell, 1899). The complete correspondence has been transcribed by Kinns and Abell (2009). Original wooden components were replaced with new parts made from metal. These included the ball, which had used a zinc skin on a wooden frame, the shaft and the external mast. The diameter of the cylinder used to provide the air spring had also been increased to accommodate the increased weight of the ball, shaft and piston. Later changes include provision of an electric motor to raise the ball to complement the windlass. The new ball made of Muntz metal had been painted black, but the colour was changed to yellow for the Millennium celebrations in 2000 and has been retained since. The apparatus remains in operation. Figure 22.2 shows the existing time ball and tower in Sydney.

It has been shown that the design for Sydney was later used for Lyttelton, New Zealand (Kinns, 2009). The 1873 apparatus for Lyttelton did not include Russell's changes and is effectively a manifestation of the original apparatus for Sydney (see Chap. 23).

Fig. 22.2 Sydney time ball in 2007. (Author's photograph)

2.2 *Newcastle*

The first time ball at Newcastle was introduced in 1871, as well as a time gun. A larger ball was installed in 1874. The early balls were made of wickerwork. The time ball on the clock tower at the Customs House was the third at Newcastle. The Russell modifications to the Sydney apparatus were included in the design of the new apparatus, which used a rack and pinion mechanism. The apparatus was made by Potter and Sons of Sydney in 1877 and was the first of its type to be made in New South Wales. The ball operated for the first time in February 1878.

The history of the Newcastle signals was included in a conservation plan for the third time ball at the Newcastle Customs House (McDonald, 2000)

> The first time ball was erected on a mast 150 feet high at the rear of the Telegraph Office at the corner of Watt and Hunter Street. The ball was made of black painted wickerwork and dropped at 1.00 pm each day. A signal gun was not provided by the government and there was general dissatisfaction with this state of affairs. Eventually the government agreed to provide a six pounder gun and necessary ammunition, which operated with the time ball from 1 May 1871. A larger, more conspicuous ball replaced the original time ball on 24 July 1874. This time ball operated until the existing time ball was erected at the top of the clock tower of the new Newcastle Customs House. The Customs House time ball first dropped on 21 February 1878. The time ball operated without major overhaul for 63 years and it was 94

> years before the elements took their toll when, on 7 July 1972, an 85 mile per hour gale broke off the top portion of the mast. The time ball had been inoperative since 7 November 1941. The wreckage of the time ball was salvaged during late 1972 and early 1973. The ball was provided with a new inner steel framework and was re-erected in a fixed position on a shorter mast on 14 March 1973. The original weathervane was placed at the top of the mast.

The restored apparatus is believed to be operable, but is not presently in use after various changes of tower ownership. The photograph in Fig. 22.3 was taken in 2008.

3 Queensland

3.1 Brisbane

The only Queensland time ball was at Brisbane. An unreliable early time ball was erected in 1861 and discontinued in 1866 (The old windmill, 2023). It was replaced by a time gun which was not listed by the Admiralty in 1880 and ceased operation in 1891. A new time ball became operational in 1895 and was discontinued in 1930 (The time ball in Brisbane, 1894; Observatory time ball, 1930). It was listed by the Admiralty throughout that period and is known to have used the Devonport principle (Lewis, 1910). The black ball was installed "On top of the Signal tower, Wickham Terrace" and had a diameter and drop height of 1.5 m and 3.3 m. It is shown in Fig. 22.4.

4 South Australia

The time balls in South Australia were dropped at noon Standard time after 1927 while those in Queensland, New South Wales, Victoria, Western Australia and Tasmania all continued to be dropped at 1 pm.

4.1 Port Adelaide: Semaphore

The time ball at Semaphore near Port Adelaide was designed and commissioned by Charles Todd who had seen the Sydney apparatus under construction in London before his departure for Adelaide in 1855. His post at Greenwich had given him responsibility for the time balls at Greenwich and Deal (Kinns & Abell, 2009; Abell & Kinns, 2010; Bateman et al., 2014). Todd saw the need for an Adelaide time ball service from the time of his appointment and would have been able to exploit his earlier experience with Maudslay mechanisms. He is best-known for his work on the overland telegraph between Darwin and Adelaide that transformed international

Fig. 22.3 Newcastle time ball in 2008. (Author's photograph)

Fig. 22.4 Brisbane time ball. (Courtesy Klaus Hülse Collection)

communications with Australian cities in 1872 (Taylor, 1980; Thomson, 1999). Alice Springs is named after his wife.

Todd developed a design that used electrical triggering in 1867 and presented a working model in January 1869, but he was still unable to obtain necessary funding. It was only after completion in 1872 of the telegraph link between Adelaide and London, which caused a further increase in shipping movements, that a time ball service was provided at last. The Semaphore apparatus is likely to have been a development of Todd's 1867 design and was manufactured locally (Abell & Kinns, 2010).

The Semaphore mechanism, which became operational on 2 August 1875, differed considerably from that at Sydney, reverting to a chain hoist as in the original 1833 Greenwich mechanism. The original drawing is shown in Fig. 22.5. This would have been cheaper and easier to manufacture than the Sydney rack and pinion arrangement. The reported cost of the mechanism and ball was less than half that for Sydney. Unusually, an array of vertical segments was used to simulate a sphere from a distance. The tower at Semaphore was restored in 2007 using a modern apparatus to raise and lower the ball (Kinns & Abell, 2009). The original and restored tower and time ball are shown in Figs. 22.6 and 22.7.

4.2 Port Adelaide

By 1920, the time ball at Semaphore had been obscured from shipping by new buildings and a supplementary time ball was installed at the Adelaide Harbour Board Office in Port Adelaide. It is shown in Fig. 22.8. It was 14″ South and 1′ 30″ East of Semaphore, a separation of 2.3 km. The two balls were dropped at the same time. The ball at Port Adelaide was raised 3 minutes before the signal while that at Semaphore was raised half-way at 5 minutes and close-up at 3 minutes before the signal (1922 and 1930 lists). Both time balls were discontinued in 1932.

4.3 Port Pirie

Port Pirie is located on the east side of Spencer Gulf, about 220 km north of Adelaide. It continues to be an important port for exportation of minerals and grain.

A time ball was erected at Port Pirie in 1917 and continued after those at Port Adelaide had been withdrawn. It was still listed in 1935 but had been discontinued before 1939. The time ball was located on "Flagstaff on harbour master's residence". As at Semaphore, the ball was raised half-way at 5 minutes and close-up at 3 minutes before the signal.

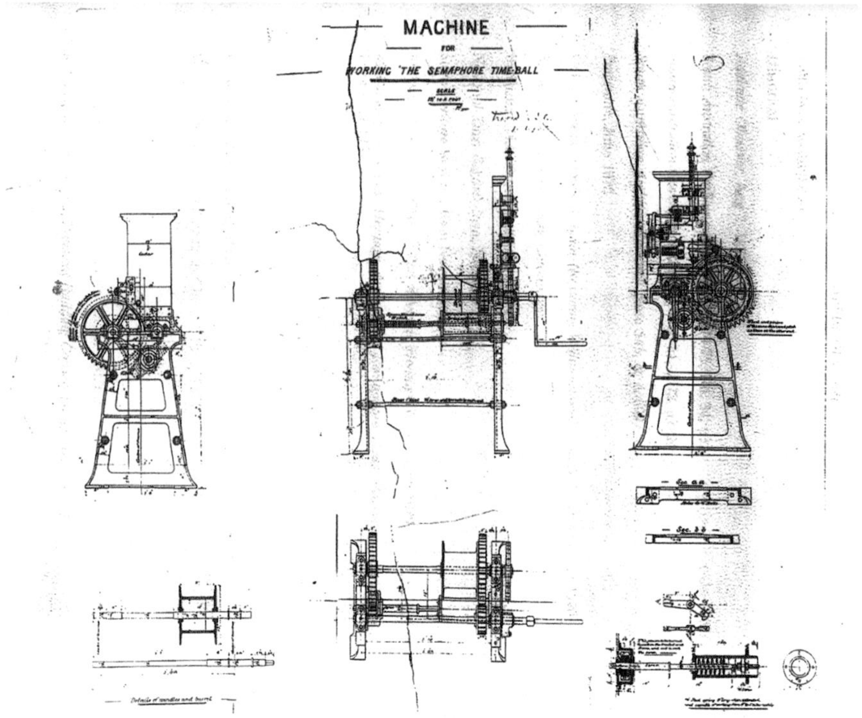

Fig. 22.5 The 1875 mechanism at Semaphore. (G.T. Light, Architect-in-Chief's Office, 1874, supplied by McDougall and Vines, Norwood, SA)

5 Tasmania

5.1 Hobart

Discussion at the Royal Society in Hobart in 1865 and acoustic experiments in 1868 led to a combined time ball and time gun service at Battery Point in Hobart from March 1875 (Kinns, 2011). Complaints from residents led to relocation of the gun a month later. It was then fired from Queen's Battery in the Domain for half a century. The change of location in 1875 had not been recognised in the 1880 Admiralty list but was corrected later. The drop of the ball at Battery Point was always the master signal; the gun was fired when the ball was seen to drop.

A large house called Lenna, now a hotel, was erected in the 1870s after the time ball had been introduced. The building features in almost all known photographs that show the time ball. An example is shown in Fig. 22.9. This and other images are described by Kinns (2011).

The time ball service had a remarkable history, having been provided initially by Francis Abbott (1799–1883), a private citizen, using his own transit telescope. Abbott had been deported as a convict from England to Tasmania in 1845 before gaining his freedom and establishing himself as a respected citizen (Orchiston,

Fig. 22.6 Semaphore time ball, c1877. (Courtesy SLSA B2408)

Fig. 22.7 Semaphore time ball in 2006. (Photograph by Marion Kinns)

Fig. 22.8 Port Adelaide time ball. (Courtesy Klaus Hülse Collection)

Fig. 22.9 Hobart time ball in 1908. (Courtesy Martin George)

1992). Ill-health forced his retirement in 1880 and the time service was provided until 1886 by his son Charles. Responsibility for it was then transferred to Hobart Observatory, but there were ongoing concerns about signal accuracy and the extent of manual intervention required to drop the ball. These led to independent enquiries into time service provision in 1894 and 1908 (see Kinns, 2011). During February 1910, the source of the telegraph signal was changed from Hobart to Melbourne Observatory, but the service was still unreliable and there was pressure to re-equip Hobart. Finally, automatic dropping of the time ball by telegraph from Melbourne was introduced in November 1910. The time ball service ended in February 1927.

The time gun had probably ceased to operate by the end of 1923, but there were sometimes long gaps in the time gun service before that date.

5.2 Devonport

A lesser-known time ball was installed at Devonport on the northern coast of Tasmania in 1907. Devonport was the usual destination of ferries operating between Melbourne and Tasmania. The time ball "Near post office" was withdrawn in 1928.

6 Victoria

6.1 Port Phillip Bay

Port Phillip is the large bay that shields Melbourne from the Bass Strait. The principal time signal was established at Williamstown, close to Melbourne near the mouth of the River Yarra. Queenscliff is located at the narrow entrance to the bay and Geelong is at its western extremity. Signals at Williamstown, Queenscliff and Geelong were all regulated by Melbourne Observatory from 1863.

6.1.1 Williamstown

The earliest time ball in Australia was established at Gellibrand Point, Williamstown in 1853 by Robert Ellery (1827–1908), later Government Astronomer in Melbourne. His intention to erect a time ball was published in May 1853 (Ellery, 1853). Part of the first notice, issued on 30 July 1853, is reproduced below (Ferguson, 1853).

> Commanders of vessels are requested to take notice that, for the purpose of enabling them to rate their chronometers, a time ball, painted black, will be dropped daily (Sundays excepted), from the top of the flagstaff at Gellibrand's Point, at the instant of one o'clock mean solar time there, corresponding to 9h 41m 8s Greenwich time. …
>
> Assigned position of the time ball, lat. 37d 52m 42s S., long. 144d 55m 28s E.

Another time ball in Melbourne was introduced at the same time (Public Notice, 1853).

> A ball (painted red) will be dropped from a gaff attached to the top of the flagstaff daily (Sundays excepted) at one o'clock mean solar time. …

The drop times of both balls were published regularly in the *Melbourne Argus*. These early time balls are thought to have been made of wickerwork. Admiralty list entries from 1880 did not include time signals in the City of Melbourne, which would not have been visible to ships in harbour.

Ellery described his early work at Williamstown in a paper published in *Monthly Notices of the Royal Astronomical Society* (Ellery, 1855). Initially, the time service was maintained using a small transit telescope, two chronometers and an astronomical clock at Williamstown. Ellery sought to improve the accuracy of time signals by purchasing new instruments. In March 1854 a new Telegraph Office was opened there and the time ball was transferred to a mast on the roof. The time ball had been moved to the top of the Old Lighthouse by 1862. This remained its location until the service was withdrawn in September 1926.

Williamstown had a deteriorating environment during the 1850s. In 1860 the government appointed a Board of Visitors to oversee Ellery's astronomical work at Williamstown and Georg Neumayer's magnetic and meteorological researches at Flagstaff Observatory. The Board recommended that the observatories should be merged at a new site in Melbourne (Williamstown Observatory, 2022). This led to the establishment of the new observatory next to the Botanic Garden in 1863 (Melbourne Observatory, 2022). A gun for signalling time to Melbourne citizens was introduced in Melbourne in 1865.

The Admiralty list entries for Williamstown all stated that the ball was on the staff of the Old Lighthouse, 68 feet above ground, and that the drop was 11 feet (3.3 m). The ball was listed as having a diameter of 5 feet (1.5 m) in 1898. The ball was still black in 1908, but the lower half of the ball was red from 1911 onwards. It was an accurate signal, the ball being dropped electrically from Melbourne Observatory. It was stated in Admiralty lists that errors of more than ⅓ second would be posted in the following day's newspapers. The Williamstown tower and apparatus were restored in 2009 as a famous landmark and the ball can now be operated with a slow descent. The tower and ball are shown in Fig. 22.10.

6.1.2 Queenscliff

A flag time signal was used at Queenscliff Signal Station for the whole of the period from 1880 to at least 1928. The flag "was dipped in coincidence with the time ball signal".

Fig. 22.10 Williamstown time ball in 2011. (Author's photograph)

6.1.3 Geelong

Another time ball was established at the Telegraph Office in Geelong in 1862. It was listed in 1880 and 1898 but had been replaced by a flag signal before 1908. There had been an intent to install a time ball on the Geelong Post Office tower (Geelong time ball, 1907), but the flag signal was still listed in 1916. This too was discontinued and a new time ball was eventually installed at the end of Moorabool Pier in 1921 or 1922. It was still listed in 1934. A replica time ball has been installed at the original Telegraph Station site and was operated for the first time in 2006. It is shown in Fig. 22.11.

7 Western Australia

7.1 Fremantle

A time ball was introduced at Fremantle in 1900, after William Cooke (1863–1947) had been appointed Government Astronomer in Western Australia. He had previously worked with Sir Charles Todd in Adelaide who advised him on selection of a time ball design (Kinns & Abell, 2009). It was listed until at least 1922 as a black

Fig. 22.11 Geelong time ball, after restoration at its original site. (https://www.workingheritage.com.au/places/6-geelong-telegraph-station)

ball having a diameter of 5 feet and a drop height of 10 feet on the flagstaff at Arthur Head, dropped by electricity from Perth Observatory. By 1927, its indicated location had been changed to "Tower of Harbour Trust Office, Cliff Street" with almost identical latitude and longitude. The same entry was retained up to 1939 but was removed from the 1940 list.

A display at the time ball site (Fremantle, 2018) indicates that the time ball had been moved from its original location to the base of the then recently decommissioned lighthouse at Arthur Head in 1903. This is the location shown in Fig. 22.12. The Federal Government reclaimed land at Arthur Head and required the time ball to be moved again. The tower of the Harbour Trust Office was the chosen site, but the move appears to have taken place before 1910. Also, the time ball and gun signal had been withdrawn before 1938. It appears that the Admiralty lists may not have been updated promptly.

It was also noted in Admiralty lists from 1927 that on weekdays a gun was fired when the ball was seen to drop. This gun was never included as a primary signal and was not mentioned at all in 1922 or earlier lists. Press announcements on display at Fremantle show that a time gun was first fired in October 1902, much earlier than its first mention in Admiralty lists. It also appears that gelignite was exploded in place

Fig. 22.12 Fremantle time ball. (Courtesy Klaus Hülse collection)

of firing a cannon in later years (Fremantle, 2018). The modern arrangement in Fremantle includes a cannon and a representative ball dropped from a yard arm.

Acknowledgements My interest in Australian time signals was stimulated by my search for time ball apparatus in Australia that might have been supplied or influenced by Maudslay, Sons & Field, visiting the time ball at Semaphore and working initially with Lesley Abell on the archives of Sir Charles Todd in Adelaide. Nick Lomb in Sydney was most helpful in providing access to the Sydney apparatus and in introducing me to Wayne Orchiston. My participation in a conference at James Cook University in March 2011 about the history of Australian astronomy led to further investigations into the time signals in Hobart and Williamstown, thanks especially to Martin George and Jenny Andropoulos. I am also grateful to Nicole Kessissoglou at UNSW in Sydney for her support and interest when I was working there as a Senior Visiting Fellow in mechanical and production engineering, and to Ross Ewington of the Tasmanian Philatelic Society for introductions to historians in Tasmania. Staff at libraries throughout Australia were always helpful and I am glad to acknowledge the extent to which digital access to newspapers using "Trove" has facilitated my research.

References

Abell, L., & Kinns, R. (2010). 'Telegraph' Todd and the semaphore time ball. *Journal of the Historical Society of South Australia, 38*, 42–57.
Bateman, D., Abell, L., & Kinns, R. (2014). Time balls at Greenwich and Adelaide – A direct personal connection. *Antiquarian Horology, 35*(3), 964–972.
Ellery, R. (1853). Letter headed "Time" to the Editor of the *(Melbourne) Argus* dated 3 May 1853. Published in the *Argus* on 7 May 1853, 9.
Ellery, R. (1855). Account of operations associated with the advancement of commercial astronomy in Australia. *Monthly Notices of the Royal Astronomical Society, 15*(5), 153–156.
Ferguson, C. (1853). Notice to Commanders in Hobson's Bay. Published in the *(Melbourne) Argus*, 4 August 1853, 4; and the *Sydney Morning Herald*, 8 August 1853, 4. [Ferguson was the Port and Harbour Master at Williamstown.]
Fremantle Display. (2018). *The story of the time ball and time gun at Fremantle*.
Geelong time ball. (1907). Article in the *Sydney Morning Herald*, 28 October 1907, 8.
Kinns, R. (2009). Time keeping in the Antipodes: A critical comparison of the Sydney and Lyttelton Time Balls. *Journal of Astronomical History and Heritage, 12*(2), 97–107.
Kinns, R. (2011). The Hobart time ball and time gun: A critical review. *Journal of Astronomical History and Heritage, 14*(2), 145–164.
Kinns, R., & Abell, L. (2009). The contribution of Maudslay, Sons & field to the development of time balls in Australia. *International Journal for the History of Engineering & Technology, 79*(1), 59–90.
Lewis, T. (1910). *Notes for the Astronomer Royal dated 21–25* April 1910. Royal Greenwich Observatory Archives RGO 7 257, Papers of William Christie, Cambridge University Library.
McDonald, B., & Associates. (2000). *Newcastle Customs House: A plan for its conservation and management*.
Melbourne Observatory 1863–1944 (2022). https://collections.museumsvictoria.com.au/articles/1634. Accessed on 14 September 2022.
Observatory time ball (1930). *The Brisbane Courier*, 3 October 1930, 12.
Orchiston, W. (1992). The contribution of Francis Abbott to Tasmanian and Australian astronomy. *Vistas in Astronomy, 35*, 315–344.
Pickett, C., & Lomb, N. (2001). *Observer and observed: A pictorial history of Sydney Observatory and Observatory Hill*. Powerhouse Publishing.
Public Notice (1853). Published under "Notices to mariners (from the Victoria Government Gazette, August 2)" in the Colonial Times (Hobart), 13 August 1853.
Russell, H. C. (1899). Letter to Charles Todd, 21 March 1899.
Taylor, P. (1980). *An end to silence*. Methuen of Australia.
The old windmill. Official notice erected near the windmill tower in Brisbane. http://3.bp.blogspot.com/-blHziLYWSkM/VViOCaQ9x-I/AAAAAAAABiI/3YTWJVCgHUE/s1600/DSCF5066.JPG. Accessed on 26 September, 2023.
The time ball in Brisbane. (1894). *The Brisbane Courier*, 6 December 1894, 2.
Thomson, A. (1999). *The singing line*. Chatto and Windus.
Williamstown Observatory 1853–1863 (2022). https://collections.museumsvictoria.com.au/articles/1629. Accessed on 14 September 2022.

Chapter 23
New Zealand

Roger Kinns

Abstract Time signals in New Zealand ranged from the 1864 time ball at Wellington to time lights at Wellington and Auckland that were withdrawn in 1937. The Wellington time ball was relocated in 1888 following a long period of harbour development and burned down in 1909, to be superseded by time lights in 1912. Various unsuccessful attempts were made to establish a reliable time ball service at Auckland. A final, short-lived time ball was replaced by time lights in 1915.

The first time ball service in South Island was established in 1867 at Port Chalmers, using local observatory facilities up to 1877. It was then re-established in 1882 as a weekly service with regulation from Wellington until its withdrawal in 1909. The 1876 time ball at Lyttelton became the principal signal in South Island and was in service until 1934. It was regulated by telegraph from Wellington, so did not require a nearby observatory.

1 Introduction

New Zealand is almost as far away from Greenwich on the Earth's surface as it is possible to be, so it was exceptionally difficult to establish the exact longitudes of time signal locations and associated drop times for Pacific navigators in the pre-wireless era. That was a particular challenge for the observatory at Wellington in North Island. The longitude of Wellington had been corrected in 1874 and was corrected twice more before 1908.

The history of visual time signals in New Zealand has been established in greater depth than almost any other country, using official sources and newspaper articles (Kinns, 2009, 2017a, b). It demonstrates some of the difficulties in providing accurate signals for navigators when there were major developments in the built environment and budgets were limited.

R. Kinns (✉)
The Maudslay Society, Pembroke College, Cambridge, England

R. Kinns (ed.), *The World of Visual Time Signals for Mariners*, Historical & Cultural Astronomy, https://doi.org/10.1007/978-3-031-57334-7_23

1.1 Overview

The wider history of New Zealand astronomy has been explored by Orchiston (2016a, b). The visual time signal history ranges from the 1864 time ball at Wellington to time lights at Wellington and Auckland that were withdrawn in 1937. The first transit telescope was erected in an observatory next to the time ball in Wellington and regulated time signals before harbour development in the 1880s. The observatory then changed location before a new observatory was built in Wellington Botanic Garden during 1907. This Dominion Observatory was known until 1925 as the Hector Observatory (after Sir James Hector, 1834–1907) and played an important role in the provision of time signals throughout New Zealand.

The Wellington time ball was relocated in 1888 following a long period of harbour development and burned down in 1909, to be superseded by time lights in 1912. The first time ball service in South Island was established in 1867 at Port Chalmers, using local observatory facilities up to 1877. It was then re-established in 1882 as a weekly service with regulation from Wellington until its withdrawal in 1909. The 1876 time ball at Lyttelton became the principal signal in South Island and was in service until 1934. It was regulated by telegraph from Wellington, so did not require a nearby observatory.

Various unsuccessful attempts were made to establish a reliable time ball service at Auckland following provision of signals by local citizens in the 1860s. An official time ball was finally introduced in 1901, discontinued in 1902, redesigned in 1903 and then discontinued again in 1904. Another time ball was erected on a different building and operated for a short period in 1915 before it was replaced by time lights that operated until 1937.

1.2 Entries in Admiralty Lists

Table 23.1 shows entries in Admiralty lists between 1880 and 1947. Their locations are shown in Fig. 23.1.

Only time balls at Wellington and Lyttelton were included in the 1880 Admiralty list. The 1898 list also included the Otago time ball at Port Chalmers, near Dunedin, which was not operating in 1880. The 1898 list formed the basis of a study by the New Zealand Historic Places Trust to show the recent status of time signals and to provide additional information about their origin (Wright, 2007). The 1904 Admiralty list was the only one to have included a time ball at Auckland.

Table 23.1 Signals in New Zealand

Country	Location	1880	1898	1904	1911	1915	1919	1922	1928	1934	1939	1947
New Zealand	Auckland Harbour Bd.			TB								
	Auckland Ferry Building						Light (w,r,g)	Light (w,r,g)	Light (w,r,g)	Light (w,r,g)		
	Wellington First location	TB (r&w)										
	Wellington Wool Shed		TB	TB								
	Wellington Observatory					Light (w,r,g)	Light (w,r,g)	Light (w,r,g)	Light (w,r,g)	Light (w,r,g)		
	Lyttelton	TB	TB	TB	TB	TB	TB	TB	TB	TB		
	Otago Port Chalmers		TB	TB								
Signal type totals		1880	1898	1904	1911	1915	1919	1922	1928	1934	1939	1947
Time ball, including collapsible		2	3	4	1	1	1	1	1	1	0	0
Gun, including powder flash		0	0	0	0	0	0	0	0	0	0	0
Lights		0	0	0	0	1	2	2	2	2	0	0
Other: disc, drum, flag, etc.		0	0	0	0	0	0	0	0	0	0	0

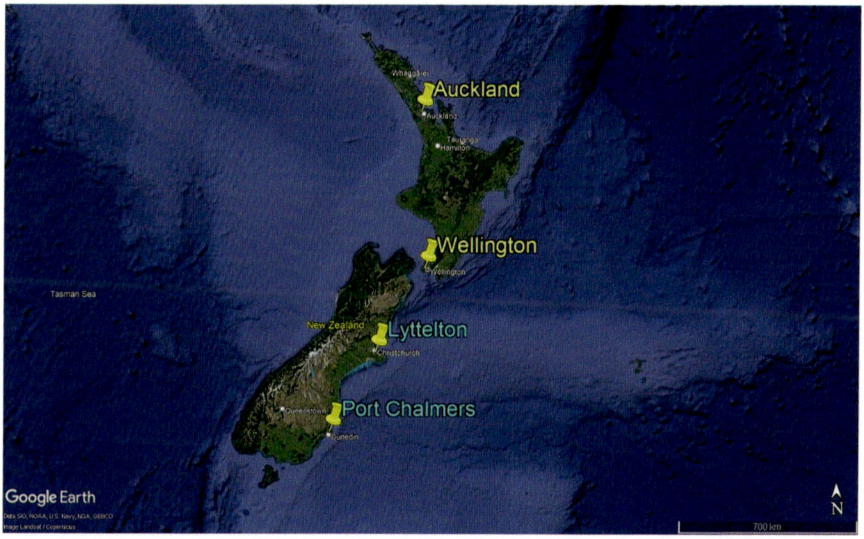

Fig. 23.1 Signals in New Zealand (Google Maps)

1.3 Entries in New Zealand Nautical Almanacs

The later history of time balls and other time signals can be traced through successive editions of the *New Zealand Nautical Almanac and Tide Tables*, which was first issued in November 1902 for the following year and superseded less comprehensive publications. It was then published annually. Each *Almanac* included statements about time signals that were expected to be available during the year. The only entries concerning time balls were for Wellington, Auckland, Lyttelton and Otago, many indicating that the signal had been withdrawn either temporarily or permanently.

By 1928, the entries in the *Almanac* had changed to include a section entitled "Dominion Time-service Arrangements for Chronometer-rating Purposes", as well as entries for individual ports. This included a list of the different types of signal that were then available throughout New Zealand and the means available for rating chronometers. The following sub-headings were used.

> Day wireless time-signals;
> Night wireless time-signals;
> Chronometer-rating time-signals by lights;
> Chronometer-rating time-signals by time-ball;
> Charging for telegraphing time-signals;
> Rating chronometers at Wellington;
> Dominion Standard Mean Time.

The following note in the 1928 almanac indicated that no manual intervention was required for signal transmission.

> Transmission of the time-signals is free of manual interference. The lights are extinguished, galvanometer deflected, wireless signals transmitted, and time-ball dropped automatically by direct communication from the Observatory clock at Wellington.

Although the time ball at Lyttelton was dropped automatically by 1928, manual intervention was still required to raise the ball and set the release triggers.

2 Wellington

The origin of the first time ball in Wellington is known from correspondence relating to a time ball at the Cape of Good Hope.

2.1 *The First Wellington Time Ball*

A new time ball apparatus for the Cape Observatory in South Africa was described in a notice published by Sir Thomas Maclear, Astronomer Royal at the Cape (Maclear, 1863a). It had been manufactured by Sandys & Co. in London and featured in two letters from Maclear to George Airy, the Astronomer Royal at Greenwich, which show that a similar apparatus had been ordered for New Zealand (Maclear, 1863b, c).

The ball had a diameter of 5 feet 6 inches (1.65 m) and a drop height of 18 feet (5.5 m). The Wellington ball was stated to be red with a white band in the 1880 Admiralty list, but its colour may have been changed more than once to maximise its visibility against the local background. A ball with this colour scheme had been supplied for the Strand in London (*The Times*, 1852).

Maclear was told by Sandys that they been awarded a contract for a New Zealand time ball (Sandys & Co, 1863). Maclear wanted Airy to check its design and operation before it was shipped, having found defects in the machinery delivered to the Cape. The letters were uncertain about the precise destination in New Zealand, but it must have been Wellington. Maclear wrote in the first letter (Maclear, 1863b):

> I suspect that our late excellent Governor – Sir Geo^{e} Grey, now Governor of New Zealand, originated the order, & am exceedingly anxious both on private & public grounds, that the machine should be sent out complete – in a state as perfect as the construction will admit of. My reasons for caution will appear presently ....
>
> My reason for the precaution is the fact that he sent out the Cape machine unfinished. Because of the great fall (18 feet) & the danger of the wood of the upper or slotted shaft twisting in this climate, it is covered outside with plate iron, & inside knee bent plate iron in the corners ....
>
> As the New Zealand Machine is to be on the plan of the Cape one, it will be found a heavy concern to manage. The weight of the Time Ball, gun metal rack rod & wood behind it, & metallic piston, come to about 350 or 360 pounds. 160 turns of the windlass are required to raise the ball, which on the average occupies 3 minutes. But the fault if any is my own. I wanted a great fall because of the low position & distance from the Anchorage.

The second letter, written 2 days later, was concerned with the lack of telegraph and observatory facilities in New Zealand (Maclear, 1863c). Although Airy was willing to inspect the apparatus and Sandys would have agreed, the letters arrived too late. The apparatus had already been shipped to New Zealand by the time Airy contacted the manufacturer (Sandys & Co, 1863).

The location of the time ball at the Customs House (also known as the Custom House or Customhouse) is shown in Fig. 23.2. The large drop of the ball is consistent with MacLear's description of the apparatus at the Cape. It indicates that the ball had a uniform dark colour at the time of the photograph, seen also in an engraving (Customs House, 1870s).

2.1.1 Announcement in Wellington

The time ball apparatus had arrived in Wellington by January 1864. The following brief account was published soon afterwards (1864).

> Many improvements have of late been made in the way of buildings, &c. and one of the chief objects that catches the eye from the wharf, is a time ball erected on the top of the Custom House, the pole passing through the centre of the building. It is on the same principle as the Greenwich time ball, and will fall every day at 12 o'clock. The clocks in connection with the works will be under the management of S. Carkeek, Esq., Collector of Customs ....

A fulsome description of the new time ball apparatus was published in the *New Zealand Spectator and Cook's Strait Guardian* (The Time Ball, 1864). Extracts from that article are reproduced below. It gives due credit to Stephen Carkeek (1815–1878), the leading astronomer in New Zealand at that time (Orchiston, 2016a).

Fig. 23.2 Wellington time ball at first location (Alexander Turnbull Library, H.N. Murray Collection, Ref: PAColl-0824-1)

Saturday last ought to be marked as a red letter day in all future editions of New Zealand almanacks, for at 3 o'clock the Wellington Time Ball was dropped by electricity ....

To Mr. Carkeek belongs the chief credit of this work; as most certainly it would never have been thought of had we not possessed one so thoroughly capable of directing the setting up of the somewhat complicated gear which carries and works the ball. Saturday's experiment proved that all the machinery was in working order, but the ball will not, we understand, be dropped regularly each day for at least a month. The observatory has yet to be built at the side of the Custom house; the transit instrument to be set up; the two clocks to be fixed and rated; all this done, each day at 12 o'clock the ball will drop, shewing with perfect accuracy true time ....

The necessity of thus checking the chronometers of ships is so well known in England, that a ball is dropped at Greenwich, at Deal, and at another point of the Channel. All three balls are dropped by the same current of electricity, and all at the same instant precisely, although so far distant from one another ....

The time ball is of zinc, weighing about two hundredweight. It is carried by an iron rod, which rod at its lower end is attached to a piston. The rod and piston are fitted into an iron cylinder resting on a foundation built up carefully from the rock below the Customhouse. The cylinder is packed at the bottom with India-rubber, forming an elastic cushion to deaden the blow of the piston when the ball is dropped ....

When the ball is wound up by the rack work the bottom of the piston is caught by a small piece of steel, which locks it securely. This trigger forms part of a most beautiful and delicate system of levers, which work one upon the other. The last of these, when the whole are set, needs but the slightest touch to release the trigger supporting the piston, and to drop the ball. Each day, when the ball is wound up to the top of the mast, these levers must be set by the assistant ....

... We hope that the Provincial Council will supplement the work of the ball, by voting funds for powder, &c, for a cannon, which, when fired, will carry the tidings, it is 12 o'clock, to the Hutt and to those parts of the town where the fall of the Time Ball is not visible.

The time ball dropped for the first time as an official signal on 9 March 1864, as indicated by the following notice in the *New Zealand Spectator and Cook's Strait Guardian* (Monthly Summary, 1864).

Notice is hereby given, that on and after Wednesday next, the 9th instant, the Time-ball at the Custom House will be dropped on each and every day, Sunday's excepted. The Ball will be hoisted half-mast high at ten minutes before 12, to the mast head at five minutes before 12, and will fall precisely at 12 o'clock at noon, Wellington mean time.

Notices about the time ball location and drop time were published regularly in newspapers. The following notice is typical (Wellington Independent, 1864).

The Time Ball at Wellington is situated in latitude 41° 17′ 01″ S., and longitude 174° 49′ 15″ E. It is dropped at noon every day (Sunday's excepted), or at 12h. 20m. 43s. past noon at Greenwich of the previous day.

Carkeek carried out the appropriate astronomical observations to determine its latitude and longitude (Thomson & Jackson, 1871). Amateur astronomer and minister of religion, Archdeacon Arthur Henry Stock (1823–1901), was responsible for the day-to-day operation of the time ball.

Announcements of the time ball drop were published in the *New Zealand Times*, usually every 1 or 2 weeks. The ball was dropped daily with a local newspaper announcement when star transit observations had allowed increased precision for chronometer-rating purposes. Otherwise, the drop was regulated using observatory clocks, with time extrapolated from preceding transit observations. The usual accuracy was better than 1 second. The service had to be withdrawn from time to time when repairs became necessary.

2.1.2 Introduction of New Zealand Mean Time

Stock published a letter to the editor of the *Wellington Independent*, explaining the change that would occur on 2 November 1868 when New Zealand mean time was introduced (Stock, 1868):

Sir, The Time Ball will drop on Monday at 12 o'clock New Zealand mean time. This time is 9 minutes 17 seconds slower than Wellington mean time, as the longitude of the Time Ball is 174° 49′ 15″. This longitude differs from that generally given, but it was calculated by Mr Carkeek from several observations taken by the transit instrument of the Time Ball observatory. The master of H.M.S. Esk told me that he was aware of this error of the chart longitude. If true time is wanted for setting sun-dials, or for any other purpose, 9 minutes 17 seconds should be added to the Time Ball time. Clocks and watches should be put back 9½ minutes on Sunday night.

The longitude quoted in the letter is that given in 1865 notices about time ball operation, which appears to have differed from contemporary Admiralty charts. The letter stated the need to readjust clocks and watches to take account of the introduction of New Zealand mean time (NZMT) in 1868, which was 11½ hours ahead of GMT and was legal time throughout New Zealand. Remarkably for the period, the longitude of the meridian 175° East of Greenwich was used to define legal time in New Zealand, not the longitude of the Observatory.

The reliability of the time ball apparatus started to become a problem in the early 1870s. A notice in April 1871 stated that the time ball must not be used for rating chronometers until further notice (Stock, 1871). The problems had become more serious by November 1873, when there was a need for clarification about responsibility for repairs (Stock, 1873). Problems with decay in the time ball mast had become significant by September 1874, and the ball had ceased to operate regularly (Evening Post, 1874). The winding mechanism was again faulty in the following year (Evening Post, 1875).

The last notice of a time ball drop at the Customs House location appears to have been in February 1882, prior to a long period of waterfront redevelopment (Shipping, 1882a).

2.1.3 Longitude Correction

There was ongoing controversy about the longitude of the time ball and revised calculations led to a time signal correction on 1 April 1874 (Longitude of Wellington, 1874). It was stated that the drop had previously been 6.15 seconds too early, corresponding to a longitude error of 1 minute, 32 seconds of arc. In effect, the previously assumed meridian was too far East. The revised longitude of the time ball was estimated to be 174° 47′ 43″, almost exactly that appearing in the 1880 Admiralty list. There was, however, ongoing doubt about longitude accuracy. Another correction of 3.8 seconds had to be made to the longitude of Wellington in 1903, moving the meridian still further West. It was changed again between 1905 and 1908 by introducing another correction of 0.8 seconds, again to the West.

2.1.4 Relocation of the Wellington Time Ball

The harbour in Wellington underwent large changes between 1882 and 1888 which necessitated moving the time ball and astronomical apparatus.

The Harbour Master at Wellington offered his views about future time ball location to the Harbour Board (Harbour Master, 1882). He referred to the poor visibility of the time ball against the background of houses and hills at its former location and argued that it should be positioned on Mount Victoria. Discussion continued for years afterwards. The Board engineer concluded in a memorandum that the best location would be on the brow of Mount Wellington, next to an existing telegraph line (Wellington Harbour Board, 1885). He noted:

> … as the time ball is completely smashed, the question of size is not an element to be considered as the new ball could be made as large as might be deemed necessary …

The matter was referred to Dr. James Hector, Director of the Colonial Museum of New Zealand, for his opinion. Hector (September 1886) indicated that he had received various papers about the time ball and that the favoured location was by then "one of the tees of the wharf". He also noted that:

> … I think it rather absurd that we should have the time observatory in Wellington and no time ball, nor any means by which the public can check the time that it is distributed through the telegraph and railway clocks, while the time ball apparatus is lying idle.

It appears that the original time ball apparatus had been saved and that a new mast and ball would allow it to be re-used. It was to be another 2 years before a Wellington time ball was again in operation.

2.2 *The Second Wellington Time Ball*

Notice of a new Wellington time ball service appeared in newspapers with the statement that "The time ball is to be erected once more for the benefit of shipping." (Wellington, 6 November 1888). The 1898 Admiralty list indicated that it was located on "Staff on square tower at inner end of Railway Wharf". The title of the drawing of a square tower issued to potential contractors by the Wellington Harbour Board was "Hydraulic Accumulator House at Waterloo Quay Woolshed (Wool Store J)" (Hydraulic, 1888).

A contract to erect the new building was duly let (Wellington Harbour Board Contract, 1888). The specification included the statement "At the eastern corner provision is to be made as shown on Sheet No 2 and in detail on Sheet No 4 for the subsequent erection of a time ball mast and apparatus". Two elements of Sheet No 4 are shown in Fig. 23.3. The outline drawing for the time ball and mast indicates that the ball drop would be 13 feet (4.0 m), with a ball diameter of 4 feet (1.2 m). Both dimensions had changed from the first Wellington time ball in Fig. 23.2 and changed again in the final installation. An arrangement of gears and a capstan can be seen in the left-hand drawing. An offer to supply a new time ball for £5 10s was made on 31 August 1888, using a design supplied by the Harbour Board (Luke & Sons, 1888). A drawing of the Lyttelton ball, which had a diameter of 1.5 m, was probably given to the ball manufacturer. This was the diameter specified in a later report (Wellington Harbour Board Report, 1910). No evidence has been found that tenders were sought for a replacement apparatus, so it appears that the original 1863 apparatus was re-used with a new ball.

The photograph in Fig. 23.4 shows the time ball at its new location. Comparison of the photograph with the drawing of the building indicates that the ball diameter was 1.5 m and that the drop height was about 5.2 m, close to the original drop height. Figure 23.4 shows that the second ball had a coloured central band.

The first notice in the *New Zealand Times* concerning drop of the time ball at its new location at the head of Waterloo Quay was in January 1889 (Late Shipping, 1889a):

> CHRONOMETER-RATING NOTICE. The time-ball may be used to-day for rating chronometers. The ball will drop precisely at noon, New Zealand mean time, equivalent to 11hr 30min Greenwich mean time. Any difference on 11hr 30min shown by a chronometer at the moment of the signal will be the error, plus or minus, of the chronometer. Colonial Observatory, Wellington.

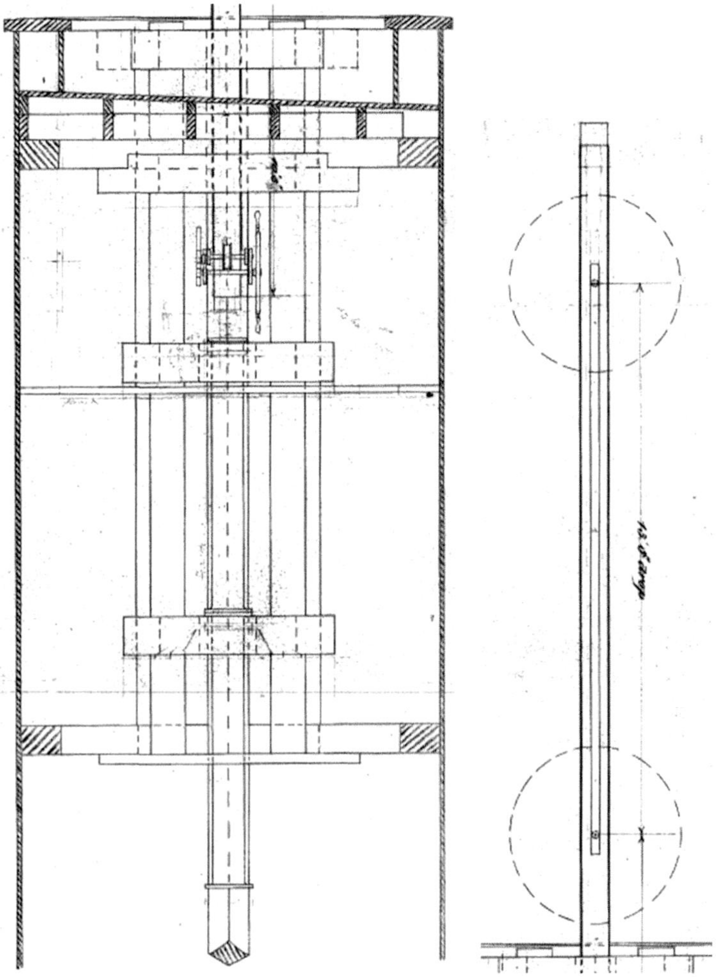

Fig. 23.3 Extracts from 1888 installation drawing for the Wellington time ball apparatus (Wellington Harbour Board archive, Drawing Office No. 3511)

Fig. 23.4 Wellington time ball at second location after 1888 (Wellington City Archives 2012/2:6725)

This notice followed star transit observations and did not mention that it was a new time ball service, so the ball may have been in use for some time before that. The newspaper announcement was usually repeated every 1 or 2 weeks. The early notices implied that noon, NZMT was equivalent to 11 hours 30 minutes, GMT. This should have been 12 hours 30 minutes (0.30 am GMT) At that time, GMT was based on noon at Greenwich for astronomical purposes.

The incorrect notice re-appeared on several occasions, but was corrected during April (Late Shipping, 1889b):

> COLONIAL OBSERVATORY. Chronometer-rating Notice.
>
> The time-ball may be used to-day for rating chronometers. The ball will drop at noon precisely, at which hour a chronometer set to Greenwich Mean Time should show 12h. 30m. Any difference will be the error of the chronometer, fast or slow on Greenwich Mean Time.

2.2.1 Entries in New Zealand Nautical Almanacs

The entry for Wellington in the *New Zealand Nautical Almanac* for 1903 is particularly informative, including notification that the assumed longitude of the observatory was about to be changed and that the time ball would be dropped 3.8 seconds later than before. There was also a clear distinction between days when the exact time could be confirmed by astronomical observations and other days where there had to be reliance on the astronomical clock for interpolation.

> There is an astronomical observatory at Wellington, and approximately correct time may be obtained from daily signals which are given by a time ball situated on the tower of "J" Shed, Waterloo Quay, at the root of Railway Wharf. The time may be taken as absolutely correct for chronometer-rating purposes on the days when a flag is flown on the flagstaff close alongside the time ball; also, on these close-rating days a notice is inserted in the *New Zealand Times* to this effect. The ball falls at 12.30 Greenwich mean time, which is equivalent to noon in New Zealand, and this is the time which is kept throughout the colony. There is also a galvanometer, deflecting every hour, in the Public Room of the Telegraph-office, close to the Queen's Wharf, which is controlled by the same clock which drops the time ball.
>
> [NOTE. – The longitude assumed for the Observatory in calculating time is 11h. 39m. 9.13s., which corresponds to that of the coastal charts. The accepted longitude of the Observatory is 11h. 39m. 5.31s., based on the longitude of Sydney. It is probable that this longitude will very shortly be adopted in giving Greenwich time by the time ball.]

The 1905 and 1906 entries showed the expected change in assumed longitude, but it had changed again by 1908.

The entry for 1907 contained an important announcement about relocation and development of the observatory at Wellington. A temporary observatory and a reduced time signal service, including suspension of the Wellington time ball service, would be available in that year.

> The time ball has been temporarily discontinued owing to the observatory having been demolished. A new observatory is about to be erected on another site. Meanwhile the time for the colony is kept by chronometers which are checked by theodolite observations taken from a temporary observatory in the grounds of the Government Buildings by the Lands and Survey Department. Approximately correct time is given daily by galvanometer to the Telegraphic and Railway Offices, and once a week the correct time from observation is telegraphed to Auckland, Lyttelton and Dunedin. Any shipmaster in Wellington wishing to correct his chronometer should apply at the Museum to the Permanent Observer (Mr. King), who has charge of the chronometers, and gives the time to the colony.

The Admiralty lists for 1898 and 1904 gave the same coordinates of 41° 16′ 50″ S, 174° 46′ 55″ E for the signal at "Staff on square tower at inner end of Railway Wharf" and did not reflect the longitude change highlighted in the 1903 *Almanac*. The 1908 list specified, incorrectly, that the time ball had been moved to the "Observatory on Battery Hill" with coordinates of 41° 17′ 04″ S, 174° 46′ 07″ E. This may have been an anticipated change that was never implemented. The entry in the 1908 *Almanac* still referred to the Railway Wharf location.

The longitude assumed for the Observatory in calculating time was specified from 1908 onwards as 11 hours 39 minutes 4.48 seconds, a change of 0.83 seconds from the assumption in 1903. It then corresponded to 174° 46′ 07″ E, the longitude given in the 1908 Admiralty list.

The service had been re-established in 1908, but the tower burned down in March 1909. The outline of the time ball apparatus can be seen in Fig. 23.5. Subsequently, chronometer calibration had to be carried out by taking chronometers ashore or by using an intermediate watch (Shipping News, 1909):

> In this week's Gazette it is notified that, owing to the destruction by fire of the "J" Shed, Waterloo Quay, the time ball which was situated on the tower of the said shed is no longer available for the information of masters of vessels frequenting the port. Correct mean time may be obtained in the public room of the telegraph office, close to the Queen's' Wharf, where a galvanometer, controlled by the observatory clock, is deflected every hour.

The 1910 *Almanac* mentioned its destruction by fire.

2.2.2 Possible Replacement in 1910

Enquiries about a replacement time ball apparatus were made in London (New Zealand High Commission, 1910a). W.H.M. Christie, the Astronomer Royal at Greenwich, arranged for a reply to be drafted by a member of his staff (Lewis, 1910a, b; see Kinns, 2017a). His notes also included estimates of cost. These notes were summarised in the official reply from the Astronomer Royal (1910a).

Fig. 23.5 Wellington's Latest Conflagration—The Last of Capt. Edwin's Tower and Time Ball (New Zealand Free Lance, 13 March 1909)

In reply to your letter of April 21, R. B 21/58, relative to the supply of a Time Ball and apparatus for Wellington, New Zealand, I have to inform you that the Time Ball and hoisting apparatus can be supplied by Messrs Saxby and Farmer Ltd., Railway Signal Engineers, 53 Victoria Street, Westminster, and the Clock and Electrical appliances by Messrs E Dent & Co., Ltd., 61 Strand, W. C., or doubtless the latter firm would undertake the complete contract.

These firms have satisfactorily carried out the fitting up of Time Balls and apparatus in various parts of the world.

Tenders were sought by the New Zealand High Commission (1910b) from the two firms. A subsequent report confirmed that offers were received from both, with additional offers from Messrs Smith & Sons and Messrs Gillett & Johnston. The latter firm stated that their time ball would be similar to those supplied for Port Said and Cape Town. Despite protests from shipping companies about the lack of a time ball (Wellington Harbour Board, 1911), no order was ever placed for a new time ball apparatus.

The entry in the 1912 *Almanac* heralded the introduction of time lights at Wellington observatory, as well as final discontinuance of the time ball service in Wellington.

There is an astronomical observatory at Wellington on Battery Hill, in the Botanical Gardens, in latitude 41° 17′ 3.76″ S. and longitude 174° 46′ 7.2″ E. = 11h. 39m. 4.48s. From the tower of this observatory a time-signal by electric lights is being inaugurated (See further page 328.) The time-ball which was situated on the tower of J. Shed, Waterloo Quay, has been discontinued since the destruction of this shed by fire. Correct mean time may be obtained in the public room of the telegraph-office, close to the Queen's Wharf, where a galvanometer controlled by the observatory clock is deflected every hour.

Later editions include details of the telegraph, time lights and eventually radio signals that were provided by Wellington.

2.3 Wellington Time Lights

The first announcement including time light signals at Wellington was published in the *New Zealand Times* (Ships and Shipping, 1912).

HECTOR OBSERVATORY, WELLINGTON. Latitude 41deg 17min 3.76sec south. Longitude 11hr 39min 4.27sec east of Greenwich. Height above 1909 mean sea level, 418 feet. CHRONOMETER RATING NOTICE. At 1 p.m. to-day a galvanometer signal for rating chronometers will be sent from the Observatory to the Public Telegraph Office, and to the Dominion Museum. The needle will move at 1 p.m. exactly of New Zealand standard mean time, when a chronometer set to Greenwich mean time should show 13hr 30min. Any difference will be the error of the chronometer on Greenwich mean time. At 9 p.m. correct time will also be signalled from the Observatory by means of electric lights. A green light will be switched on at about 8.45 p.m., a red one at about 8.55 p.m., and a white one at about 8.59 p.m., and all lights will be switched off at 9 p.m. exactly of New Zealand standard mean time. The preparatory switching on of the lights must be considered as only approximately correct, and must not be used for rating chronometers. The correct time for rating will be given by switching off the lights simultaneously at 9 p.m. C. E. ADAMS, Government Astronomer.

Figure 23.6 shows the arrangement of time lights at Wellington, on a mast next to the Observatory buildings. The photograph was taken in 1913 soon after the lights had been introduced. The lowest green lamp in the vertical array was illuminated first, the middle red lamp second and the upper white lamp last, at nominated intervals. This gave three preparatory signals, later reduced to two when precise calibration using star transits was not available. Extinction of all lights was the time signal, repeated at least once.

Figure 23.7 shows an arial view of the Dominion Observatory buildings and time lights in 1935. The arrangement of lights remained unchanged from 1912. Other photographs are included in a comprehensive article about the history of structures in Wellington Botanic Garden (Tomlinson, 2015) and in a factsheet about the Observatory (Dominion Observatory, 2007).

Fig. 23.6 Wellington time lights in 1913 (Wellington City Council Archives, 2005/24-13-23, Dominion Observatory, photograph: W.F. Tibbutt)

Fig. 23.7 Kelburn, Wellington, Meteorological Office, 1930 (detail) (Evening Post Collection, Alexander Turnbull Library, Wellington, EP-2845-1/2-G)

2.3.1 Entries in Nautical Almanacs

The entry for Wellington in the 1914 *Almanac*, published in October 1913, is transcribed below. It is similar to the newspaper announcement in February 1912, showing that the timings of individual lights had not been altered.

> **Time-signal and General Service Arrangements.** – There is an astronomical observatory at Wellington on Battery Hill, in the Botanical Gardens, 418ft. above sea level, in latitude 41° 17′ 3.76″ S. and longitude 174° 46′ 4.0″ E. = 11h. 39m. 4.27s. From the tower of this observatory a time-signal by electric lights has been established.
>
> At 1 p.m. on chronometer rating-days a galvanometer signal for rating chronometers is sent from the observatory to the Public Telegraph Office, Customhouse Quay, Wellington, and to the Dominion Museum, Wellington. The needle moves at 1 p.m. exactly of New Zealand standard mean time, when a chronometer set to Greenwich mean time should show 13h. 30m. Any difference will be the error of the chronometer at Greenwich mean time.
>
> At 9 p.m. on chronometer rating-days correct time is also signalled from the observatory by means of electric lights. A *green* light is switched on at about 8.45 p.m., a *red* one at about 8.55 p.m. and a *white* one at about 8.59 p.m., and all the lights are switched off at 9 p.m. exactly of New Zealand standard mean time. The preparatory switching-on of the lights must be considered as only approximately correct, and must not be used for rating chronometers. The correct time for rating is given by switching-off the lights simultaneously at 9 p.m. The lights are now shown from a flagstaff in a vertical line one over the other; the lower light *green*, middle light *red*, and upper light *white*.

Changes in the signal timings were included in the 1915 *Almanac*. The Wellington time light signals were now being repeated on four successive hours from 8 pm to 11 pm. Also, the switching on of the individual lights had been changed from 15, 5, 1 minutes to 50, 10, 5 minutes before the hour. A flag was introduced to show that the signal times had been derived from transit observations during the previous night, rather than by extrapolation using astronomical clocks.

> Time Service. – 1. Accurate Time Signals. – On days when accurate time signals are given, the flag T of the international code will be hoisted on the Observatory flagstaff about midday.
>
> (a) When the flag is flying, chronometers may be compared with a galvanometer in the public telegraph office, Featherstone Street. The galvanometer is controlled by the Observatory clock, and is deflected every hour of New Zealand mean time.
>
> (b) Time signals are given by three vertical electric lights erected on the Observatory flagstaff. The bottom light is green, and is 30 ft. above the ground; the middle light is red, and is 36 ft. above the ground; the top light is white, and is 42 ft. above the ground. The green light is shown at 50 minutes, the red light at 10 minutes and the white light at 5 minutes to the hour; all three lights are extinguished simultaneously at the hour. The switching-on of the lights must be considered as only approximately correct; the correct time is given by switching off the three lights. This signal is given at 8, 9, 10 and 11 p.m. of New Zealand mean time ….

There were other timing changes during the life of the signal. In 1918, the signal was made three times, at 7.30 pm, 8.30 pm and 9.30 pm NZMT. The 1922 Admiralty list indicated that the signal was made at 8.30 pm (9 am GMT) only, with illumination of the green, red and white lamps at 50, 10 and 5 minutes before 8.30 pm. The 1922 notice also included the qualifications below:

> … The green light is only used on Tuesdays and Fridays, New Zealand date.
>
> When owing to bad weather or other causes accurate time signals cannot be given, approximate ones will be given on application being made to the Observatory, but in such cases the green light will not be shown.

The 1930 Admiralty list indicated that the signal continued to be made at 8.30 pm, but the lamps were then switched on at 20, 10 and 5 minutes before 8.30 pm.

The final entry concerning time light signals was in the 1937 *Almanac*. The green light was no longer used at Wellington and the service was fully automated. The additional entries were:

> … The lights are *white* and *red*, shown vertically, 6 ft. apart, with *white* uppermost, 42 ft. above the ground. The *red* light is shown 10 minutes and the *white* 5 minutes before the signal. Simultaneous extinction of the lights at 9h., G.M.T. is the time-signal. The lights are operated automatically, but as they are not supervised, failures sometimes occur through unforseen causes.

There were no entries concerning time lights in the 1938 *Almanac*, confirming that the service had been terminated during 1937.

3 Lyttelton

Lyttleton, near Christchurch in South Island provided a time ball service between December 1876 and November 1934. It was the only listed time ball in New Zealand after 1915 and was never replaced by time lights.

3.1 The Lyttelton Time Ball

Booklets about the Lyttelton time ball were published by The New Zealand Historic Places Trust in 1979. Two versions have been published, both entitled *The Lyttelton Time-Ball Station.* The preliminary version included the manufacturer's drawing. The second, longer version included several photographs and also a three-dimensional sketch of the mechanism in place of the drawing.

Siemens Brothers shipped the apparatus for Lyttelton from London in July 1874, following an order in March 1873. Siemens had become a principal supplier of telegraphic equipment, with heavy commitments to supply and installation of telegraph cables at the time (see Kinns, 2009). Lyttelton time ball operation had to await completion of the necessary tower and it became operational on 23 December 1876. It was restored faithfully during the 1970s and was a much-admired New Zealand landmark. The apparatus was severely damaged in the 1911 earthquakes which devasted Christchurch and caused the time ball tower to collapse.

The following quotations from the booklets show how the project came to fruition:

> Canterbury's time-ball project was backed by two Australians in the Provincial Council, who had a long-term interest in shipping. Businessman and farmer J. T. Peacock was the son of a Sydney shipowner who had a fleet of ships in Lyttelton, and Henry Webb, Provincial Secretary, had come to Lyttelton initially to take control of the Peacock shipping business. The old Sydney time-ball would have been a familiar sight to both of them.
>
> At a Council session in November 1870, Peacock proposed: "That a respectful address be presented to His Honour the Superintendent requesting him to place upon the Supplementary Estimates a sum sufficient to erect a time-ball and tower at the port of Lyttelton." The motion was passed, and Webb contacted the Telegraph Department in Wellington who requested a quotation from London for a ball and its machinery and also a special clock: the ball was to drop automatically, released by an electric current which was to be switched on at the correct time by an astronomical clock. Electricity was one of Webb's many interests – he had demonstrated the first electric light in Lyttelton.

Webb's interests in lighting and telegraphy would have made him aware of the important contributions by Siemens Brothers in these areas.

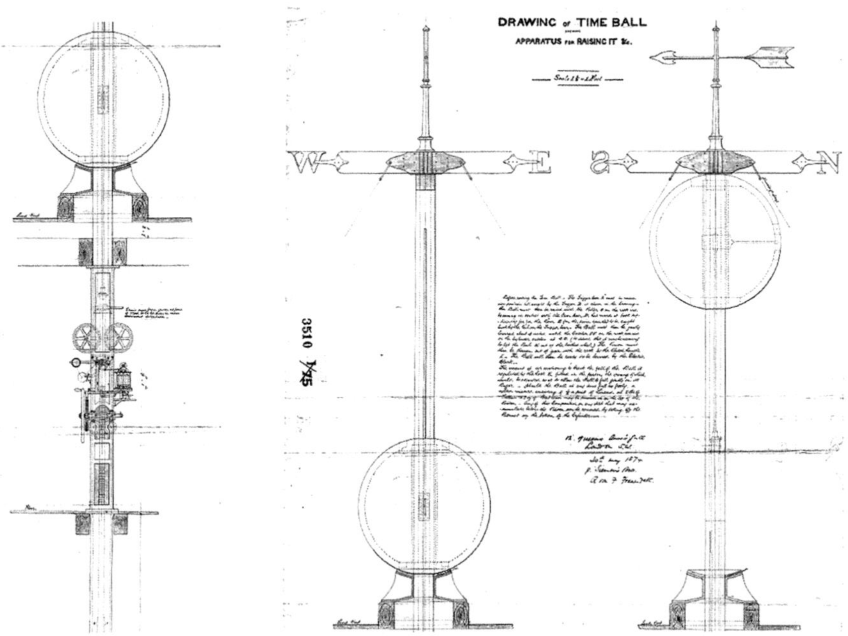

Fig. 23.8 1874 drawing of Lyttelton time ball apparatus (Wellington Harbour Board archive, Drawing Office No. 3510)

3.1.1 Cost of the Time Ball Apparatus

The Telegraph Department in Wellington ordered "one time-ball with all necessary apparatus and one astronomical clock" from London on behalf of the Canterbury Provincial Council on 14 March 1873. The complete apparatus was supplied by Siemens Brothers, as shown in the invoice that is reproduced in Fig. 23.8. "One Time Ball with all necessary apparatus including electrical parts" cost £650. The original quotation had been for "about £500". Interestingly, that is Sir Charles Todd's recollection of the cost of the 1855 mechanism for Sydney, supplied by Maudslay, Sons & Field (Todd, 1899). Siemens' invoice was sent to Canterbury's London agent who was shocked at the price, but paid promptly, nevertheless.

3.1.2 Design of the Time Ball Apparatus

It has been demonstrated in Chap. 5 that the apparatus for Lyttelton is actually a replica of the 1855 design for Sydney, NSW, despite a long interval between their dates of supply (Kinns, 2009). The evidence is summarised below.

Maudslays built the time ball apparatus for Sydney in 1855, using a rack and pinion mechanism to hoist the ball that had been developed from the design for Edinburgh and Deal (Kinns & Abell, 2009). Henry Russell, the NSW Government

Astronomer, modified this apparatus during the 1870s, but most principal features were retained. He identified many of his changes in a letter to Sir Charles Todd in Adelaide (Russell, 1899). Detailed comparison of the Sydney and Lyttelton time ball arrangements showed that Lyttelton used the unmodified 1855 design (Kinns, 2009). The only surviving note in Maudslays records about an 1873 time ball indicates provision for the Cape of Good Hope and an association with the nearby firm of Siemens Brothers (Sells, 1842–1883). No evidence has been found that an apparatus built by Maudslays was ever delivered to Cape Colony. Also, no Siemens records showing supply of time ball apparatus to a location other than Lyttelton have been found. It was deduced that a single 1873 apparatus was built by Maudslays for Siemens Brothers who added electrical equipment and shipped the complete system to Lyttelton in 1874. Extracts from the drawing supplied by Siemens Brothers are shown in Fig. 23.9. The ball diameter was 5 feet (1.5 m). The time ball drop was 10 feet, not 16 feet as stated in the 1898 Admiralty list.

3.1.3 Time Ball Colour

The time ball colour was not stated in Admiralty lists but is thought to have been red with a central white band during its early life, as in Fig. 23.10. This was the scheme specified for Wellington in the 1880 Admiralty list.

3.1.4 Time Ball Location

The Lyttelton time ball was always at the signal station, often described as the time ball tower. It was given incorrectly as the "Custom House" in 1880, which was actually the location of the first time ball in Wellington. It was then given as "The Observatory" in 1898 and lists up to at least 1911, but there was never an astronomical observatory at Lyttelton. The 1916 and later Admiralty lists gave the location as "At the time ball tower".

The entry for Lyttelton in *New Zealand Nautical Almanacs* from 1903 until at least 1918 also gave the time ball location as the Observatory, perpetuating the anomalous description given in Admiralty lists. The description of the location given in *Almanacs* was later changed to the "signal-station".

3.1.5 Time Ball Drop Time

The drop time was 1 pm NZMT (1.30 am GMT) from signal inception in 1876 to 1917. The ball was dropped daily during that period. The *Almanac* entry for 1918 showed a change of drop time from 1 pm to 3.30 pm NZMT and a reduction in the service to Tuesdays and Fridays only, but still referred to "the Observatory".

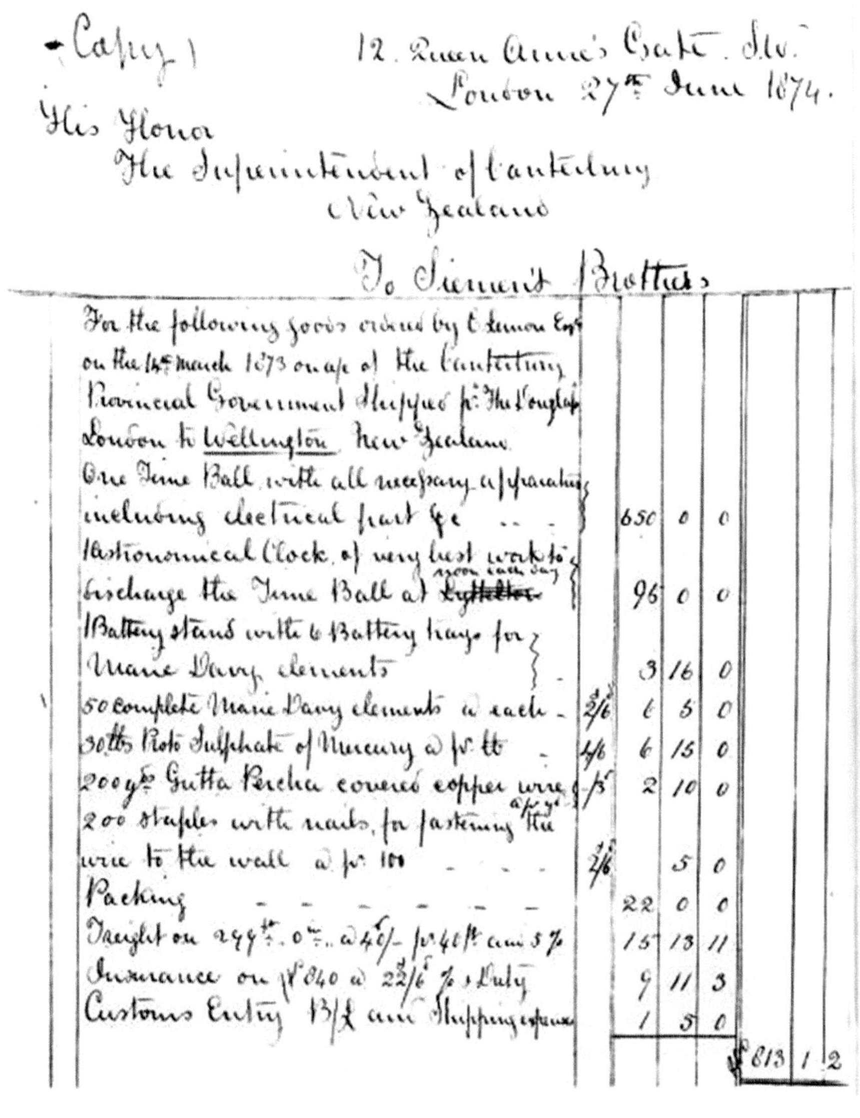

(Copy) 12. Queen Anne's Gate. S.W.
London 27th June 1874.

His Honor
The Superintendent of Canterbury
New Zealand

To Siemens Brothers

For the following goods ordered by C. Lemon Esq on the 14th March 1873 on a/c of the Canterbury Provincial Government shipped per "The Douglas" London to Wellington New Zealand					
One Time Ball with all necessary apparatus including electrical part &c. . . .		650	0	0	
1 Astronomical Clock of very best work to discharge the Time Ball at ~~Lyttelton~~ noon each day		96	0	0	
1 Battery stand with 6 Battery trays for Marie Davy elements		3	16	0	
50 complete Marie Davy elements @ each	2/6	6	5	0	
30 lbs Proto Sulphate of Mercury @ per lb	4/6	6	15	0	
200 yds Gutta Percha covered copper wire @ per yd	/3	2	10	0	
200 staples with nails for fastening the wire to the wall @ per 100	2/6		5	0	
Packing		22	0	0	
Freight on 24 … @ 40/- per 40 ft and 5%		15	13	11	
Insurance on £840 @ 22/6 % & Duty		9	11	3	
Customs Entry B/L and Shipping expenses		1	5	0	
					£813 1 2

Fig. 23.9 The 1874 invoice from Siemens Brothers (NZ Historic Places Trust)

At the Observatory, on the east side of Lyttelton, on days when accurate time signals are given a ball is dropped at 3h. 30m. p.m. New Zealand standard mean time, or 16h. 0m, Greenwich mean time. The ball is hoisted five minutes before the signal. The time signals will be sent on Tuesdays and Fridays, provided satisfactory observations have been obtained. If time signals are required on other days or at other exact hours of Greenwich mean time, application for them should be made to the Observatory, Wellington.

Fig. 23.10 Painting of Lyttelton time ball by Crane, 1880 (NZ Historic Places Trust)

The 1919 Admiralty list entry indicated that the timing change had been introduced in 1917. The twice-weekly service continued as the only official time ball in New Zealand until it was rendered obsolete by radio time signals and had ceased by the end of 1934 (New Zealand Historic Places Trust, 1979). It was listed as "Unreliable" after 1927.

The tower and apparatus were restored to working order in the 1970s and became an important tourist attraction as well as a memorial to an earlier age. The ball was painted black with a central red band, as shown as in Fig. 23.11.

3.1.6 Damage in 2011

Tragically, the tower collapsed during the earthquakes and aftershocks in 2011 that devastated Christchurch. The extent of the damage is shown in Fig. 23.12. The time ball itself can be seen in the lower left hand corner of Fig. 23.12. The mechanism

Fig. 23.11 Lyttelton time ball station before 2011 (https://nzhistory.govt.nz/media/photo/lyttelton-timeball-station)

Fig. 23.12 After collapse of the Lyttelton Time Ball Tower, 18 June 2011 (www.stuff.co.nz/national/christchurch-earthquake/5145289/June-13-earthquakes-Morale-costs-get-a-shake-up)

was severely damaged, but many key components have survived and it may be possible to restore it to working order. A new tower has been constructed and the restored original ball with its original red and white colour scheme is now operated using a modern control system.

4 Port Chalmers

The prospect of a time gun for Otago, in preference to a time ball, was introduced in March 1864 (Harbor [*sic*] Department, 1864). The report was published in the *Otago Daily Times* on 9 April. Extracts are transcribed below.

> … A flagstaff has been erected at Port Chalmers, and a signal master appointed, who will enter upon the duties of his office immediately …. I would now recommend that an astronomical clock be procured for this station, to assist in keeping accurate time, which will also be computed every clear day by observation; and that a small signal gun be got to be fired at eight p.m., to enable ship masters to correct their chronometers – an accommodation much prized by them. A gun would be much preferable to a time ball, as in the case of the latter the time would necessarily have to be signalled at noon, an hour when ship masters are frequently on shore, and when the duties of the day would much prevent their giving attention to it. The gun would also indicate the time to the inhabitants of the town, many of whom are not within sight of the signal station, and to the river steamboat masters, who are desirous of keeping correct time with a view to secure punctuality in arriving and departing.

The idea of a time ball for Dunedin was raised again in a letter to the editor of the *Otago Daily Times*, with an emphasis on the need for improved clock accuracy in the City (Time Ball for Dunedin, 1865).

> Sir – Now that the telegraph has been established throughout the Provinces, a very great addition, and one of much value to the public in Dunedin, would be the erection of a ball over the Telegraph Office, and some regular time adopted for the town ….

4.1 Port Chalmers Time Ball Under Local Control

The imminent provision of a time ball for Port Chalmers, near Dunedin, was announced in the *Otago Daily Times* (Shipping, 1867). It used local observatory facilities.

> Active preparations are being made for the working of the long needed time ball at the Flagstaff, Port Chalmers. A conductor has been laid from the Observatory to the Flagstaff, which being attached to the haulyards by a simple arrangement, the ball is dropped instantaneously by the signal master. The whole of the arrangements are expected to be completed in a week, when the great want of a time ball to the shipping community will be numbered with the things of the past.

The service actually started on Saturday, 1 June 1867, 2 months later. This still preceded establishment of New Zealand Mean Time. The time ball was dropped every day, except Sundays (Customs Entries, 1867).

> The time ball at the Port was dropped, for the first time, on Saturday, and it will in future be dropped daily (Sundays excepted) at 1 p.m., Port Chalmers mean time or 1h 37m. 23.5s a.m., Greenwich mean time.

The following notice is typical of newspaper entries that appeared regularly in the *Otago Daily Times* until 5 October 1877 (Port Chalmers Observatory, 1877). Curiously, the stated drop time after 1868 continued to be 1 pm Port Chalmers mean time, not 1 pm NZMT.

> PORT CHALMERS OBSERVATORY. Latitude, 45.48.55 south; longitude, 11h. 22m. 37s. east. Time ball drops daily (Sundays excepted) at 1 p.m. Port Chalmers mean time, or 1h. 37m. 23s. a.m., Greenwich mean time.

The time ball service is believed to have been discontinued soon afterwards but action was required in response to protests from shipmasters (Time Ball, 1881). After considerable discussion concerning previous inaccuracies, location of the time ball and methods for its control, "... it was resolved to inform the memorialists that a time-ball would be erected shortly".

4.2 Port Chalmers Time Ball Under Wellington Control

It was announced (*Evening Star*, 1881) that the time ball was to be regulated in future by telegraph signal from Wellington.

> On receipt this morning of intimation from the Wellington Telegraph Office that the time ball at Port Chalmers would in future be worked in connection with the telegraph, notice thereof was sent round to the masters of vessels by the harbor authorities. The time ball was dropped at the instant the current moved the needle of the galvanometer. The time given is mean time at longitude 11h 30min 00.3 [*sic*] sec east.

The drop time was to be 1 pm NZMT, not local astronomical time at Port Chalmers.

4.2.1 Change to a Weekly Service

The lack of announcements during the following 12 months suggests that the new service was only introduced in April 1882, when the signal was offered once per week (Shipping, 1882b).

> TIME BALL AT PORT CHALMERS. New Zealand mean time at noon, calculated for the meridian of longitude, in time 11 hours 30 minutes east of Greenwich, will be signalled once a week by the time-ball dropping at the instant of mean noon. A blue flag will be hoisted at the mast-head, Port Chalmers signal-station, on the forenoon of the day when the time-signal will be given.

Regular time ball announcements in the *Otago Daily Times* re-appeared in April 1882 after the change to a weekly service. The last newspaper notice appears to have been in 1906 (Time ball at Port Chalmers, 1906), although nautical almanacs indicated that the service was operating in 1908 and 1909, following development of the new Wellington observatory during 1907. It was also included in the 1908 Admiralty list. The signal was withdrawn in 1909.

The entry in the 1903 *New Zealand Nautical Almanac* was similar to those published in the late nineteenth century.

> A time ball is dropped about once a week from the flagstaff at Port Chalmers at noon New Zealand standard mean time, or 12. 30m. 0s. Greenwich time. When the ball is to be dropped a blue flag is hoisted on the signal-staff at about 10h. 0m. 0s. a.m. *New Zealand mean time.*

The same notice was repeated up to 1906, but the entry for 1907 was changed to "The time ball has been temporarily discontinued." This is likely to have been caused by construction of the new observatory at Wellington during 1907, when only a temporary observatory was available. The original notice reappeared in the 1908 and 1909 editions. That changed in 1910 to "The time-signal ball dropped from Port Chalmers Signal-station has been discontinued." The same notice appeared in editions up to at least 1918.

The Otago time ball service had long ceased by 1919, but the demand for an accurate signal was still present (*Otago Daily Times*, 12 July 1919). The Port Chalmers time ball service was never re-established.

4.3 Port Chalmers Time Ball Arrangement

Figure 23.13 shows the arrangement in 1905, which can also be seen in photographs of sailing ships as early as 1873, so it remained the same throughout (see Kinns, 2017a). The design was different to others in New Zealand. The ball clearly slides up and down the mast, but there are no cross-trees or compass arms above the ball to define its position when raised. It is believed to have been made locally. An 1875 engraving showing the time ball is included in *Colonial New Zealand* (Hargreaves & Hearn, 1988). A replica arrangement, externally similar to the original, is now operating using a modern control system.

Fig. 23.13 Port Chalmers time ball in 1905 (Ref: 1/1-002534-G. Alexander Turnbull Library, Wellington)

5 Auckland

5.1 Williams 1864 Time Ball

The first time ball at Auckland was established by Captain Williams at Smales's Point. He offered a Sunday service only at 9 am, Auckland time, with a newspaper notice on the following Monday to state the time of the drop. The new service started on Sunday, 7 August 1864 (Williams, 1864). Its announcement was republished 50 years later (New Zealand Herald, 1914):

> To the Editor of the New Zealand Herald. A time ball for the benefit of the city and shipping will be dropped every Sunday morning about 9 a.m., and will be hoisted at the mast head, Smale's Point, some five minutes before it is let fall. The corresponding Auckland, as also Greenwich mean time, at which the ball was dropped, will be published in this journal on the Monday morning.

A typical Monday notice is transcribed below (Williams, 1865):

> The mean time when the Ball dropped yesterday morning.
> Auckland … 8h. 59m. 48s. Greenwich … 9h. 20m. 37s.

The local mean time was estimated to be 11 hours, 39 minutes and 11 seconds ahead of Greenwich. There is no known photograph of the time ball at Smale's point. Figure 23.14 shows the view in January 1864 from close to its probable location later that year.

Fig. 23.14 Auckland Harbour from Smale's Point, 29 January 1864 (Sir George Grey Special Collections, Auckland Libraries, 4-1166)

The service appears to have ceased on Sunday 10 June 1866, less than 2 years after it started, with a final notice on the following day (Williams, 1866). It is not known how Captain Williams determined mean time or how he designed and constructed a time ball apparatus.

5.2 *Bartlett's 1866 Time Ball*

The apparent demise of Williams' time ball is likely to have been associated with the provision of a time ball by Arthur Gregory Bartlett (1818–1909). He described himself in advertisements as a "Chronometer, Watch and Clock Maker" with a business on the corner of Shortland-street and High-street. He gave his credentials as "23 years Examiner to Messrs. Crockland & Atkins of 7 Cowper's Court, Cornhill, London". Cornhill was the location of many chronometer makers in England. The new service was advertised in the *New Zealand Herald* (Bartlett, 1866).

> A.G.B. begs to inform his patrons and the public generally that for the convenience of the Town, he has completed arrangements in connexion with his Observatory, at his residence, Coburg-street, Barrack-hill, to drop a Time Ball as formerly, at 1 o'clock daily.

Although Bartlett's time ball may have been useful to local businesses, it fell short of providing a viable service to mariners. Published correspondence in 1874 showed further interest in erection of a new time ball for mariners.

5.2.1 Proposal for a Time Ball in 1874

The Telegraph and Post Office had burned down in 1874 and there were suggestions that it should be rebuilt with a time ball. Extracts from an editorial in the *Auckland Star* (Editorial, 1874) are shown below:

> The renovation of the lately burnt down Post-cum-Telegraph office is approaching completion, and the freshened aspect of its facade is quite imposing ....
>
> As the support for a time ball, which is surely a necessity in every commercial port, as a means of signalling any approaching gales to the shipping in harbour, in accordance with the new system inaugurated, and as an embellishment to a city that has but few objects of architectural pretension, the tower once intended and since neglected is deserving of an effort on the part of our citizens.

The response was ambivalent and no action appears to have been taken (Public Works, 1874).

> The Mayor wished first to direct the Commissioner of Customs' attention to the desirability of completing the Post-office building, which had been handed over to the General Government, by the erection of a tower with the addition of a time-ball, which was of very great use to an important shipping port like Auckland. The tower could be utilized for many purposes, and the clock would not only be useful but ornamental.
>
> Mr Reynolds said that the matter was one which came within the Department of the Minister of Public Works .... He should certainly not be in favour of erecting a time-ball there, as it would not be of any use; it would be better to erect it in a position nearer the harbour ....
>
> Captain Daldy said if the Government intended to carry out the system of storm signals the time-ball would be required in connection with the telegraph. It was desirable that the time-ball should be connected with the Telegraph Department, which he believed would occupy that building.

No evidence of further activity has been found before 1883. Possible provision of a time ball was discussed on several occasions in the years up to 1900.

5.2.2 Public Discussion Between 1883 and 1900

The possibility of progress was noted in 1883 (*Auckland Star*, 4 May).

> We have been favoured with a glance at the conditions for the guidance of architects sending in competitive designs for the new Harbour Board offices. The building is to be situated on the reserve bounded by Little Queen, Quay, and Albert-streets, and Mr Goldie's property, and is to be three stories high, and provision is to be made for a time ball tower.

Still nothing happened and a discouraging note appeared in 1885 (Harbour Board, 1885).

> The Works and Tariffs Committee's recommendations were: – (1) That the Board decline to expend £700 or £800 in providing a reliable time ball apparatus.

Another editorial in support of time ball procurement and the required operational budget was published in 1886 (New Zealand Herald, 1886). It was in support of a correspondent and showed that the design of the Victoria Arcade building had made provision for a time ball apparatus.

> ... When the Victoria Arcade was in course of erection, representations were made to the directors of the New Zealand Insurance Company of the suitability of the projected clock tower of their new building for the purposes of a time ball, and urging the importance of one being fixed on it. On these representations, the directors caused the plans of the building to be altered, and the tower strengthened in order to render it available for the proposed object. After this expense has been incurred, we learn that the Government refuses to sanction the very small expenditure that would be involved in placing and maintaining the tower in electric contact with the telegraph office, and thence with Wellington, in order to secure the desired end. This is, we think, a matter on which strong representations might be made by the Chamber of Commerce and the Harbour Board.

Again, nothing happened. Another attempt to establish a time ball was made in December 1893 (Auckland Star, 1893).

> ... Mr Crowther will move (1) "That the next ordinary meeting of the Board be hold on Tuesday the 9th January, 1894, at 2.30 p.m." Mr Porter will move (2) "That the Secretary be instructed to communicate with Sir J. Hector, Colonial Museum, Wellington, with the view of obtaining information as to time ball for harbour purposes, and estimated cost of fitting and erecting same on the offices of this Board; time ball to be of a similar description to those in use at Lyttelton and Wellington."

It was at least recognised that the apparatus would have to be of similar quality to those at Wellington and Lyttelton, to be of value to mariners. Progress was still snail-like and there was no time ball during the last 20 years of the nineteenth century. That certainly explains the lack of an entry for Auckland in the 1880 and 1898 Admiralty lists.

Another strongly worded petition for a time ball was published in April 1900 (Auckland Harbour Board, 1900). The lack of a time ball was an embarrassment to such an important port.

> Time Ball: A petition signed by over 60 shipowners, agents, and shipmasters, was read, drawing attention to the great inconvenience caused by the absence of a time ball at the port. Auckland, said the petition, was the only port in Australasia where such a necessary adjunct to navigation was absent. The petition was referred to the Works and Tariff Committee.

5.3 *Auckland Harbour Board Building Time Ball*

A time ball for Auckland was finally under detailed discussion in August 1900, with support from the Admiralty (*Auckland Star*, 1900).

> Machinery, Time Balls, etc. – Messrs. W. and A. McArthur wrote that the Admiralty considered it better that shop tools and iron ladders should be made in Auckland, and enclosing sketch thereof. Information with regard to the time ball, core former, sheer legs, etc., was also forwarded. – Referred to the Works and Tariff Committee to report. – The Acting-Chairman said the work could be done here for a small sum of money. It was at first thought it would cost a lot.

Tenders for construction of the time ball were invited with a closing date of 30 April (Tenders, 1901). The possibility of a time gun was also debated (Auld Reekie, 1901).

> 1 see by your paper that, we have got a time ball for Auckland, by which to regulate our clocks and watches, but like many others 1 have no opportunity of seeing such a thing from my point of view, and without undervaluing the time ball I think that (if practicable) the electric wire might be extended to the Albert Park and connected with one of the guns there, and at the same moment that the current frees the ball it will also discharge the gun, which could be heard all over the city, and so greatly extend its usefulness, a plan adopted in Edinburgh.

Mr A.G. Bartlett, who had provided his own time service in 1866 and was over 82 years old in 1901, was given a minimal budget of £30 per annum to operate the time ball (Auckland Star, 1901).

> Works and Tariff Committee. – The report of this committee was adopted … that Mr A. Bartlett be requested to take the necessary steps to cause the time ball to be dropped at 12 noon every day, excepting Sundays and holidays, payment not to exceed £30 per annum; ….

5.3.1 Design of the 1901 Apparatus

The time ball can be seen on top of the Harbour Board Building in Fig. 23.15. The time ball is clearly offset from a tall mast and was guided by a pair of wires, using eyes at the sides of the ball. The arrangement was dissimilar to those used at Lyttelton or Wellington and is likely to have used the "Devonport Principle". This

Fig. 23.15 Auckland time ball on the Harbour Board Building in 1901 (Auckland Libraries Heritage Collections, 4-2938)

type of design was outlined in correspondence relating to a new time ball for Portsmouth (Astronomer Royal, 1910c).

> In the more recent form of apparatus, there are two vertical guides (wire) and when once the ball is hoisted it is held by a small rope. Everything else being loose, the ball drops about a foot before the thick rope used in hoisting comes into real action.

The Auckland time ball finally became operational in September 1901 (Time Ball, 1901).

> Harbour Board Offices, September 6th, 1901. Notice is hereby given that the Time Ball erected at the Offices of the Board, Quay-street, will be Hoisted every day (Sundays and holidays excepted) at 11.50 a.m. and Dropped at 12 Noon, Colonial Mean Time (same time as observed in the town).

So far, so good, but other expenditure was needed (Harbour Board, 1901).

> Mr A.G. Bartlett wrote asking for a clock, cost £50, to assist in keeping the time ball correct – The letter was referred to the Works and Tariff Committee.

Regular notices about the time ball drop appeared in the *Auckland Star* and continued up to August 1902 in one of two forms (Time Ball, 1902a, b):

> The time ball drops at the Harbour Board Buildings at noon daily. New Zealand mean time, equivalent to 12 hours 30 minutes Greenwich mean time.
>
> Harbour Board Offices, August 2nd, 1902. Notice is hereby given that the TIME BALL erected at the Offices of the Board will be hoisted every day (Sundays and Holidays excepted) at 11.50 a.m., and dropped at 12 noon, Colonial mean time (same time as observed in the town).

The service did not last long and appeared to have stopped on 5 August 1902. Not long afterwards, it appeared to have been abandoned altogether (Local and General News, 1903).

> The Harbour Board, at its meeting yesterday, decided on the recommendation of the Works and Finance Committee to incur no further expenditure in regard to the time ball over the Board's buildings. The chairman (Mr. J. T. Julian) said the time-ball had become the laughing stock of the port on account of its unreliableness, and Mr. J. A. Walker said that the visiting naval officers, as well as shipping masters, had ceased to take any notice of the thing for the reason stated by the chairman. Mr. W. J. Philson said he thought a correct timeball was essential for any sub-naval station, but other members regarded the matter of adjusting the present instrument as apparently hopeless, and adopted the committee's report.

There was, in fact, an attempt to re-establish the time ball in 1903.

5.3.2 Revised Arrangement in 1904

The Admiralty list for 1904 included a time ball at Auckland but with the qualification "Reported unreliable". It was not mentioned in the 1903 Almanac and was described as "temporarily discontinued" in the 1905 and later almanacs. The following letter was published in the *New Zealand Herald* and indicated that the service had ceased by September 1904 (Time Ball, 1904).

Fig. 23.16 Auckland time ball on the Harbour building in 1901 and after 1904 (Auckland Libraries Heritage Collections, 4-2938 and 4-889)

> A few years ago the Harbour Board, at some considerable expense, imported and erected a time ball for the benefit of vessels frequenting this port, but, as the antics of the said ball became very erratic, further expense was incurred by an additional pole. An astronomical clock was next to arrive, followed by an individual with electrical knowledge: still the ball remains a fixture ….

The change to a pole can be seen in subsequent photographs which show the disused time ball in its lowered position. It was probably removed at some time before 1912, when another time ball was erected on the new Ferry Building. Figure 23.16 shows a comparison of the 1901 arrangement using guide wires with the 1904 arrangement using a pole. The second photograph, showing United States flags, was taken during a 1908 visit of the US fleet to Auckland.

5.3.3 Proposals to Re-establish a Time Ball Between 1905 and 1912

There were further attempts to re-establish the service from 1905, still involving Mr. Bartlett (Harbour Board, 1905).

> Time Ball Wanted. – Mr Dacre moved. "That it is in the interest of the port that a correct, time ball be kept in use." He found there was a great need of this, and it was a necessity for any up-to-date port. – Mr Napier, seconding, asked why it was not gone on with. – The Chairman said it was because of complaints that it was inaccurate. It cost £50 a year to maintain, and he thought it was unnecessary expense. When the matter was gone into two years ago Mr Bartlett told them that to ensure correct time it would be necessary to erect an observatory in the Albert Park – The matter was referred to the Works and Tariff Committee.

This was followed by another announcement in the *Auckland Star* (Commercial, 1905).

> The Harbour Board has instructed its engineer to submit an estimate of the cost of putting an observatory into the Albert Park, and requested Mr A. G. Bartlett to quote the cost of maintaining the time ball at accurate time.

Yet another plan was published in the following year (Advertisement, 1906), whereby electro-magnetic triggering was to be used, with control from Wellington. Progress appeared to have been made with a decision by the Harbour Board to erect a new time ball (Auckland Star, 1907).

> The fortnightly committee meeting of the Auckland Harbour Board was held yesterday afternoon. In reply to requests from Captain McNellar, of the SS Kaikoura, it was decided to state that a time ball would be erected on top of the proposed offices in the corner of Queen-street as soon as convenient.

Inclusion of an observatory was suggested after another year had elapsed (Harbour Board Committee, 1908).

> Time Ball. – In answer to a letter from J. Hamilton, of Wellington, re a time ball, it was decided on the motion of the Chairman to inform the writer that the Board contemplated erecting a building on the water front, which would be fitted with an observatory.

Bartlett, who had long been regarded as the Auckland authority on time determination, died in 1909 (Obituary, 1909). The basic problems of chronometer calibration without a time ball were well-described in a letter to the *Auckland Star* (Chudley, 1911) but there was still no time ball (A Time Ball for Auckland, 1911).

5.4 Auckland Ferry Building Time Ball

The 1912 *Almanac* included the statement "When the Ferry Buildings, now in the course of erection, are completed the time-ball will be placed at the summit of the tower." The statement in the 1914 *Almanac* was "The time ball has been erected on the tower of the Ferry Building and will shortly be ready for use. It will be dropped at noon each day, except Sundays." That was changed again in the 1915 *Almanac* to "The time ball has been erected on the tower of the Ferry Building, and is now awaiting satisfactory connections with the Government Observatory at Wellington." The 1916 *Almanac* confirmed the final demise of the Auckland time ball: "The time ball has been discontinued and a system of night signals has been adopted. Not less than two guaranteed time signals per week will be given at 9 p.m."

The time ball on the new Ferry Building should have been operational in 1912 but there had been a problem in connecting the time ball to the Post Office (The Time Ball, 1912). The unfortunate saga was exposed in the *Auckland Star* (Time Ball: A Department's Inaction, 1913).

> The Harbour Board engineer reported to the Board this afternoon that although the time-ball had been ready for many months and connection with the post office had been promised in April last, nothing whatever had been done by the Department, and the gear was simply rusting from inaction. In connection with this matter a "Star" reporter, who made

Fig. 23.17 Auckland time ball on the Ferry Building in 1913 (Auckland Libraries Heritage Collections, 1-W1555)

> inquiries in official circles, was informed that the Department had never been made aware that the time-ball was ready for connection. The Department would fix it up at once when the Board was ready, but no intimation of any kind had been received, nor any suggestion of a conference made. When the connection is made it is understood that the exact mean time will he made available from Wellington at certain defined periods, possibly twice a week.

Figure 23.17 shows a photograph of the Ferry Building in April 1913, taken before the clock had been installed. A small diameter ball can be seen on the mast above the tower. A notice 2 years later (Auckland Star, 1915a, b) suggested that a new time ball service was at last imminent.

> The time ball on the Ferry Buildings will in future be dropped daily at l. p.m., while on the day of the week on which a time signal is flashed from Wellington a red flag will be displayed from the building.

The plan was changed in August 1915 to use of night-time signals, the time ball having proved to be unreliable yet again (Time Signals, 1915).

5.5 *Auckland Ferry Building Time Lights*

The following notice was published in the *Auckland Star* in August 1915 (Time Signals, 1915):

> Another attempt probably will be made by the Auckland Harbour Board to provide time signals for the harbour. At to-day's meeting of the Board a letter was received from the acting Government astronomer approving of the proposal for two time signals per week at 9 p.m. Usually signals of guaranteed accuracy were provided on three nights a week at Wellington, and he felt fairly safe in promising two for Auckland.
>
> The harbourmaster (Captain Sergeant) recommended that the signals should be given only on those nights when the observatory could guarantee them. The Government Astronomer should notify the Board by telegram at 2 p.m. on the days when he could guarantee the signals. When such telegram was received a red flag should be hoisted on the tower of the Ferry Buildings from 4 p.m. to 4.30 p.m., notifying shipping that a guaranteed time signal would be given that night. At least two guaranteed signals per week should be given at 9 p.m. and the following lights shown from the flagstaff of the Ferry Buildings. – Green light switched on at 8.10 p.m., red light at 8.50 p.m., white light at 8.55 p.m. All lights switched off at 9 p.m.
>
> The Harbourmaster further recommended, that the present time ball-signals should be discontinued at once owing to their irregularity, and the night signals commenced as soon as the mechanism had been installed.

The Auckland time lights were operating by the end of 1915. Figure 23.18a shows the time ball on the Ferry Building, resting on supports while the building awaited

Fig. 23.18 Auckland time ball in 1913 and time lights in 1915 (Auckland Libraries Heritage Collections, 1-W1555 and 1-W1321)

installation of the clock. Figure 23.18b shows the time lights at the end of 1915. They were attached to the mast that had previously been used to support the time ball. The same arrangement of lights can be seen in photographs of the Ferry Building up to 1937.

Notices concerning the time lights at Auckland were included in *New Zealand Almanacs*. As at Wellington, there were changes in timing over the years, using the same arrangement of lamps throughout.

The initial notice in 1915 is reproduced below.

> **Time Signals.** – The time ball has been discontinued and a system of night signals has been adopted. Not less than two guaranteed time signals per week will be given at 9 p.m.
>
> On those days when such signals are to be given, a red flag will be hoisted on the tower of the Ferry Buildings from 4 p.m. to 4.30 p.m.
>
> The following lights will be used on the flagstaff of the Ferry Buildings when such signals are given:-
>
> Green light switched on 8.10 p.m.
> Red light switched on 8.50 p.m.
> White light switched on at 8.55 p.m.

All lights switched off 9 p.m. – 21h. 30m. 0s. G.M.T. These lights will be in a vertical line, 6 ft. apart, the green light being the lowest.

The lights were switched on 50, 10 and 5 minutes before the signal 9 pm NZMT. The entry for Auckland in the 1917 *Almanac* included an additional statement:

> When the Time Signal is expected, but does not come through, the red light will remain burning till 9h. 5m. p.m., thereby notifying shipping that the signal has not been received.

The signal time had been changed to 8.30 pm NZMT in the 1918 *Almanac*. The 1928 entry showed concern about accuracy of the signal:

> Auckland. – From flagstaff atop Ferry Buildings. Should telegraphic transmission from Wellington fail, the red light will continue until 8.35 pm. (This time-signal is considered unreliable.)

The same sequence was still being used in the 1930s, with the following entry in *Almanacs* up to 1936 (9 am GMT corresponded to 8.30 pm NZMT):

> ... The lights are extinguished by direct signal from the Observatory, Wellington, on Tuesdays and Fridays.

The final entries in 1936 and 1937 included an additional statement about accuracy:

> ... Tests recently carried out by the Auckland Harbour Board indicate that the time taken by the telegraph signal operating the lights to reach Auckland from Wellington is less than one-tenth of a second.

5.5.1 Withdrawal in 1937

An article was published in the *New Zealand Herald*, anticipating withdrawal of the Auckland time light service on 31 July 1937 (Visual Time Signals, 1937a). It was actually decided to continue the service for another 3 months and it operated for the last time on 29 October, when the article was reprinted in a modified form (Visual Time Signals, 1937b). Both newspaper articles contained a number of significant errors concerning time signal provision at Auckland (Kinns, 2017b).

6 Other New Zealand Time Signals

Time signals existed at various other places in New Zealand, but they were used mainly to provide time signals for local inhabitants.

6.1 Time Balls at Timaru and Wanganui

Wright (2007) noted those at Wanganui and Timaru. A Wanganui newspaper editorial on 16 October 1874 expressed the frustration that occurred when a local time ball was not maintained or operated properly (Wanganui Herald, 1874).

> The irregular manner in which the time ball has been attended to of late has been the cause of great complaints being made by the public, especially the working men .... If Wanganui has a time ball, we do not see why it should, not be attended to properly, and if anything goes wrong, the cause should be ascertained and remedied. As at present attended to, it is quite useless, as it misleads some, while others take no notice whatever of it.

Most references to a time ball at Wanganui are in 1886 or later, when a time ball was located on a mast at the Harbour Board's signal station on Durie Hill. There appear to have been persistent difficulties in operating the station with limited staff and budgets (The Signal Station, 1899). It was reported that the ball had once blown down in a gale (Local and General, 1914). The Wanganui time ball was not mentioned in any edition of the *New Zealand Nautical Almanac* from 1903 onwards, so it was not regarded as an official signal for mariners.

6.2 Time Guns

No time gun in New Zealand was included as an official signal in either Admiralty lists or *New Zealand Almanacs*. A time gun at Nelson was fired at noon on Saturdays from 11 September 1858 (Nelson Town Improvements, 1858). There were occasional incidents concerning safety (New Zealand Herald, 1878). Time guns were used at other locations, including Dunedin and Auckland, but were often controversial because of disturbance, concerns about safety and the cost of maintaining the service. A gun was installed on Mount Victoria in Wellington at the end of 1877 but ceased to operate in 1880 when the supply of cartridges was discontinued on grounds of cost (News and Notes, 1880).

Acknowledgements I shall always be grateful to the late Bruce Carr (1930–2023), for his demonstration of the Lyttelton apparatus that sparked my interest in visual time signals over 20 years ago, and to Ken Philpott, an old friend and expert photographer who was my frequent host during visits to Lyttleton.

The work was based initially on notices, articles and correspondence that appeared in New Zealand newspapers from the 1860s onwards. This material was accessed online using "Papers Past", a free service provided by the National Library of New Zealand, complemented by research in Cambridge University Library.

Jan Titus of Heritage New Zealand Pouhere Taonga in Christchurch provided much encouragement for the work concerning the Lyttelton time ball, before and after the 2011 earthquakes. I am grateful to Ayla Koning-Thornton of Wellington City Council for her support in accessing archives of the Wellington Harbour Board, which helped to clarify the complicated history of Wellington time ball provision. I also thank staff at the Auckland Maritime Museum for their help in providing access to the extensive collection of New Zealand almanacs in the Bill Laxon library. Staff at the Macmillan Brown Library, University of Canterbury, helped me to access almanacs held there. Phil Tomlinson, a senior guide at Wellington Botanic Garden and author of articles concerning observatory buildings, provided important background information.

References

A Time Ball for Auckland. (1911, July 7). *Auckland Star*, 2.

Advertisement. (1906, 21 February). *Auckland Star,* 10.

Astronomer Royal. (1910a). *Letter to the New Zealand High Commissioner*, dated 25 April 1910. Papers of William Christie, Cambridge University Library, RGO 7/257.

Astronomer Royal. (1910c). *Letter to the Admiral Superintendent, Portsmouth*, dated 7 December 1910. Papers of William Christie, Cambridge University Library, RGO 7/257.

Auckland Harbour Board. (1900, April 4). *New Zealand Herald*, 3.

Auckland Star. (1883, May 4), 2.

Auckland Star. (1893, December 16), 5.

Auckland Star. (1900, August 22), 6.

Auckland Star. (1901, September 4), 2.

Auckland Star. (1902a, February 15), 4.

Auckland Star. (1902b, August 2), 8.

Auckland Star. (1907 June 8), 4.

Auckland Star. (1915a, April 21), 4.

Auckland Star. (1915b, August 5), 4.

Auld Reekie. (1901, July 5). Letter to the editor of the *Auckland Star*, 8.

Bartlett, A. G. (1866, June 22). Advertisement in the *New Zealand Herald*, 1.

Chudley, C. (1911, June 9). Correct times: Timeball required locally. *Auckland Star*.

Commercial. (1905, July 12). *Auckland Star*, 6.

Customs Entries. (1867, June 3). *Otago Daily Times*, 4.

Customs House. (1870s). *Engraving of Customs House* by Swan & Wrigglesworth. https://commons.wikimedia.org/wiki/File:Customs-House-Wellington.jpg

Dominion Observatory. (2007). Factsheet published by the Department of Conservation, Te Papa Atawhai.

Editorial. (1874, October 6). *Auckland Star*, 2.

Evening Post. (1874, September 10), 2.

Evening Post. (1875, March 2), 2.

Evening Star. (1881, March 15), 2.

Harbor [sic] Department, No. III. (1864, April 9). Report to the Provincial Council, dated 14 March 1864, published in the *Otago Daily Times*, 4.

Harbour Board. (1885, March 4). *Auckland Star*, 2.

Harbour Board. (1901, November 27). *Auckland Star*, 3.
Harbour Board. (1905, June 14). *Auckland Star*, 6.
Harbour Board Committee. (1908, June 5). *Auckland Star*, 2.
Harbour Master. (1882). *Letter to Wellington Harbour Board*, dated 18 October 1882. Wellington City Archives, AC020:559.
Hargreaves, R. P., & Hearn, T. J. (1988). *Colonial New Zealand: Engravings of the Victorian period*. John McIndoe Ltd.
Hector, J. (1886). *Letter to Wellington Harbour Board*, dated 9 September 1886. Wellington City Archives, AC020:3098.
Hydraulic Accumulator House at Waterloo Quay Woolshed (Wool Store J). (1888). [Drawing]. Wellington City Archives, AC016:43.
Kinns, R. (2009). Time keeping in the antipodes: A critical comparison of the Sydney and Lyttelton Time Balls. *Journal of Astronomical History and Heritage, 12*(2), 97–107.
Kinns, R. (2017a). The principal time balls of New Zealand. *Journal of Astronomical History and Heritage, 20*(1), 69–94.
Kinns, R. (2017b). The time lights of New Zealand; yet another way of communicating time in the pre-wireless era. *Journal of Astronomical History and Heritage, 20*(2), 211–222.
Kinns, R., & Abell, L. (2009). The contribution of Maudslay, Sons & Field to the development of time balls in Australia. *International Journal for the History of Engineering & Technology, 79*(1), 59–90.
Late Shipping. (1889a, January 21). *New Zealand Times*, 4.
Late Shipping. (1889b, April 13). *New Zealand Times*, 4.
Lewis, T. (1910a). *Notes*, dated 21–25 April 1910. Papers of William Christie, Cambridge University Library, RGO 7/257.
Lewis, T. (1910b). *Note to W. H. M. Christie*, dated November. Papers of William Christie, Cambridge University Library, RGO 7/257.
Local and General. (1914, October 6). *Wanganui Herald*, 4.
Local and General News. (1903, October 14). *New Zealand Herald*, 4.
Longitude of Wellington. (1874, April 1). *Wellington Independent*, 3.
Luke & Sons. (1888). *Memorandum to Wellington Harbour Board*, dated 31 August 1888. Wellington City Archives, AC020:4546.
Maclear, T. (1863a). *New time-ball at the Royal Observatory*, notice signed by Thos. Maclear. Papers of George Airy, Cambridge University Library, RGO 6/615, File 7.
Maclear, T. (1863b). *Letter to G. B. Airy*, dated 16 June 1863. Papers of George Airy, Cambridge University Library, RGO 6/615, File 7, Leaves 145–146.
Maclear, T. (1863c). *Letter to G. B. Airy*, dated 18 June 1863. Papers of George Airy, Cambridge University Library, RGO 6/615, File 7, Leaf 147.
Monthly Summary. (1864, March 12). *New Zealand spectator and Cook's Strait guardian*, 2.
Nelson Town Improvements. (1858, September 14). *Colonist*, 2.
New Zealand Herald. (1878, August 17), 4.
New Zealand Herald. (1886, March 4), 4.
New Zealand Herald. (1914, August 1), 11.
New Zealand High Commission. (1910a). *Letter to the Astronomer Royal*, dated 21 April 1910. Papers of William Christie, Cambridge University Library, RGO 7/257.
New Zealand High Commission. (1910b). *Letter to the Astronomer Royal*, dated 26 April 1910. Papers of William Christie, Cambridge University Library, RGO 7/257.
New Zealand Historic Places Trust. (1979). *The Lyttelton Time-ball Station: An account of its history, operation and restoration*. New Zealand Historic Places Trust.
News and Notes. (1880, April 19). *The Press*, 3.
Obituary, Bartlett, A. G. (1909, February 22). *Auckland Star*, 3.
Orchiston, W. (2016a). *Exploring the history of New Zealand astronomy: Trials, tribulations, telescopes and transits*. Springer International Publishing.
Orchiston, W. (2016b). James Hector, Arthur Stock and early astronomy in Wellington. *Journal of the Royal Society of New Zealand, 47*(1), 88–93. Available on-line http://www.tandfonline.com/toc/tnzr20/47/1

Port Chalmers Observatory. (1877, October 5). *Otago Daily Times*, 2.
Public Works in Auckland. (1874, October 13). *Auckland Star*, 2.
Russell, H. C. (1899). *Letter to C. Todd*, dated 21 March 1899. Transcribed as Appendix 2 in Kinns and Abell, 2009.
Sandys & Co. (1863). *Letter to G. B. Airy*, dated 31 August 1863. Papers of George Airy, Cambridge University Library, RGO 6/615, File 7, Leaves 151–152.
Sells, C. (1842–1883). Notebooks: (1) 1842, also 1847; (2) April 1842 to 1849; (3) January 1849 to June 1856 (*sic*); (4) June 1853 to December 1855; (5) January 1856 to March 1859; (6) March 1859 to November 1861; (7) November 1861 to May 1865; (8) May 1865 to December 1867; (9) December 1867 to January 1871; (10) January 1871 to May 1875; (11) May 1875 to April 1878; (12) April 1878 to March 1883.
Shipping. (1867, March 30). *Otago Daily Times*, 4.
Shipping. (1882a, February 28). *New Zealand Times*, 2.
Shipping. (1882b, April 3). *Otago Daily Times*, 2.
Shipping News. (1909, July 16). *Dominion*, 7.
Ships and Shipping. (1912, February 22). *New Zealand Times*, 8.
Stock, A. (1868, October 31). The Time Ball (article dated 29 October 1868). *Wellington Independent*, 5.
Stock, A. (1871, April 25). The Time Ball. *Wellington Independent*, 2.
Stock, A. (1873, November 26). The Time Ball, letter to the editor. *Evening Post*, 2.
Tenders. (1901, April 27). *New Zealand Herald*, 3.
The Electric Time Ball in the Strand. (1852, August 19). *The Times*.
The Signal Station. (1899, April 18). *Wanganui Herald*, 2.
The Time Ball. (1864, February 10). *New Zealand spectator and Cook's Strait guardian*, 2.
The Time Ball. (1912, September 10). *Auckland Star*, 7.
The Time Ball. (1901, September 7). *Auckland Star*, 8.
Thomson, J. T., & Jackson, H. (1871). *Report relative to measurement of longitude between Wellington and Otago by means of electric telegraph*. Government Printer. Cited by Wayne Orchiston in *Exploring the history of New Zealand astronomy: Trials, tribulations, telescopes and transits*, 2016.
Time Ball. (1881, January 28). *Evening Star*, 4.
Time Ball. (1901, September 7). *Auckland Star*, 8.
Time Ball. (1902a, February 15). *Auckland Star*, 4.
Time Ball. (1902b, August 2). *Auckland Star*, 8.
Time Ball. (1904, September 19). Letter to the editor of the *New Zealand Herald*, 7.
Time Ball at Port Chalmers. (1906, September 27). *Otago Daily Times*, 6.
Time Ball for Dunedin. (1865, June 26). *Otago Daily Times*, 5.
Time Ball: A Department's Inaction. (1913, July 29). *Auckland Star*, 2.
Time Signals: A New Proposal. (1915, August 17). *Auckland Star*, 4.
Todd, S. C. (1899). *Letter to H.C. Russell*, dated 29 March 1899. SRSA GRG 31/1 0001, vols. 1–5, 345–346. Transcribed as Appendix 4 in Kinns and Abell, 2009.
Tomlinson, P. C. (2015). *Structures in the garden: Wellington, New Zealand observatories*. Friends of the Wellington Botanic Garden. http://www.friendswbg.org.nz/newobservatories.html
Visual Time Signals. (1937a, June 23). Shipping facility, system to be abandoned, displaced by wireless. *New Zealand Herald*, 14.
Visual Time Signals. (1937b, October 29). Shipping facility, system to be abandoned, last service tonight. *New Zealand Herald*, 12.
Wanganui Herald. (1874, October 16), 2.
Wellington. (1864, January 14). Article dated 11 January 1864 in *The Nelson Examiner and New Zealand Chronicle*, 3.
Wellington Harbour Board. (1885). *Engineer's memo re time ball*, dated 11 June 1885. Wellington City Archives, AC020:1956.
Wellington Harbour Board. (1888). *Contract No 43*, signed 15 March 1888. Hydraulic Accumulator House at Waterloo Quay Woolshed [Wool Store J]. Wellington City Archives, AC016:43.

Wellington Harbour Board. (1910). *Report on proposed new time ball*, dated 30 November 1910. Wellington City Archives, AC020:32094.

Wellington Harbour Board. (1911). *Letter to Marine Department – Copy of letter from Shipowners urging that timeball should be re-established*, dated 2 August 1911 [with enclosures dated 30 November 1910 and 21 July 1911]. Wellington City Archives, AC020:33271.

Wellington Independent. (1864, March 19), 2.

Williams, S. J. (1864, August 2). Letter to the editor, *New Zealand Herald*, 4.

Williams, S. J. (1865, May 8). Notice in the *New Zealand Herald*, 4.

Williams, S. J. (1866, June 11). Notice in the *New Zealand Herald*, 3.

Wright, G. (2007). *A compilation of world time signals, past and present, based on the list of time signals (5th edition, Lords of the admiralty, 1898)*, compiled for the New Zealand Historic Places Trust – Lyttelton Time Ball Station, 14 April 2003. Revised 29 June 2003 and 12 November 2007.

Chapter 24
Canada, Newfoundland and Bermuda

Roger Kinns

Abstract A time ball on the Citadel in Quebec City was erected in 1852. Another at Saint John was erected in 1870 and later changed location, followed by a time ball at Halifax in 1904. A gun at Halifax had been used previously and complemented the time ball afterwards. Several time balls in eastern Canada used a caged arrangement so the time ball could be operated in harsh winter conditions. Many were seasonal. A time gun at Vancouver was first listed in 1904, followed by a time ball at Victoria in 1916. Canadian time balls had all been withdrawn before 1947, leaving four time guns.

The only signal in Newfoundland was a time gun on Signal Hill at St. John's. This was listed by the Admiralty throughout the period from 1880 to 1947, with a warning that it was not suitable for rating chronometers. A time ball at the dockyard in Bermuda was listed from 1880 to 1904.

1 Introduction

Canada was established as a Dominion within the British Empire in 1867, but the island of Newfoundland was then governed separately. It had acquired Dominion status of its own in 1909, but it reverted to being a Crown Colony in 1933 after severe economic problems. Newfoundland became a province of Canada in 1948.

Bermuda is a compact archipelago of 181 islands, many interconnected by bridges. It is now a British Overseas Territory in the North Atlantic, further north than islands in the West Indies. It was a principal port for the Royal Navy with large dockyard facilities. Bermuda is located east of North Carolina in the USA.

Table 24.1 show the signals for Canada, Newfoundland and Bermuda in Admiralty lists between 1880 and 1947. The signal locations are shown in Fig. 24.1.

R. Kinns (✉)
The Maudslay Society, Pembroke College, Cambridge, England

R. Kinns (ed.), *The World of Visual Time Signals for Mariners*, Historical & Cultural Astronomy, https://doi.org/10.1007/978-3-031-57334-7_24

Table 24.1 Signals in Canada, Newfoundland and Bermuda

Area	Location	1880	1898	1904	1911	1915	1919	1922	1928	1934	1939	1947
Newfoundland	St. John's	Gun	Gun	Gun	Gun	Gun	Gun	Gun	Gun	Gun	Gun	Gun
Canada	Saint John	TB (bl, g.)	TB (bl, g.)	TB (bl, g.)	TB (bl, g.)	TB (bl, g.)	TB (bl, g.)	TB (bl, g.)	TB (bl, g.)	TB (bl, g.)	TB (bl, g.)	
	Quebec	TB (black)	TB (black)	TB (black)	TB (black)	TB (black)	TB (black)	TB (black)	TB (black)	TB	TB	
	Montreal	TB (black)	TB (black)	TB (black)	TB (black)	TB (black)	TB (black)	TB (black)	TB (black)	TB (black)	TB (black)	
	Halifax				TB (black)	TB (black)	TB (black)	TB (black)	TB (black)	TB (black)	TB (black)	
			Gun	Gun	Gun	Gun	Gun	Gun	Gun	Gun	Gun	Gun
	Vancouver			Gun	Gun	Gun	Gun	Gun	Gun	Gun	Gun	Gun
	Victoria						TB	TB	TB	TB	TB	
												Gun
Bermuda	Ireland Island	TB	TB	TB								
Signal type totals		1880	1898	1904	1911	1915	1919	1922	1928	1934	1939	1947
Time ball, including collapsible		4	4	4	4	4	5	5	5	5	5	0
Gun, including powder flash		1	2	3	3	3	3	3	3	3	3	4
Lights		0	0	0	0	0	0	0	0	0	0	0
Other: disc, drum, flag, etc.		0	0	0	0	0	0	0	0	0	0	0

Fig. 24.1 Locations of signals in Canada, Newfoundland and Bermuda. (Google Maps)

2 Canada

2.1 *Quebec City*

The first visual time signal in Canada was erected on the Citadel in Quebec City in 1852 or soon afterwards. It was implemented by Edward David Ashe (1814–1895). His distinguished early career in the British Royal Navy and his subsequent work as an astronomer in Canada have been described by Jarrell (1975, 1990). He was invalided out of the Navy after a shipboard accident in 1849 and was offered the directorship of a proposed new observatory in Quebec. He arrived in Quebec in 1850 to supervise development of the observatory, using instruments supplied from Greenwich. The exact date of time ball introduction is uncertain but the apparatus was supplied by Messrs Forrester in Liverpool (Airy, 1871). It was erected in July 1852, soon after its delayed arrival (Jarrell, 1988). The following notes are on display at the restored Ball House, which includes a static representation of the time ball arrangement.

> In 1857, the exact longitude of Quebec City was determined by linking it to the longitude of Harvard based on an exchange of astronomical clock beats using the telegraph network rather than the signals of star transits. … the longitudes of Montréal, Toronto and Kingston were determined based on the longitude of Quebec City.
>
> Ball House was the most accurate source of time available to navigators for a number of years. … During the navigation season from May 13th to November 29th the time ball was used every day except Sunday. It was raised to the half-way point at 12:45 p.m., was at the top by 12:55 p.m., and was dropped at 1:00 p.m. local time.
>
> In 1874, the Citadel's observatory and its instruments were moved to a new building located in Bonner pasture on the Plains of Abraham. The time ball remained at the Citadel and was converted to run on electricity. …

Ashe had erected an equatorial telescope at the Plains site in 1865 but other instruments, including the transit telescope, were retained at the original site until 1874, when additional buildings had been completed (Jarrell, 1975).

The 1880 Admiralty list included the following details, which are consistent with the above notes after the observatory relocation. It is, however, possible that the time ball arrangement had been changed in 1874 (see later).

> At the Citadel, 1,370 yards from the Observatory.
> 355 feet above high water.
> 46 feet above ground.
> (Drop 8 feet)
> Ball dropped (by electricity from the Observatory) at 1h 0m 0s Quebec mean time.

The latitude and longitude of the time ball were given as 46° 48′ 23″ N, 71° 12′ 17″ W. Quebec mean time at noon corresponded to 4 h 44 m 50.3 s GMT in 1880.

The 1898 list included the same details about the location, except that the ball was dropped at 6h 0m 0s GMT, or 1h 15m 11.0s local mean time. Quebec mean time had been changed to 5 hours behind Greenwich by 1898. There had been a local mean time adjustment of 1.3 seconds at some time after 1880, or a longitude correction of 18 arc-seconds. The following details about preparatory signals were included in the 1898 list. These are consistent with the Ball House display.

> Ball hoisted half way up as preparatory 15 minutes before signal.
> Ball hoisted close up 5 minutes before.
> Ball dropped (by electricity from the Observatory) at 1h 00m 00s mean time for 75° W. long., or 5h 00m 00s from the meridian of Greenwich. …
> [*Note.* – Signal not made on Sundays.]

The above details were repeated until the signal was withdrawn, apart from the change to civil time based on zero hours at midnight in 1925. By 1908, the stated longitude of the "Signal Station" had been adjusted to 71° 12′ 35″ W: the local mean time was then consistent with the stated longitude of the signal. It became common after the introduction of time zones to indicate the latitude and longitude of the signal location to the nearest arc-minute: the precise time of the signal was the critical parameter.

There is a mystery about the time ball installation itself. An 1870 photograph is shown in Fig. 24.2a (Ball House, 2007). The arrangement then used a caged ball, which was later used at several other locations in eastern Canada. The modern static arrangement at the restored Ball House includes a representation of this original design (Fig. 24.2b) but the associated display also shows another design, where an uncaged ball was dropped from a yard-arm (Fig. 24.2c). This design may have been introduced after instruments had been relocated in 1874. All the Admiralty lists from 1880 onwards included the same time ball parameters.

Fig. 24.2 Quebec Citadel time balls: (**a**) 1870 photograph (Ball House, 2007); (**b**) modern restored arrangement at the Ball House; (**c**) arrangement using a yard-arm. (Courtesy Klaus Hülse collection)

2.2 *Other Locations in Eastern Canada*

Several time balls in eastern Canada used a caged arrangement so the time ball could be operated in harsh winter conditions (Bartky, 1983).

2.2.1 Fredericton, NB

The first astronomical observatory in Canada was established at Fredericton, New Brunswick in 1851, thanks largely to the persistence of William Brydone Jack (1817–1886) who was a professor and amateur astronomer at King's College, Fredericton (Kennedy, 1982). Jack and colleagues determined the longitude of Fredericton in 1854 by exchanging clock signals and star transit timings with Harvard College Observatory in Cambridge, Massachusetts, using a telegraph link. The Harvard longitude had been determined by exchange of numerous chronometers with Greenwich, there being no transatlantic telegraph cable at that time. This in turn allowed Ashe to re-determine the longitude of Quebec which was the meridian reference for other locations in British North America. It was almost two decades before a time ball was established at Saint John, New Brunswick. The time ball at Montreal followed in the 1860s, but a time ball at Halifax, Nova Scotia was not listed until the twentieth century.

In 1880, time balls were listed at Saint John, New Brunswick and at Montreal, as well at Quebec City. The Saint John time ball was distinguished by its gold band, the others being black. It had been established at Fort Howe in 1870 and later moved to the Customs House, continuing with the caged arrangement, as shown in Fig. 24.3. Another black time ball, also caged, was first listed for Halifax, Nova Scotia in 1908. It had been erected in 1904 (Brooks, 2022). Figure 24.4 shows the arrangement at Halifax. These time balls in eastern provinces were still listed in 1939 but not in 1947.

A time gun at Halifax was listed between 1898 and 1947. A description of its operation was published in 1891 which also included a short history of earlier signals at Halifax which had not been included in the 1880 Admiralty list and appear to have been of dubious accuracy.

> At Halifax, Nova Scotia, a signal gun is fired from the citadel each day at noon and at 2.30 p.m. Twenty years ago the gunner in charge regulated his watch at a famous jewelry establishment. A flag signal system was subsequently adopted, and that in turn was succeeded by a time ball signal system. The interference with the signals by fogs made a new and more reliable arrangement necessary. A cable has been laid from a clock-maker's establishment to the citadel, and is connected with a clock which automatically fires the gun. ...

This was clearly an improvement on earlier arrangements, but the 1898 Admiralty list included the following statement.

> Gun fired at noon mean time of the 60th meridian.
> [Note. This gun is fired for local purposes only, and is not to be depended upon for rating chronometers.]

Fig. 24.3 Saint John, NB time balls. (Courtesy Klaus Hülse collection)

Fig. 24.4 Halifax Citadel time ball. (Courtesy Klaus Hülse collection)

Fig. 24.5 Montreal time ball. (Courtesy Klaus Hülse collection)

There was no mention of the gun in the 1908 Admiralty list which included the time ball for the first time, while the 1911 to 1922 lists included the statement "A gun is fired by hand" when the ball dropped at 1 pm. This had been changed by 1926, when the gun was "Fired electrically from St. John's (*sic*), New Brunswick." It was still operating in 1947 when the time ball was no longer in service.

The black time ball at Montreal was dropped by electricity from Montreal Observatory and was located on the tower of the Harbour Commissioner's building, shown in Fig. 24.5. The Observatory at McGill University was operational by 1863 and a time ball was introduced before 1870 (Bignell, 1962). A larger transit telescope was acquired 1n 1879. As at Quebec, the ball had a drop of 2.4 m. The comment "Signal may always be relied upon within half a second." was repeated in all Admiralty lists from 1880 to 1937. The time ball was only operated during the navigation season when the St Lawrence seaway was free of ice. Clearly, the signal was highly regarded for its accuracy.

2.3 *Western Canada*

A time ball on the Pacific coast was established at Victoria on Vancouver Island in 1916. This time ball was operated by "The superintendent of the Meteorological Observatory at Gonzales Heights" throughout its life. It was positioned on the Belmont building.

A time gun at Vancouver, British Columbia was first listed in 1904. It was "fired electrically from McGill College, Montreal, at 9 am Standard time" and located at Brockton point. It was still listed in 1947. Another gun at Victoria was listed in 1947, but not in 1939 or earlier. It was located at Work Point barracks.

3 Newfoundland

The only signal in Newfoundland was the 32-pounder time gun on Signal Hill at St. John's. This was listed throughout the period from 1880 to 1947. The 1880 list included the comment "No preliminary signal is given.". Lists from 1898 onwards included the unusually strong statement "The signal has been reported to be useless for rating chronometers".

4 Bermuda

A time ball at the Ireland Island dockyard in the British colony of Bermuda was listed from 1880 to 1904. The dockyard is shown in Fig. 24.6 but the photograph may have been taken after the time ball service had been discontinued. The same location was stated in 1880 and 1898, but the latitude and longitude of the signal had been corrected from 32° 19′ 19″ N, 64° 51′ 34″ W to 32° 19′ 23″ N, 64° 49′ 35″ W in 1898. The signal was made on Saturdays only.

The ball was described as having a diameter of only 0.5 m with a drop height of 9 m. It was positioned on "Flagstaff on western jetty in front of the main building of the Dockyard", 25 m above ground in its raised position (1898 list). An ensign "usually on the flagstaff" was hauled down and the ball hoisted as a preparatory signal 5 minutes before the signal at noon, Bermuda mean time (4 h 19 m 18.3 s behind

Fig. 24.6 Royal Navy dockyard in Bermuda. (Courtesy Klaus Hülse collection)

GMT) in 1898. Bermuda mean time had been corrected by 1898 from 4 h 19 m 26.1 s behind GMT in 1880, corresponding closely to the correction in longitude of almost 2 minutes of arc. Also, the preparatory signal had been 15 minutes before the time signal in 1880.

References

Airy, G. B. (1871). *Letter to John Lilley & Sons*, dated 5 September 1871. RGO 6/617, File 2, Leaves 54–55, Papers of George Airy, Cambridge University Library.

Ball house and observatory (*Ancien observatoire et tour de la boule*), 2007. Notice on the Quebec Citadel that celebrates the first time ball in Canada.

Bartky, I. R. (1983). *Time Dissemination, Sky with Ocean Joined*, Proceedings of the Sesquicentennial Symposia of the U.S. Naval Observatory, December 5 and 8 1980, Washington D.C, edited by Dick, S.J. and Doggett, L.E., US Naval Observatory.

Bignell, N. (1962). *McGill Observatory through 100 years*. McGill News.

Brooks, R. C. (2022). Research notes: A Canadian time ball. https://journals.lib.unb.ca/index.php/MCR/article/view/17871/22106.

Jarrell, R. A. (1975). Origins of Canadian Government Astronomy. *Journal of the Royal Astronomical Society of Canada, 69*(2), 77–85.

Jarrell, R. A. (1988). *The cold light of dawn: a history of Canadian astronomy*. University of Toronto Press.

Jarrell, R. A. (1990). Ashe, Edward David. In *Dictionary of Canadian Biography* (p. 12). University of Toronto/Université Laval. Accessed 24 June 2022. http://www.biographi.ca/en/bio/ashe_edward_david_12E.html

Kennedy, J. E. (1982). Jack, Edward Brydone. In *Dictionary of Canadian Biography* (p. 11). University of Toronto/Université Laval. Accessed 26 June 2022. http://www.biographi.ca/en/bio/jack_william_brydone_11E.html

Chapter 25
The United States of America, Including Hawaii

Roger Kinns

Abstract This chapter is based strongly on earlier research by authors in the USA, complemented by additional information in lists by the British Admiralty. The first time ball in the USA was at Washington, DC in 1845. It was operated by the Naval Observatory, but it was intended as a signal for public use. The US Navy had a policy of purchasing only high-quality chronometers and rating them carefully before issue to the fleet. Later, long-distance electric telegraphs were developed to provide signals for mariners along an extended coastline. In 1880, only the signals in Boston, New York and Washington were intended specifically for mariners, but the number increased greatly after 1900, with signals around the Pacific and Atlantic coasts, the Gulf of Mexico and the Great Lakes. Unusually, the official time signal for Honolulu in Admiralty lists from 1898 until the 1920s was a whistle, close to the main berthing facilities.

1 Introduction

A comprehensive discussion of time signals in the USA has been presented by Bartky (1983), who included images of the many different time balls that used to exist around the coasts of the United States. Many of the images in Bartky's paper are included in the collection assembled by Hülse (2023) and are reproduced in this paper.

Wauchope had proposed the use of time balls for rating chronometers in the USA in his 1830s correspondence, but the required infrastructure for accurate measurement and distribution of time with such a large coastline took a long time to develop. Time balls at Washington DC, New York City, and Boston were the only entries for the USA in the 1880 Admiralty list, although many other time balls had been erected for public use across the USA by that time. The history of these three time balls will be considered later in more detail. The history of the Washington time ball, the first

R. Kinns (✉)
The Maudslay Society, Pembroke College, Cambridge, England

R. Kinns (ed.), *The World of Visual Time Signals for Mariners*, Historical & Cultural Astronomy, https://doi.org/10.1007/978-3-031-57334-7_25

to be introduced in North America, has been described by Bartky and Dick (1982). The 1981 and 1982 papers by Bartky and Dick have been reprinted in Dick (2020).

The Hawaiian Islands had become a US Territory in 1898 and a US State in 1959. They were described initially as the Sandwich Islands.

Entries in Admiralty lists are shown in Table 25.1, including Honolulu. Figure 25.1 shows the locations of signals in mainland United States. Figure 25.2 shows the location of the signal at Honolulu in the Hawaiian Islands, using a much larger scale.

Table 25.1 Signals in the United States, including Hawaii

Area	Location	1880	1898	1904	1911	1915	1919	1922	1928	1934	1939	1947
Great Lakes	Duluth Minnesota			TB (red)	TB (red)	TB (red)						
	Sault St. Marie Michigan			TB (black)	TB (black)	TB (black)						
	Chicago Illinois				TB (red)	TB (red)						
	Cleveland Ohio				TB (black)	TB (black)						
	Buffalo New York			TB (black)	TB (black)	TB (black)						
USA (East Coast)	Boston Mass.	TB (copper)		TB (black)	TB (black)	TB (black)	TB (black)	TB (black)	TB (black)			
	Newport Rhode Island		TB (black)	TB (black)	TB (black)	TB (black)						
	Woods Hole Mass.		TB (black)									
	New York	TB (black)	TB (black)	TB (black)	TB (black)		TB (black)	TB (black)	TB (black)	TB (black)	TB (black)	TB (black)
	Philadelphia Pennsylvania			TB	TB	TB (black)	TB (black)	TB (black)				
	Baltimore Maryland		TB (black)	TB (black)	TB (black)	TB (black)	TB (black)	TB (black)	TB (black)	TB (black)		
	Washington DC	TB (black)	TB (black)	TB (black)	TB (black)	TB (black)	TB (black)	TB (black)	TB (black)	TB (black)	TB (black)	
	Newport News Virginia				TB (black)	TB (black)	TB (black)					
	Norfolk Virginia			TB (black)	TB (black)	TB (black)	TB (black)	TB (black)	TB (black)	TB (black)		
	Savannah Georgia		TB (black)	TB (black)	TB (black)	TB (black)	TB (black)	TB (black)	TB (black)	TB (black)		
	Key West Florida				TB (black)	TB (black)	TB	TB	TB	TB		
USA, (Gulf of Mexico)	New Orleans Louisiana		TB (black)	TB (black)	TB (black)							
	Galveston Texas			TB	TB	TB	TB	TB				
USA (West Coast)	San Francisco California		TB	TB	TB	TB	TB	TB	TB	TB	TB	
	Mare Island California		TB (black)	TB (black)	TB (black)	TB (black)	TB (black)	TB (black)	TB (black)	TB (black)	TB (black)	
	Portland Oregon				TB	TB (r, w)	TB (r, w)	TB (r, w)	TB (r, w)	TB (r, w)	TB (r, w)	
Hawaii	Honolulu		Whistle	Whistle	Whistle	Whistle	Whistle	Whistle				
Signal type totals		1880	1898	1904	1911	1915	1919	1922	1928	1934	1939	1947
Time ball, including collapsible		3	9	15	20	18	13	12	10	9	5	1
Gun, including powder flash		0	0	0	0	0	0	0	0	0	0	0
Lights		0	0	0	0	0	0	0	0	0	0	0
Other: disc, drum, flag, etc.		0	1	1	1	1	1	1	0	0	0	0

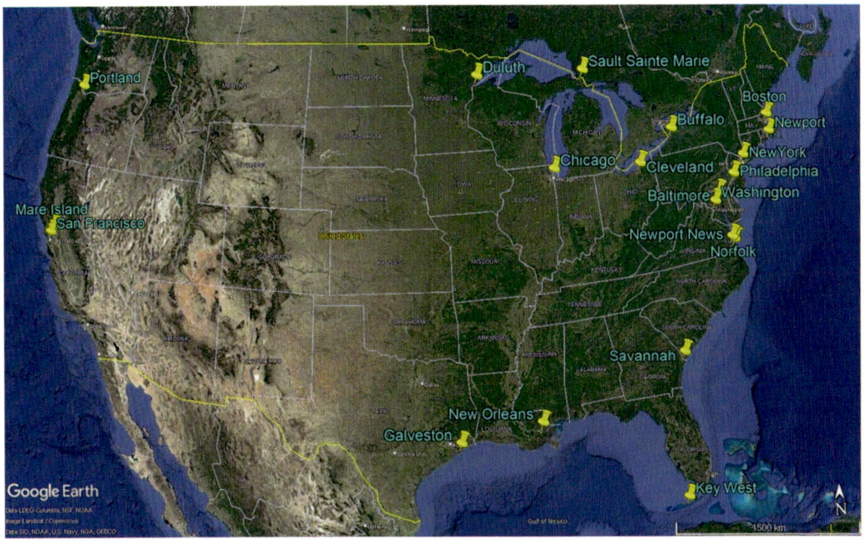

Fig. 25.1 Locations of signals in mainland USA

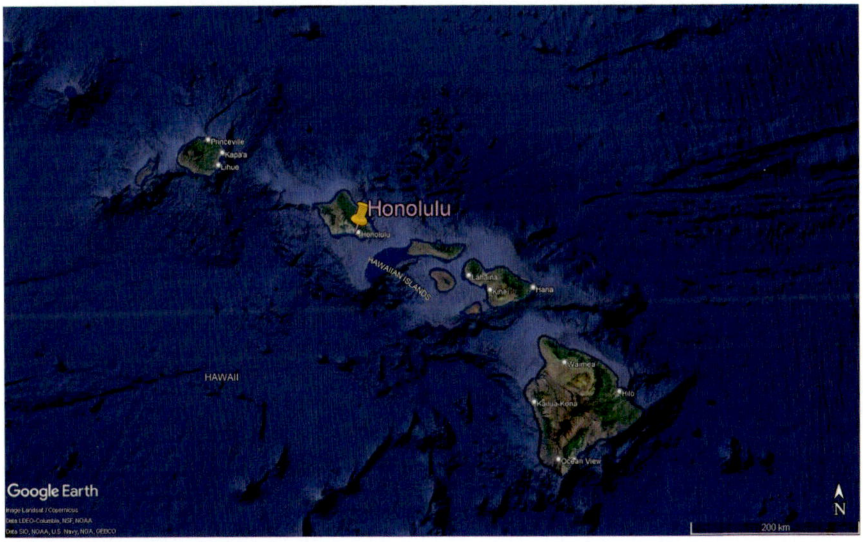

Fig. 25.2 Location of signal in Hawaii

2 Time Balls for Chronometer Rating in the USA

Two time balls for rating chronometers in the USA were identified in an 1877 report prepared for the Secretary of the Navy (Rodgers, 1877). These were at Washington DC and New York City. The author of the report was Rear Admiral John Rodgers

(1812–1882), who was Superintendent of the Naval Observatory from 1877 to 1882 (Dick, 2003, 168).

The time ball at New York became operational in September 1877. It was anticipated prematurely that another at Baltimore would soon be erected. The failure rate of the time-ball signal at Deal in England, which was dropped by direct telegraph signal from Greenwich, was also noted. Extracts from the report for 1877 are shown below (Rodgers, 1877).

> … In September arrangements were perfected for dropping a time-ball in New York City at exact New York noon; and on the 10th of that month the ball was dropped from the chronometer room for the first time. It has been dropped at New York mean noon daily, except Sundays, to the present time, without a single failure.
>
> It may be of interest to remark that the time-ball at Deal, England, which is dropped by signal from the Royal Observatory at Greenwich, is subject to about fifteen failures annually …
>
> This New York time-ball is already used … in rating the ship's chronometers.
>
> The Baltimore Board of Trade contemplates the establishment of a similar time-ball at Baltimore, to be dropped at Washington mean noon daily, except Sundays; and it is to be hoped that similar time-balls may be established gradually at various other ports.

The corresponding report for 1878 (Rodgers, 1878) stated that there had been eight failures of the New York time ball drop since its introduction in September 1877, for a variety of reasons. The Baltimore time ball was not mentioned in the 1878 report.

Time balls for marine chronometer calibration were adopted enthusiastically after 1880. Bartky described how time was disseminated by telegraph around the United States, initially using observations from different observatories but later using telegraph signals from Washington, DC throughout the Nation. By 1888, time balls at Woods Hole, Massachusetts; Newport, Rhode Island; New York City; Philadelphia, Baltimore; Hampton Roads; Savannah; and New Orleans were regulated by telegraph signals from Washington (Dick, 2003, 189). Bartky (1983) included a chart showing approximate dates of operation. Time ball apparatus was changed and moved between buildings in various American cities. These changes are not shown in Table 25.1. Dick (2003, 46) noted that 19 time balls were dropped using telegraph signals at principal ports in 1905. Many had been introduced in the early years of the twentieth century. The Admiralty lists for 1908 and 1911 included 20 time balls in the USA, while only 9 had been listed in 1898 and 16 in 1904.

Ironically, the best known time ball of them all, dropped slowly at Times Square in New York to announce the arrival of each New Year since 1908, with the exception of 1942 and 1943, had nothing to do with chronometer rating.

Most, but not all, time balls in the USA had small diameters and light weights relative to most of those used in the British Isles, where a diameter of 1.5 m had been preferred after visibility trials at Greenwich that preceded erection of the 1833 time ball. Many had diameters of 1.2 m, but some were much smaller. Even the Washington time ball had a diameter of less than 0.9 m.

The following summary concerning the first time signals in Washington is derived primarily from the authoritative history of the Naval Observatory in *Sky and Ocean Joined* (Dick, 2003). The Observatory was the home of the first visual time signal in the United States in 1845, but the signal was not promoted initially as a means for rating chronometers.

3 Washington, DC

The Depot of Charts and Instruments had been founded in 1830. Its functions included selection and calibration of chronometers for the US Navy, using a transit telescope to rate chronometers and to check rate stability before issue to ships. There was no provision for visual time signals that could be seen outside the Depot. The organisation developed into the Naval Observatory which became operational in 1844. Matthew Fontaine Maury (1806–1873) was Superintendent from 1844 to 1861 (Dick, 2003, 60). The Observatory was located initially at Foggy Bottom in Washington close to the Potomac River. It was relocated in 1893.

Secretary of the Navy John Mason (1799–1859) wrote to Lt. Maury, on 10 December 1844 "You will be pleased to devise some signal by which the mean time may be made known every day to the inhabitants of the city of Washington…". Maury responded quickly by ordering a flagstaff, stays, halyards and a black ball with a diameter of 3 feet (0.9 m) from the Navy agent in Washington (Dick, 2003, 85). The time ball service was announced in the *Daily National Intelligence* on 2 January 1845, but the nature of the ball and its dropping arrangements appear to have been subject to subsequent experimental development (Dick, 2003, 85). Maury wrote to the New York agent for Charles Goodyear, the inventor of vulcanised rubber, in February 1845 "Be pleased to make and send … four air tight balls of Gumelastic Composition capable of being inflated into spheres. They are wanted for signal balls…. Let two of the balls be at least 4 feet in diameter when inflated one of 3 feet and one of 18 inches…". These balls may have been for various purposes, not necessarily for signalling time. A letter in May 1845 to another supplier in New York included the statement "What will you charge for that copper ball? I may want it for a time ball…".

The ball was dropped at noon each day, except Sunday, above the dome of the 9.6 inch refractor at the observatory's original site and served as the first public time service in the USA. The early arrangement is shown in Fig. 25.3 (Gideon, 1846; Bartky, 1983; Dick, 2003). It was a far cry from the sophisticated 1833 arrangement at Greenwich, England.

The Washington time balls were of light construction (Dick, 2003, 154; from Gilliss, 1863)

> In 1863, the supply of time balls having been exhausted, Gilliss ordered three others to be constructed. As in the past, they were constructed of hoops of hickory three feet in diameter, covered with canvas painted black, and fitted with a loop at each extremity of diameter, the latter undoubtedly for attachment to the flagstaff atop the Observatory dome.

James Melville Gilliss (1811–1865) succeeded Maury as Superintendent in 1861 and saw the Observatory through the Civil War (Dick, 2003, 45).

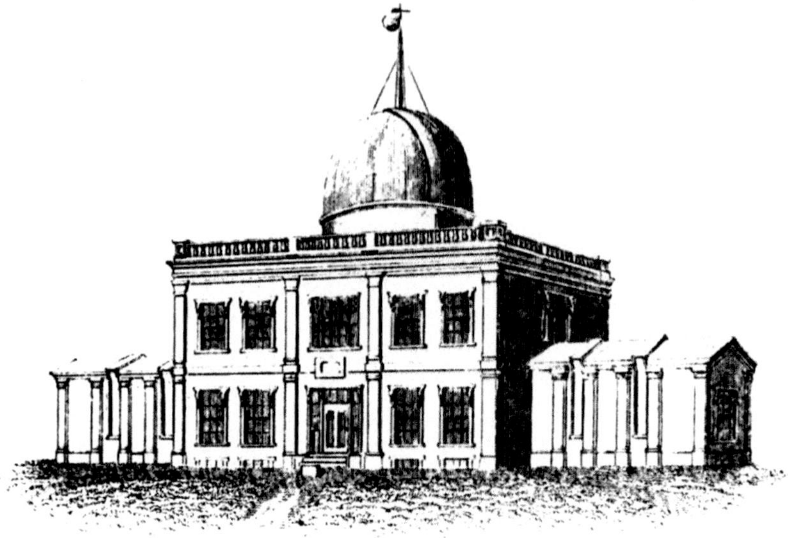

Fig. 25.3 The first time ball at the Naval Observatory in Washington (Bartky, 1983)

The time ball was dropped manually until some time in the 1850s (Dick, 2003, 186). It may have been operated like a flag signal for at least part of that period, using halyards to lower the ball. It was subsequently dropped by a clock observer who pressed a key to release electromagnetic triggers. This was changed to automatic operation in 1879. Halyards were used for hoisting the ball throughout its existence, rather than a mechanism for pushing the ball up a slotted mast. According to the 1880 Admiralty list, the black ball had a diameter of only 2 feet 8 inches (0.8 m) and a drop height of 12 feet (3.6 m). It was 162 feet above high water and 68 feet above ground, located at the "Marine Observatory" with latitude and longitude 38° 53′ 39″ N, 77° 3′ 6″ W. It was dropped at noon local time, 5 hours 8 minutes 12.4 seconds behind GMT.

The time ball was relocated in 1885 to the nearby State, Navy and War building where it would be more visible (Dick, 2003). However, automatic ball release was discontinued. The Naval Observatory itself was relocated further north to Massachusetts Avenue in 1893 but the time ball remained at its 1885 location. The time ball description had changed significantly in the 1898 Admiralty list when its location was given, misleadingly, as the "Naval Observatory". The Observatory was the source of the signal but was not the time ball location in 1898. The stated time ball latitude was unchanged from 1880 but the longitude had increased slightly to 77° 3′ 8″ W. There were significant changes to other parameters after 1880. The ball diameter was changed slightly to 2 feet 10 inches but the drop height had more than doubled to 25 feet by 1898, while the heights above high water and ground had been increased to 205 feet and 157 feet respectively. The drop time was then 5 hours

Fig. 25.4 Washington time ball on the Executive Office building (Bartky, 1983)

GMT or 23 hours 51 minutes 47.5 seconds local time, based on zero hours at noon. These details were repeated until at least 1911. The principal change after that was in the description of the building, rather than the time ball location. It was given as the State, Navy and War building in Admiralty lists after 1915. There were some minor changes to the principal dimensions after 1919. In the 1922 list the drop height was reduced to 21 feet, while the heights above high water and ground had been reduced to 197 feet and 150 feet respectively. The ball diameter was then given as 2 feet 7 inches. These parameters were still the same in lists up to 1946. Dick indicated that the Washington ball was dropped until 1936 (Dick, 2003, 468), so such late Admiralty list entries are surprising. It had been deleted from the 1947 list.

Figure 25.4a shows the time ball at its location on the State, Navy and War building, specifically the Executive Office building. Figure 25.4b shows the lowered time ball in a tub with the halyards used for hoisting the ball (Bartky, 1983). A new time ball at the United States Naval Observatory was erected in 1999 for the Millennium celebrations (Dick, 2020, Fig. 26.5).

4 East Coast Time Balls

Time balls on the East Coast extended from Boston in the north to Key West in Florida. A time ball at Woods Hole, Massachusetts was listed in 1898 but not in 1904 or later. There was no time ball at Hampton Roads in Admiralty lists: they referred instead to an accessible regulated chronometer. The Admiralty lists for 1908, 1911 and 1915 included ten east coast time balls, many at naval dockyards. These were at: Boston, Newport RI, New York, Philadelphia, Baltimore, Washington DC, Newport News, Norfolk, Savannah and Key West. The time balls at Newport RI and Newport News were no longer listed in 1922. The time ball at Philadelphia had been delisted by 1928.

Fig. 25.5 New York time ball at the Custom-house (The Electric Time Ball, 1878)

The black time ball at Newport RI had an unusually large diameter of 6 feet (1.8 m) and a 24 feet drop height after it was moved from the machine shop to a wireless mast (Notice, 1911). Its diameter and drop height were given as 3.5 and 16 feet in Admiralty lists up to 1911, which had been prepared in the previous year. The revised parameters were listed after 1911. The Newport time ball service was discontinued in 1918 when the wireless mast was removed (Notice, 1918).

Another time ball had been established at Hampton Roads (Bartky, 1983) but this was not listed by the Admiralty in 1898 or later. The service was "suspended temporarily" because the apparatus was out of order in 1896 (Notice, 1896) so there may have been longer term reliability problems.

Those at New York and Boston will be discussed in more detail. The east coast time balls remaining in the 1939 Admiralty list were at New York and Washington.

4.1 New York

The first time ball in New York City was on the Custom-house at 55 Wall Street, which had been built in 1842. It had been operated using a telegraph signal from the "Dudley Observatory, at Albany" (The Electric Time Service, 1878). It was stated to

Fig. 25.6 New York time ball design suggestion (The Electric Time Ball, 1878)

have been in operation 20 years before 1878, so its approximate year of introduction was 1858. Bartky (1983) gave its date of introduction as 1859/1861. It is shown in Fig. 25.5. This time ball was for public use rather than chronometer rating.

A new time ball in New York was erected in 1877 for chronometer rating (Rodgers, 1877, 1878). The 1880 Admiralty list described its location as "Staff on tower of Western Union Telegraph Office". Its diameter was 3.5 feet (1.05 m). Lists up to 1911 gave similar details and specified that the ball was black with a drop of 25 feet (7.5 m). This is consistent with the description "It was to be made of copper-wire netting painted black, and 3.5 feet in diameter" (Dick, 2003, 186). The alternative design shown in Fig. 25.6 used ten segments at 36° intervals around a vertical axis to simulate a spherical ball. It was published in *Scientific American* (Electric Time Ball, 1878; Bartky, 1983), but may have been just an illustrative arrangement. A segmented time ball was used at Semaphore, South Australia and is still present on the restored tower there (Abell & Kinns, 2010).

The Western Union announced in 1914 that their office building at 195 Broadway was to be demolished and that "No provision has been made for a time-ball service" at their new location at 15 Dey Street (Notice, 1914). It had been replaced by a time ball at the "Seaman's Church Institute" in 1913, about 1.6 km away. This was described as a black ball with a diameter of 4 feet (1.2 m) and a drop of 16 feet (4.8 m) in Admiralty lists from 1916 to 1947, but was omitted from the 1915 list. These time balls were entirely separate from the Times Square arrangement that has welcomed the New Year since 1908.

4.2 Boston

The Boston Time-Ball became operational in May 1878 on top of the Equitable building (Lumbricht, 2015). The ball was made of rolled plate copper and had a diameter of 4 feet (1.2 m). It was an unusually heavy time ball and weighed 180 kg (Bartky, 1983). It had a drop height of 4.8 m. A friction brake mechanism was used to stop the ball. Figure 25.7 shows the ball in its rest position. The time signal to release the ball was relayed from the standard clock at Harvard College Observatory in Cambridge by telegraph wire to the ball location (Lumbricht, 2015). These details are consistent with the 1880 Admiralty list which gave the time ball location as "located on a mast on a building near the Harbour Signal Office".

There was no Admiralty list entry in 1898, so the original ball appears to have ceased operation by that time. A new time ball was erected in 1902 on the roof of the Ames building, then the highest building in Boston (Lumbricht, 2015). A black time ball with the same 1.2 m diameter as the 1878 ball was listed by the Admiralty between 1904 and 1932. It was removed from the 1933 Admiralty list and was never reinstated.

Fig. 25.7 Boston time ball (Bartky, 1983)

5 Gulf of Mexico

5.1 New Orleans, Louisiana

A black time ball at New Orleans was listed from 1898 to 1911. It had a diameter of 0.75 m, a drop height of 6 m and was located on the roof of the "Sugar Refining building".

5.2 Galveston, Texas

Another at Galveston, Texas was introduced in 1901 (Notice, 1901) and was listed from 1904 to 1922. It had a diameter of 1.2 m, a large drop height of 8 m and was located on the roof of the "Levy building". The building was best known as a clothing store, but it also housed the offices of the US weather bureau.

6 West Coast

6.1 Mare Island, California

There were no visual time signals for the Pacific coast of North America in the 1880 Admiralty list. The Naval Observatory at the Mare Island naval shipyard, north of San Francisco, distributed time to locations on the Pacific coast from 1884 (Bartky, 1983; Dick, 2003, 192–193). Time balls were soon established at Mare Island and San Francisco. A new longitude determination by telegraph in 1908 showed that there was a small error in Mare Island time, corresponding to a longitude error of about 5 arc-seconds.

In 1898, the latitude and longitude of the Mare Island time ball were given as 38° 5′ 53″ N, 122° 16′ 16″ W. The black ball had a diameter of 0.9 m and a drop height of 4.5 m. In 1908, the coordinates had been adjusted to 38° 5′ 57″ N, 122° 16′ 6″ W. There was a further small adjustment in longitude to 122° 16′ 9″ W by 1922, but no changes thereafter.

6.2 San Francisco, California

A time ball on Telegraph Hill was established in 1885. It had a diameter of 0.9 m and a drop height of 4.5 m, as at Mare Island. The location was changed to the Ferry Building (Notice, 1898). The Ferry building time ball mast was bent in the 1906 San Francisco earthquake (Bartky, 1987). The time ball was moved to its final location

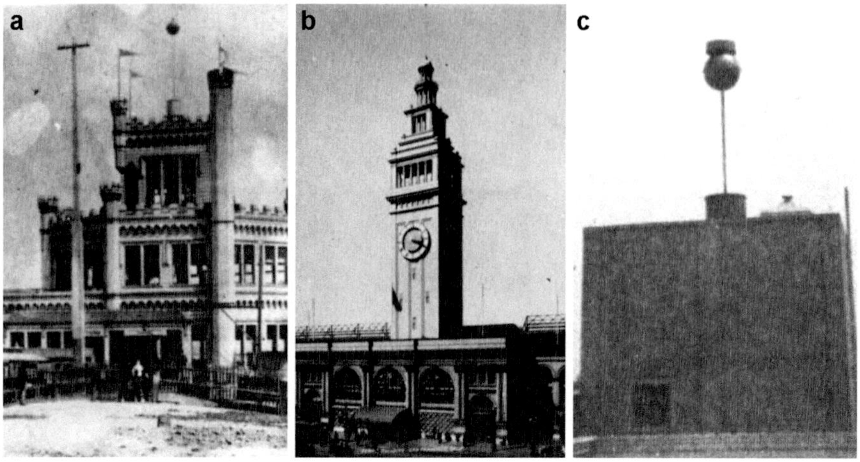

Fig. 25.8 San Francisco time balls: (**a**) Telegraph Hill; (**b**) Ferry Building; (**c**) Fairmont Hotel (Bartky, 1983)

on the Fairmount Hotel building in 1911. The three locations are shown in Fig. 25.8. Their respective coordinates were given in Admiralty lists as: 37° 48′ 5″ N, 122° 24′ 20″ W; 37° 47′ 38″ N, 122° 23′ 38″ W; 37° 47′ 27″ N, 122° 24′ 36″ W. The same diameter and drop height were used throughout. The Mare Island and San Francisco time balls were still listed in 1939 but had been withdrawn by 1947.

6.3 Portland, Oregon

The time ball at the Portland Custom House was listed from 1908 until 1939. It had a typical diameter of 1.2 m but it was noted as deep red with a white band in lists from 1911 onwards. As at other West Coast locations, it was dropped by the Mare Island Observatory at noon, Standard time of the 120th meridian, 8 hours behind GMT. Its latitude and longitude were 45° 31′ 39″ N, 122° 40′ 44″ W.

7 Great Lakes

The Admiralty list for 1908 included five time balls on the Great Lakes. These were at Duluth, Minnesota and Sault St. Marie, Michigan on Lake Superior; Chicago, Illinois on Lake Michigan; Cleveland, Ohio and Buffalo, New York on Lake Erie. They were also included in the 1911 and 1915 lists and all apart from Chicago and Cleveland had appeared in the 1904 list. The time balls at Duluth and Chicago were painted red but the others were black. Their diameters varied between 0.9 m at Sault

St. Marie to 1.3 m at Chicago and Cleveland. They were all deleted in Admiralty lists from 1916 onwards. Bartky (1983) indicated that some of the Great Lakes signals may have continued at some locations into the 1920s. Table 25.1 is otherwise consistent with Bartky's findings from independent sources in the USA.

8 Hawaii

The islands in the Hawaiian archipelago are remote from continental land masses, being 3200 km from the Pacific coast of the United States. There had been no confirmed contact with people of European origin until the 1778 visit by Captain James Cook, who named them the Sandwich Islands after his sponsor. He was killed there in 1779. The Hawaiian Islands had been an independent monarchy from 1810 until 1893, when the monarch was deposed and the islands became a republic. The islands were annexed by the United States in 1898 as a US Territory and became the 50th US State in 1959.

Unusually, the official time signal in Admiralty lists was a whistle. It was located "At Steam Planing Mill" in Honolulu (Sandwich Islands). Its latitude and longitude were given as 21° 18′ 13″ N, 157° 51′ 47″ W. The entry in the 1898 Admiralty list is shown below.

> *Honolulu.* – Whistle sounded by electric signal from the Survey Office at noon, Sandwich Island standard mean time, and repeated at 1h 30m 0s p.m. standard mean time
>
> [*Note.* – The second signal is reported to be the most accurate for rating chronometers.]

The same entry was repeated in 1908 with the qualification "Reported unreliable.". A similar entry was included in lists until at least 1922, except that "Sandwich Is." had been replaced by "Hawaiian Islands" by 1915. It was also noted that "Chronometers may be compared at the Territorial survey office."

The whistle was only about 2 km from the main berthing facilities, so the sound propagation time would have been about 6 seconds and correction for it would have been of acceptable accuracy. There was no entry for Honolulu in Admiralty lists from 1927 onwards. The only other whistle listed by the Admiralty was at Niigatako in Japan, introduced in 1924.

References

Abell, L., & Kinns, R. (2010). 'Telegraph' Todd and the Semaphore time ball. *Journal of the Historical Society of South Australia, 38*, 42–57.

Bartky, I. R. (1983). Time dissemination, sky with ocean joined. In S. J. Dick & L. E. Doggett (Eds.), *Proceedings of the sesquicentennial symposia of the U.S. Naval Observatory, December 5 and 8 1980, Washington, DC*. US Naval Observatory.

Bartky, I. R. (1987). The bygone era of time balls. *Sky and Telescope, 73*, 32–35.

Bartky, I. R., & Dick, S. J. (1981). The first time balls. *Journal for the History of Astronomy, 12*, 155–164.
Bartky, I. R., & Dick, S. J. (1982). The first North American time ball. *Journal for the History of Astronomy, 13*, 5–54.
Dick, S. J. (2003). *Sky and ocean joined: The U.S. Naval Observatory 1830–2000*. Cambridge University Press.
Dick, S. J. (2020). The first time balls and the first North American time ball. In *Space, time and aliens: Collected works on cosmos and culture*. Springer Nature.
Electric Time Service for New York. (1878). *Scientific American, 39*(22), 335–337.
Gideon, J. and G.S. (1846). *Astronomical observations (1845) of the U.S. Naval Observatory*. Washington.
Gilliss, J. M. (1863). *Letter to C.H. Davis*. 10 August 1863.
Hülse, K. (2023). *Themen-Seite Zeitsignale: Zeitbälle oder Zeitkanonen*. http://www.leuchtturm-welt.net/HTML/TIMEBPK/TIMEBALL.HTM
Lumbricht, M. J. (2015). *The Boston time ball*. Ladd Observatory (Brown University) Blog. https://blogs.brown.edu/ladd/2015/12/22/the-boston-time-ball/
Notice. (1896). *Virginia – Hampton Roads – Hygeia Hotel – Time ball temporarily discontinued*, dated 3 February 1896. Superintendent, United States Naval Observatory.
Notice. (1898). *San Francisco – Time ball station moved*. US Hydrographic Office Notice to Mariners No. 86 of 1898.
Notice. (1901). *Galveston – Time signal established*. US Hydrographic Office Notice to Mariners No. 52 of 1901.
Notice. (1911). Rhode Island – Narragansett Bay – Newport Harbor – Torpedo Station – Time ball moved. *Nautical Magazine, 3*, 1911 (US Hydrographic Office Document, No. 45901-187).
Notice. (1914). (2049) New York – Time ball to be discontinued. *Nautical Magazine, 25*, 1914 (US Hydrographic Office Document, No. 93067-187).
Notice. (1918). Rhode Island – Narragansett Bay – Eastern Passage – Goat Island – Radio, weather signal and time ball discontinued. *Nautical Magazine, 43*, 1918 (US Hydrographic Office Document, 161679-72291).
Rodgers, J. (1877). Rear-Admiral USN, 1877. *Report of the Secretary of the Navy: Chronometers, 1877*, 117.
Rodgers, J. (1878). Rear-Admiral USN, 1877. *Report of the Secretary of the Navy: Chronometers, 1878*, 162.
The Electric Time Service. (1878). Article in *Harper's New Monthly Magazine, 56*, 665–671.

Chapter 26
Central America, The West Indies and South America

Roger Kinns

Abstract The southernmost time ball was at Montevideo. The northernmost signals were in Guyana and Suriname. The first known time ball in the Americas was at Valparaiso in 1843. A small time ball at Port La Plata in Argentina was replaced by simple time lights in 1911. A large drum at Buenos Aires was replaced by complex time lights in 1915. Brazil used an inflatable drum at Rio de Janeiro, replaced by time lights in 1920. A combined flag and magnesium flash signal was provided on Fiscal Island from 1921. Peru used a time ball at Callao.

A flag time signal existed in the US Virgin Islands during the 1840s. Time balls were later used at Port of Spain, Castries, Bridgetown and Port Royal, with more extended time signal provision in Havana and at Curaçao. There were time balls at Guaymas in Mexico and at Balboa in Central America.

1 Introduction

Visual signals in Central and South America were sparsely distributed relative to North America. The first known time ball in the Americas was, however, at Valparaiso in 1843. The southernmost time ball was at Montevideo in Uruguay on the Atlantic coast.

Table 26.1 shows the evolution of signals in Central America and the West Indies between 1880 and 1947, excluding Bermuda. Figure 26.1 shows their locations. Figure 26.1 also shows the locations of signals in Guyana and Suriname which are in South America, but north of the Equator. Bermuda is further north than islands in the Caribbean and has been included in Chap. 24.

Table 26.2 shows the evolution of signals in South America, arranged in alphabetical order of countries. Figure 26.2 shows locations south of the Equator,

R. Kinns (✉)
The Maudslay Society, Pembroke College, Cambridge, England

R. Kinns (ed.), *The World of Visual Time Signals for Mariners*, Historical & Cultural Astronomy, https://doi.org/10.1007/978-3-031-57334-7_26

Table 26.1 Signals in the Central America and the West Indies

Area	Location	1880	1898	1904	1911	1915	1919	1922	1928	1934	1939	1947
Central America	Mexico Guaymas										TB	
	Panama Balboa						TB	TB				
West Indies	Havana NO		TB (black)	TB (black)								
	Havana Obs.				TB	TB	TB	TB	TB	TB	TB	
	Havana MC				Sema-phore.							
						TB	TB	TB	TB	TB	TB	
	Barbados				TB							
	St. Lucia		TB	TB	TB							
	Curaçao Guard Ship	Flag	Flag	Flag	Gun	Gun	Gun	Gun	Gun	Gun	Gun	
	Curaçao Fort				Flag	Flag	Flag	Flag	Flag	Flag	Flag	
	Trinidad		TB	TB	TB							
Signal type totals		1880	1898	1904	1911	1915	1919	1922	1928	1934	1939	1947
Time ball, including collapsible		0	3	3	4	2	3	3	2	2	3	0
Gun, including powder flash		0	0	0	1	1	1	1	1	1	1	0
Lights		0	0	0	0	0	0	0	0	0	0	0
Other: disc, drum, flag, semaphore, etc.		1	1	1	2	1	1	1	1	1	1	0

2 Central America

2.1 *Mexico*

A time ball at Guaymas in Mexico was listed in 1939. It was on "mast on roof of harbour office". The 1939 list indicated that it had been introduced in 1926, but it did not appear in Admiralty lists up to at least 1934. Modern port facilities were built at Guaymas in 1925 for the Mexican navy, which are likely to have included a time ball.

2.2 *Panama*

A time ball at Balboa was near the southern entrance to the Panama Canal. It had been introduced in 1916 and was located on "Mast over centre of signal station, Sosa Hill". It was dropped by wireless signal from Washington, DC at 1 pm Standard time, 5 hours behind GMT, and was an accurate signal: "the lag in the time of

Fig. 26.1 Signal locations in Central America, the West Indies and northern South America (Google Maps)

dropping the ball is half a second; the actual time of the dropping of the ball on each occasion will be published in the succeeding issues of all local daily newspapers" (1919 list). It appeared in the 1922 list, but not in in lists for 1927 or later.

Otherwise, telegraph signals were available at Panama for shore access.

3 West Indies

3.1 Virgin Islands

A flag time signal had been provided at St. Croix and St. Thomas, now in the US Virgin Islands, by distinguished amateur astronomer Andrew Lang during the 1840s (Spencer Jones, 2020). Lang developed a well-equipped observatory at St Croix and made an accurate determination of its longitude which needed a correction of only 17 seconds of arc (1.1 seconds of time) when telegraphic determinations were made by the US Hydrographic Office in the 1870s. No signal in the Virgin Islands was included in Admiralty lists from 1880.

Table 26.2 Signals in South America

Area	Location	1880	1898	1904	1911	1915	1919	1922	1928	1934	1939	1947
South America	Argentina Buenos Aires			Drum	Drum	Drum						
							Light	Light	Light			
	Argentina La Plata		TB (red)	TB (red)	TB (red)							
						Light	Light	Light	Light	Light	Light	
	Brazil Rio de Janeiro	Drum	Drum	Drum	Drum	Drum	Drum					
								Light	Light	Light	Light	
	Brazil Ilha Fiscal							Flash	Flash	Flash	Flash	Flash
								Flag	Flag	Flag	Flag	Flag
	Chile Coquimbo				Cone	Cone						
					Gun	Gun						
	Chile Valparaiso		TB	TB	TB	TB	TB	TB	TB			
					Gun	Gun	Gun	Gun	Gun	Gun	Gun	Gun
	Guyana Georgetown	TB (red)	TB (red)	TB (red)	TB (red)	TB (red)	TB (red)	TB (red)	TB (red)	TB (red)	TB (red)	
	Peru Callao								TB	TB	TB	TB
	Suriname Paramaribo		Discs	Discs								
		Flag			Flag	Flag						
						Gun						
	Uruguay Montevideo					TB (r, g)	TB (r, g)	TB (r, g)	TB (r, g)	TB (r, g)	TB (r, g)	
Signal type totals		1880	1898	1904	1911	1915	1919	1922	1928	1934	1939	1947
Time ball, including collapsible		1	3	3	3	3	3	3	4	3	3	1
Gun, including powder flash		0	0	0	2	3	1	2	2	2	2	2
Lights		0	0	0	0	1	2	3	3	2	2	0
Other: disc, drum, flag, etc.		2	2	3	4	4	1	1	1	1	1	1

3.2 British Colonies

Official time balls in British Colonies were under consideration in 1878. Airy was approached by T Cooke & Sons in York for advice about designs, suppliers and costs (Cooke, 1878).

> We are asked to give estimate &c for some Time Balls & trigger apparatus complete for the West Indies, & not having yet done much with these arrangements we should greatly esteem it if you could favor us with scale drawings of your apparatus / to which allusion is made / & some idea as to cost &c. We are wishful to supply the most complete & satisfactory arrangement possible & should therefore be glad if you can offer any suggestions for improvements &c.

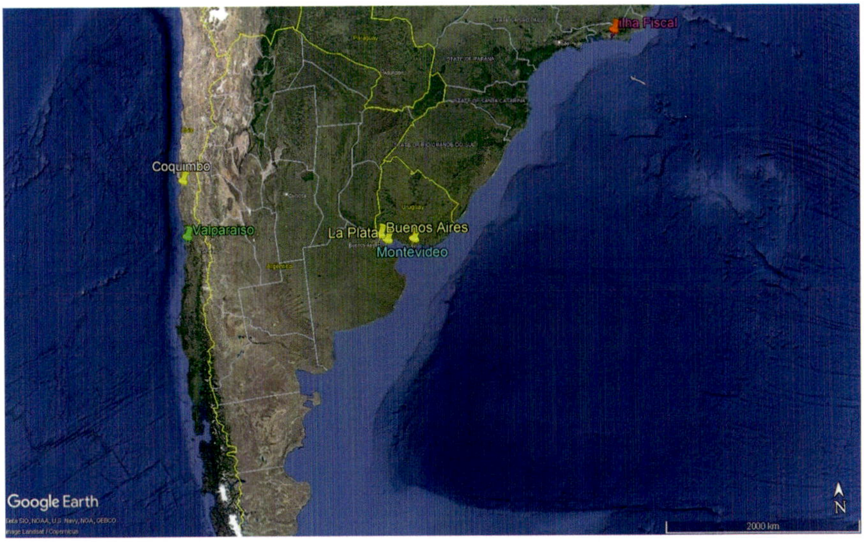

Fig. 26.2 Signal locations in southern South America (Google Maps)

As often, Airy replied promptly and recommended that Maudslay, Sons & Field should be approached (Airy, 1878):

> The time balls with which I have been concerned have all been constructed by Messrs. Maudslays & Field and you would probably find it best to consult that firm. I have no draftsman's office.
>
> If one of your firm should call here on a morning before 1 o'clock we could show him some plans….

No time balls in West Indies colonies were listed by the Admiralty in 1880, so there does not appear to have been early action. There is no evidence that Maudslays supplied any time ball apparatus after 1873, so any time ball apparatus in the West Indies would have been supplied locally or by other firms in Great Britain.

3.2.1 Barbados, St. Lucia and Trinidad

Time balls were in use at Port of Spain, Trinidad and Castries, St. Lucia before 1898. A time ball at Bridgetown, Barbados was also listed in 1911. These three in British colonies were not listed in 1915 or later. The time balls at Port of Spain and Castries were described respectively as "Signal reported of no value for rating chronometers" and as "Signal reported unreliable and as being dropped for local purposes" in 1908 and 1911. The short-lived signal at Bridgetown was dropped from a flagstaff at "Rickett's Battery". The 1911 Admiralty list contained the following qualification for Bridgetown:

> The time ball will be dropped at any arranged time, on application to Mr. Bailey, watchmaker; chronometers may also be compared at his shop. Neither method is recommended, as their accuracy is doubtful.

Early withdrawal of signals with such limited performance is unsurprising.

3.2.2 Jamaica

There had also been a time ball in Jamaica during some of the period between 1880 and 1898. It was withdrawn in 1894.

> The dropping of the Time Ball at noon daily at the flagstaff in the dock yard, Port Royal has been discontinued by order of the Admiralty.

3.3 Curaçao

A flag signal on the guard ship in Santa Anna Harbour, in the Dutch colony of Curaçao, was listed in 1880. The flag signal was changed to a gun before 1908 and a flag signal was provided at Fort Nassau as the secondary signal. The flag was hauled down when the gun fired. The gun and flag were still listed in 1939.

3.4 Havana

Cuba had been a Spanish colony until the Spanish-American War in 1898. It was then under US military rule before gaining independence in 1902. Havana had a more extended provision of time signals than other locations in the West Indies.

There was a black time ball with a diameter of 0.9 m on the flagstaff of the Naval Commandant's office by 1898. It was dropped at noon, local mean time, and was also included in the 1908 list. It had been replaced by a similar time ball at the National Observatory by 1910, also dropped at noon, local mean time. A semaphore signal at Moro Castle was dropped by hand. The 1911 list included the statement "signals reported unreliable, 1910" and the comment "Local mean time is kept throughout Cuba. Havana mean time has been adopted for meteorological and railway services only." The semaphore signal was replaced by a secondary time ball, also dropped by hand, before 1915. The 1910 comment about reliability was withdrawn. In 1939, the National Observatory and Moro Castle time balls were still operated at noon, local mean time, corresponding to 17 hours 29 minutes 23.5 seconds GMT.

4 South America

The signals in South America are described in alphabetical order, except for the northernmost signals at Guyana and Suriname which are described last.

4.1 Argentina

4.1.1 Port La Plata

A small red time ball with a diameter of 2 feet (0.6 m) at Port La Plata was listed between 1898 and 1911. The time ball was replaced by time lights on 15 March 1911 (Argentina, 1911). These lights were switched on 3 minutes before noon GMT and extinguished at noon GMT without flashing. The time lights at La Plata were still listed in 1939.

4.1.2 Buenos Aires

A large red cylindrical drum with a diameter of 1.8 m and length of 2.4 m was listed for Buenos Aires between 1904 and 1915. It was discontinued on 20 September 1915 and replaced by time lights (Argentina, 1915).

The time lights at Buenos Aires were operated in a different sequence to Port La Plata. They were switched on 2 minutes before 10 pm Cordoba mean time (4 hours 16 minutes 48.2 seconds behind GMT) then flashed on and off in a defined sequence before extinction at 10 pm. The time of the Buenos Aires signal was changed to 9.30 pm Standard time (4 hours behind GMT) after the introduction of new time zones. The signal had been withdrawn by 1931.

5 Brazil

5.1 Rio de Janeiro

Brazil used a red cylindrical drum at Rio de Janeiro from before 1880 to 1920, when it was replaced by a time light. The cylinder was inflated before the signal and had large dimensions: its diameter and drop height were 1.8 m and 2 m respectively.

The time lights at Rio de Janeiro were "on a tower of the old building of the National Observatory". The signal was made by extinction of lights at noon and 9 pm, Standard time, corresponding to 3 pm and midnight GMT. The lights were illuminated at 5 minutes before the signal and then extinguished after 1 minute. This was repeated at 3 minutes before the signal and again at 1 minute. Final extinction was the exact signal.

5.2 Fiscal Island

A combined flag and magnesium flash signal was provided on Fiscal Island in Guanabara Bay from 1921.

The magnesium flash was inside the "Central tower of building on Ilha Fiscal". A blue and white pennant was raised 5 minutes before the signal and hauled down at the same time as the flash at noon Standard time (3 pm GMT). Should the signal fail, the flag would be rehoisted and the combined flag and flash signal would be made at 12.10 pm. This unusual signal was still listed in 1947.

5.3 Bahia

An unusually short-lived time ball service was provided in Brazil at Salvador in the state of Bahia, close to the historic Fort San Marcello do Mar. It was announced in 1906 and withdrawn in the following year. It never appeared in Admiralty lists.

6 Chile

6.1 Valparaiso

6.1.1 The First Time Ball

One of the earliest time balls worldwide was established at Valparaiso in 1843. It was noted by Bartky (1987). The only known notice concerning the time ball is transcribed below (Time Ball at Valparaiso, 1843).

> We perceive by a *Valparaiso Mercury*, which has been forwarded to us, that Mr. Mouat has established a Time Ball at his Observatory in the north-east angle of the Castle of St. Joseph, at Valparaiso, for the purpose of enabling vessels in the Bay to rate their chronometers. The time of its falling is not stated.

The following translation of an article in Spanish shows how the observatory site was acquired by Mouat, who was a watchmaker by profession (History of Valparaiso Observatory).

> … a young immigrant named John Norbert Mouat arrived from Scotland at the port of Valparaíso…. Four years after his arrival, in 1840, our character purchased at a public auction a particular site that is part of the grounds of the old Castillo San José, a Spanish

> fortification that, after suffering irreparable damage in the earthquake of 1822, was decommissioned, its land subdivided and later auctioned. Strategically, Mouat purchased a lot that offered an uninterrupted view of both the bay and the customs plaza and its old dock—now Sotomayor Plaza. On this land, Mouat had his observatory built and, next to it, his family home and operations centre.

Mouat's observatory was announced on 27 February 1843 (An Observatory, 1843). Mouat was a skilled astronomer, who corrected earlier measurements of the longitude of Valparaiso and provided an accurate time service. Later determination of longitude using the electric telegraph confirmed the accuracy of his work (History of Valparaiso Astronomy, 2023).

No details of Mouat's time ball design have been found. "It could have been self-made, based on similar devices operating in his native Great Britain" (Bustamente & Bustamente, 2020).

Mouat's wider work led eventually to his bankruptcy and forced sale of his home and observatory. He died in 1871 and his time service was not continued. It had probably ceased some years previously. Castillo San José is believed to be the oldest building in Valparaiso and is now a National Historical Monument (History of Valparaiso Astronomy, 2023).

6.1.2 Time Balls in Admiralty Lists

Chile used a time gun and a time ball at Valparaiso, but neither was listed in 1880. There was then no Valparaiso time service (Hydrographic Yearbook, 1880). The time ball had been listed by 1898. The 4 feet (1.2 m) diameter ball was located on a mast in front of the Naval School and at an elevation of 90 m above high water. It had a drop height of 7 m and remained at the same location throughout its listing by the Admiralty. The ball was dropped and the gun was fired at noon, local mean time, 4 hours 46 minutes 34 second behind GMT. A time gun at Fort Covadonga was noted in 1908, but the location was changed to Fort Bueras in 1911 and later lists.

The time ball service had to be suspended because of a devastating earthquake in 1906. The 1908 list included the comment "Time ball suspended; gun is still fired, but is unreliable. Buildings were destroyed by an earthquake in 1906". The time ball was included again in the 1911 list and last listed in 1930. The gun was still listed in 1947. It continued to be fired at noon, by then changed to 4 hours behind GMT.

6.1.3 Modern Replica

There is a modern working replica of the time ball at Valparaiso which is shown in Fig. 26.3. It was erected in 2015 and is now located on top of the Naval School building.

Fig. 26.3 Valparaiso time ball replica (https://www.armada.cl/noticias-navales/proyecto-patrimonial-time-ball-de-valparaiso-es-entregado-a-la-armada)

6.2 Coquimbo

A combined cone and gun signal in Chile was first listed for Coquimbo in 1908. A black cone was raised on the mizzen mast of the training ship *Abto* in Coquimbo Bay. A gun was fired when "the ball" dropped at noon, local time. "The signal adopted" was always described as a black cone in Admiralty list, but "additional details" in the same lists included the statement "Ball hoisted half-way …". The same ambiguity has been noted at other locations. The signal was withdrawn permanently in 1916 (Chile, 1916).

7 Peru

A time ball was established at Callao in 1924. It was located on "Mast of the 'Union' near the La Punta Naval School", 85 feet above sea level and had a very large drop height of 75 feet (23 m). It was dropped using a wireless signal from Washington at noon Standard time, 5 hours behind GMT, and was still operational in 1947.

8 Uruguay

8.1 *Montevideo*

Uruguay used a 0.9 m diameter red time ball with a horizontal gilt band at Montevideo from 1912. It had a drop height of 3 m. Unusually, a flag was hoisted 10 minutes before the signal while the ball was raised 5 minutes later, giving two preparatory signals. The flag was dropped at the same time as the ball. The time ball was still listed in 1939.

9 Guyana and Suriname

The northernmost signals in South America were in British Guiana (now Guyana) and Dutch Guiana or Surinam (now Suriname), both on the Atlantic coast.

9.1 *Georgetown, British Guiana*

A red time ball at Demerara (Georgetown) in British Guiana was listed from 1880 to 1939. It was on a flagstaff near the General Post Office in Georgetown and was dropped on Wednesdays and Saturdays only. Its drop height was 4.5 m from 32 m above ground.

9.2 *Paramaribo, Suriname*

Suriname (Surinam) used a variety of visual time signals at Paramaribo. A flag signal was listed by the Admiralty in 1880 and again between 1911 and 1916. A disc signal had been used for part of the intervening period and was listed between 1898 and 1908. The disc or flag were both hoisted on the main yard of the Paramaribo guard ship. The disc was described as black and white with a diameter of 1 foot 4 inches and a length of 3 feet. It was hoisted half-way up 5 minutes before the signal and close-up 3 minutes before the signal, so it was not the rotating disc signal favoured elsewhere in the Netherlands and Dutch colonies. It might have been an unusually shaped object that was dropped like a time ball. It was actually described as a time ball in 1912 "The Netherlands Government has given notice that the time ball at Paramaribo, Dutch Guiana, is unreliable and should not be used for the comparison of chronometers" (Dutch Guiana, 1912). The latitude and longitude of the signal were the same in the 1912 notice as for the disc and flag in the Admiralty lists.

A gun was also listed at Fort Zelandia in 1915 and 1916 but not in 1917 or later.

References

Airy, G. B. (1878). *Letter from G.B. Airy to T Cooke & Sons*, 16 August 1878. RGO 6 618, File 6, Leaf 213, Papers of George Airy, Cambridge University Library.

An Astronomical Observatory in Valparaíso! (1843, February 27). Announcement in *El Mercurio de Valparaíso*.

Argentina. (1911, Match 7). Port La Plata – Buenos Aires – Time signal changed. *Avisos a los Navegantes*, Buenos Aires, also published in *The Nautical Magazine*, 15, 1911.

Argentina. (1915). Plata River – Buenos Aires – Time signal changed. *Avisos a los Navegantes*, No. 5 (130), Buenos Aires, also published in *The Nautical Magazine*, 20, 1916.

Bartky, I. R. (1987). The bygone era of time balls. *Sky and Telescope, 73*, 32–35.

Bustamente, P., & Bustamente, D. (Sociedad Chilena de Historia y Geografía). (2020). *Valparaíso y el Tiempo en el Puerto: El time ball de Juan Mouat* [in Spanish]. http://www.schhg.cl/wp-schhg/wp-content/uploads/2020/04/El-Time-Ball-de-Juan-Mouat.pdf

Chile. (1916). Coquimbo Bay – Coquimbo Harbour – Time signal discontinued. *Avisos a los Navegantes*, No. 50 (208), Valparaiso, also published in *The Nautical Magazine*, 9, 1917.

Cooke, T. (1878). *Letter from T Cooke & Sons, Buckingham Works, York to G.B. Airy*, 14 August 1878. RGO 6 618, File 6, Leaf 212, Papers of George Airy, Cambridge University Library.

Dutch Guiana. (1912). Surinam – Paramaribo – Time ball unreliable. *Bericht aan Zeevarenden*, No. 17 (130), 'sGravenhage, also published in *The Nautical Magazine*, 6, 1912.

History of Valparaiso Astronomy. (2023). [In Spanish]. https://elprimerobservatorio.cl/historia_mas/

Hydrographic Yearbook of the Chilean Navy. (1880), 328.

Spencer Jones, J. (2020). Living on an island. *Astronomy and Geography, 61*, 3.26–3.27.

Time Ball at Valparaiso. (1843). *The Nautical Magazine and Naval Chronicle for 1843*, 768.

Chapter 27
Atlantic Islands

Roger Kinns

Abstract Several Atlantic islands were important as locations on trade routes for sailing ships. Many were occupied by European powers or trading companies. One of the earliest time balls was established on St Helena in 1834 by the British East India Company, shortly before St Helena became a British Colony. It was complemented initially by a time gun and a repeater ball, following early use of rocket signals. The original observatory was closed in 1836 and a new time ball was erected at a lower level, with retention of the repeater ball. They continued operation into the twentieth century. A time ball had existed on Ascension up to at least 1865. Most islands under jurisdiction by Spain and Portugal offered telegraph time signals that could be accessed for chronometer calibration. Light signals were introduced from 1914 on three islands in the Azores. Time balls were introduced in the Cape Verde Islands in the 1920s.

1 Introduction

Several Atlantic islands were important as locations on trade routes for sailing ships. Many had strategic importance and were occupied by European powers or trading companies. The islands of St. Helena and Ascension became British Dependencies. The Azores and Madeira are autonomous regions of Portugal. The Canary Islands are an autonomous region of Spain. The Cape Verde islands were a Portuguese colony until independence as the Republic of Cabo Verde in 1975. The islands of Bermuda and the West Indies have been considered in Chaps. 24 and 26.

Table 27.1 shows visual time signals on Atlantic islands that appeared in Admiralty lists from 1880 to 1947. Figure 27.1 shows their locations, together with the location of the earlier time ball at Ascension. They covered a huge geographical range from the Azores in the North to Ascension in the South Atlantic.

R. Kinns (✉)
The Maudslay Society, Pembroke College, Cambridge, England

R. Kinns (ed.), *The World of Visual Time Signals for Mariners*, Historical & Cultural Astronomy, https://doi.org/10.1007/978-3-031-57334-7_27

Table 27.1 Signals in the Atlantic Islands

Area	Location	1880	1898	1904	1911	1915	1919	1922	1928	1934	1939	1947
St. Helena	Jamestown: Time Office	TB	TB	TB								
	Jamestown: Ladder Hill	TB	TB	TB								
Azores	Horta						Light	Light	Light	Light	Light	
	Ponta Delgada					Light	Light	Light	Light	Light	Light	
	Terceira								Light	Light	Light	
Cape Verde Islands	St Vincent Port Office							Flag	TB+ diamond	TB+ diamond	TB+ diamond	
	St Vincent Salvos battery								Gun	Gun	Gun	
	St Vincent M. Videa Fort							TB (black)	TB (black)	TB (black)	TB (black)	
Signal type totals		1880	1898	1904	1911	1915	1919	1922	1928	1934	1939	1947
Time ball, including collapsible		2	2	2	0	0	0	1	2	2	2	0
Gun, including powder flash		0	0	0	0	0	0	0	1	1	1	0
Lights		0	0	0	0	1	2	2	3	3	3	0
Other: disc, drum, flag, etc.		0	0	0	0	0	0	1	0	0	0	0

Fig. 27.1 Atlantic Island Signals (Google Maps)

2 St. Helena

2.1 A Brief History of St Helena

The island of St Helena is believed to have been discovered by the Portuguese in May 1502. The island was uninhabited, but with an abundance of trees and fresh water. They imported livestock, fruit trees and vegetables, and built a chapel and one or two houses, but did not establish a permanent settlement. The island was an important rendezvous point and source of food for ships returning from Asia to Europe. Sick mariners were frequently left on the island to recover before taking passage on the next ship to call at the island. The following historical summary is derived primarily from Saint Helena Island Info (2023).

Its existence was known only to the Portuguese until it was visited by Capt. Thomas Cavendish in 1588 on his return from a voyage around the world. The Dutch are believed to have occupied St Helena for some years before 1649 when the British East India Company ordered all of its homeward-bound vessels to wait for one another at St Helena. The Company was granted a charter to govern the island by Oliver Cromwell in 1657. A fleet commanded by Captain John Dutton arrived at St Helena in 1659 and took control of the island. Dutton became the first governor from 1659 to 1661. A fort was completed within a month and houses were built. The Company lost control of St Helena for 5 months in 1673 when it was occupied temporarily by the Dutch. It flew the Union Flag from 1687, having previously flown the Company flag.

The remoteness of St. Helena made it attractive to European powers as a place of permanent exile for the Emperor Napoleon and he was confined on the island from October 1815 until his death in May 1821. During that period the island was under the jurisdiction of the British Crown. The East India Company resumed control in 1821. A provision of the India Act of 1833 transferred control of St Helena from the East India Company to the Crown with effect from 22 April 1834. The last Company governor, Governor Charles Dallas, stayed in post after the India Act came into force, not leaving until the first Crown governor, General George Middleton, arrived on 24[t] February 1836.

Subsequent administrative cost-cutting triggered a long-term population decline as well as a reduction in observatory facilities. The latter half of the nineteenth century saw the advent of steamships not reliant on trade winds and diversion of trade from routes via the Cape of Good Hope to routes via the Suez Canal. The number of ships calling at the island fell from 1100 in 1855 to only 288 in 1889.

The extent of the population decline in St Helena after the opening of the Suez Canal was well described in an article published in the Inverness Courier (St Helena, 1900).

> Passengers to the Cape who have never taken the trouble to go up to the high ground of St Helena can have little idea (says a correspondent) of what a beautiful combination of semi-tropical and temperate growth is to be found there. The hills range up to about 2000 feet above the sea. There is no harbour, but only an open roadstead off Jamestown, sheltered

> from the south-east trade wind, which blows ninety-nine days out of the hundred. At one time, the population approached 20,000, and it managed to get along somehow by supplying provisions to homeward bound vessels returning round the Cape. But the opening of the Suez Canal led to fewer vessels calling there, and those that anchored off the island only came there to correct the Greenwich time. The salvation of the place has been the development of the South African goldfields, for though the population is now under 4000, there is no work for them except to take off watercress and cabbages to passing ships. A feature of the island is "Jacob's Ladder". A wooden staircase of 699 steps, with an average slope 39 degrees to the vertical.

Figure 27.2 shows a historic map of St Helena. Ladder Hill is located near the capital Jamestown on the NW coast of the island. Figure 27.3 shows Jacob's Ladder from the Wharf. The old Ladder Hill Observatory can be seen at the centre top of the hill.

2.2 *Ladder Hill Observatory*

The Ladder Hill observatory was commissioned by Governor Walker in the 1820s. Its foundation stone is dated 13 September 1826. It was directed by Lieutenant Manuel Johnson (1805–1859) of the St Helena Artillery, whose work in cataloguing stars of the Southern Hemisphere is still highly regarded (Johnson, 1835). The observations were made between November 1829 and April 1833. He received the Gold Medal of the Royal Astronomical Society for that work. His major

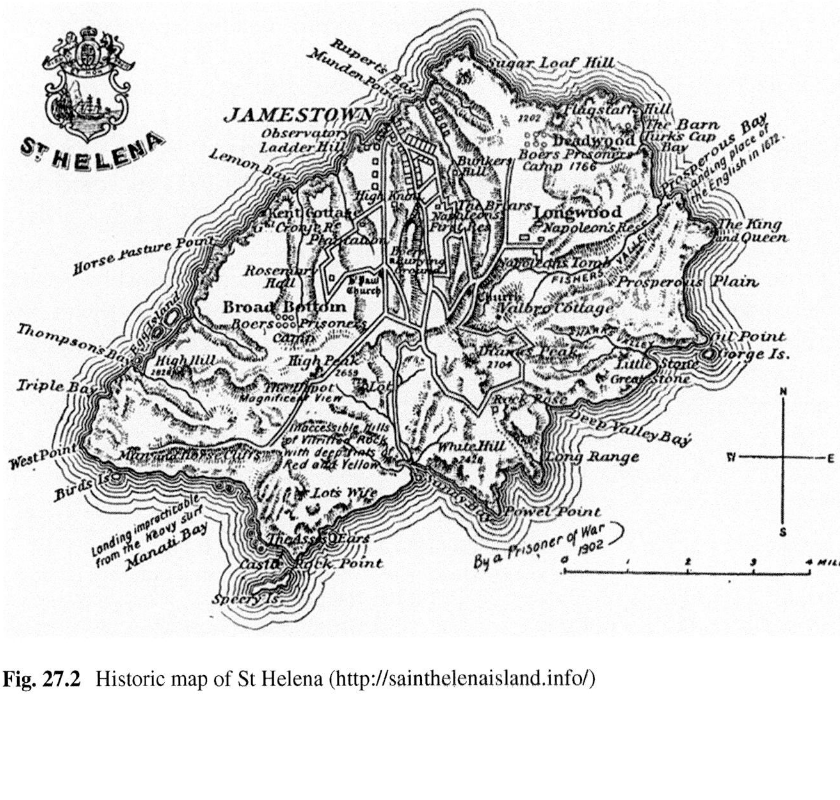

Fig. 27.2 Historic map of St Helena (http://sainthelenaisland.info/)

Fig. 27.3 Jacob's Ladder, Jamestown, St Helena (http://sainthelenaisland.info/)

contributions to astronomy in St Helena have been described by Warner (1981). He left St Helena in 1833 and became a distinguished astronomer at Oxford.

The observatory was closed on 29 February 1836 "due to its uselessness and immense annual cost of £300, the Crown Commissioners reporting that they had been unable to learn its establishment had been attended with any important result to science". That was a remarkable assertion, given Johnson's achievements. Most of the instruments were sent to Canada, although the clocks were retained. The building was then used as a mess hall for the fort and the Ladder Hill Observatory was never re-established.

2.3 *Rocket Signals*

Navigation guides for the South Atlantic included the following note about rockets in the entry relating to St Helena (Findlay, 1867, 1883):

> The practice of discharging rockets for rating the chronometers of vessels touching here has also been discontinued and a time ball has been substituted. This ball was originally hoisted at the Observatory …

The firing of rockets to indicate time had been proposed at the Cape of Good Hope in about 1826 but the plan appears to have been rejected there (Editorial, 1835, 658).

A nineteenth century photograph of the observatory building is shown in Fig. 27.4.

Fig. 27.4 The disused Ladder Hill Observatory, nineteenth century (http://sainthelenaisland.info/)

2.4 The First Time Ball at St Helena

An editorial concerning the St Helena time ball has been noted previously (Editorial, 1835). The parts of that editorial relating specifically to St Helena are reproduced below.

> We shall now extract from the remarks of H.M.S Thalia, Captain R.Wauchope, his account of the St. Helena Time-ball, on the arrival of that ship there in December last [*1834*]:–"On our arrival at St. Helena, after an eighteen days' passage from Prince's Island, we found my chronometer-signal established there. The ball drops at mean noon, St. Helena time, for the benefit of the inhabitants; and at one P.M. mean time at Greenwich, for the advantage of the shipping. The ball is hoisted half-mast high five minutes before the time, and at one minute before to the mast-head.
>
> It gives universal satisfaction both to the inhabitants and to all ships touching at the island; so much so, that I understood that several ships touched there in consequence of the signal, that otherwise would not have done so. On board the Thalia, we found the observations come out every day accurately to a second, which is much more correct than any observations that can be made by means of the sextant and artificial horizon."
>
> We shall not stop to question this latter assertion, our object being to describe the apparatus, which we may briefly do as follows:– …

The article went on describe the apparatus, using the diagram reproduced in Fig. 27.5. The description is restructured as Table 27.2, using the original wording.

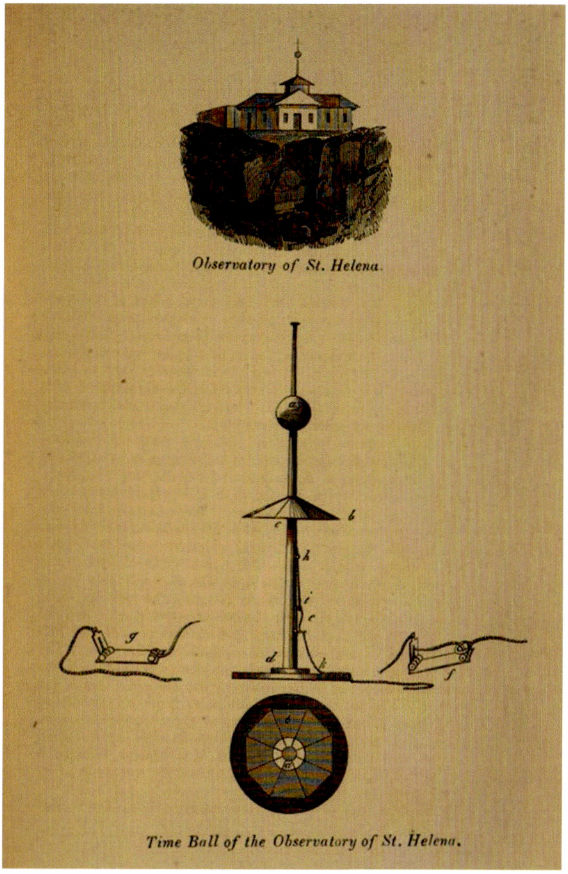

Fig. 27.5 The St Helena time ball. (From The Nautical Magazine, 1835)

Table 27.2 Features of the 1834 time ball apparatus at St. Helena

Figure identification	Description
a	The ball, half-mast high: a globe of canvas, 22 lbs weight, and 2 feet 8 inches diameter.
b and c	The covering of the building in which the mast is fixed
c–d	Imaginary, since the rope only goes from the roof to the floor of the cupola
e	The stopper fixed to the floor, by which the ball is let fall.
f and g	The figures f and g represent the stopper on a larger scale; the former as it is closed, ready for letting the ball fall; the latter, as opened after the ball is down.
h	The whole mast-ring, or the eye in the rope, through which the tongue e of the stopper is passed, when the ball is hoisted to the masthead.
i	Half mast-ring, with 8 lbs of lead attached, to check the fall of the ball when it reaches half mast high..
k	10 lbs of lead, to provide a second check when it reaches within 2 feet of the dome of the cupola, to prevent staving of the ball, or injuring of the dome.

From The Nautical Magazine (1835, 659)

The building and cupola are recognisable from the photographs in Figs. 27.3 and 27.4. The diagram suggests that the drop height was about 11 feet, given a ball diameter of 2 feet 8 inches (0.8 m). It is difficult to understand from the information in Table 27.2 how the apparatus actually worked. The ball appears to slide down a tapered mast, rather than being suspended from a gaff, but no ropes for hoisting it are shown. The ball had a mass of 22 lbs and its descent appears to have been controlled using lead weights attached to ropes. These were stated to have masses of 8 and 10 lbs. The smaller weight was lifted when the ball had fallen to the half-mast position, presumably freely under gravity, while the larger mass was lifted when the ball was only 2 feet from the cupola. The combined mass of the two lead weights was less than that of the ball, so would not even have prevented its continuing downward acceleration in the absence of friction. The arrangement clearly worked in practice, so there is great deal missing from the description.

The description of the apparatus was followed by reproduction of a notice issued by the Observatory at St. Helena that was dated 21 January 1834. It included the observation that the astronomer would allow for the delay of a fraction of a second in releasing the ball and that this applied both to the Observatory time ball and the repeater ball, which also appears to have been at Ladder Hill. A gun would also be fired. Although firing was not automatic, the gun responded to the observatory signal rather than being fired at a nominal time which might be in error by a minute or more (Editorial, 1835).

> To prevent mistakes, a *White Ball*, hoisted upon a Staff over the Observatory, will denote the time, agreeably to the following instructions:–
>
> The ball will be hoisted half mast at *five* minutes, and close up at *two* minutes before *twelve o'clock*.
>
> At the instant of the *Mean Time*, at noon of St. Helena, the ball will drop from the top of the Staff, when the gun will be fired at High Knoll.
>
> The signal will be repeated at *one o'clock*, at the instant of *Greenwich mean time*, for the benefit of the shipping.
>
> A ship wishing to correct her chronometers, and *arriving* after *one* P.M., and not likely to remain the twenty-four hours, may hoist the "Blue Peter" at the main-topgallant mast-head, when the same method will be adopted, at the next ensuing hour after the signal. Foreign ships to substitute their ensign for the "Blue Peter."
>
> Should there be any uncertainty, and the ship wishes to have the signal repeated, she will dip the flag, and rehoist it, on observing the ball half-mast. The ball will again drop, at the ensuing quarter of the last hour.
>
> Ships concealed from a view of the Observatory, will attend to the Repeating Ball at Ladder Hill, and in neither case is any allowance to be made for loss of time, since the astronomer will make the calculation of the few tenths required.

It was a controversial editorial, because it led to a protest from Wauchope that his invention of visual time signals regulated by astronomical observations had not been recognised properly (Wauchope, 1836). That was accepted by the editors of *The Nautical Magazine*.

2.4.1 The Time Gun

The time gun arrangement at St Helena was unusual during the 1830s. Firing when the time ball was released made the flash of the gun an accurate time signal. At other places, time guns were fired by the Army at a nominal time and the precise time of firing was noted by the local observatory (Madras, 1841). This time would only be available to ships on the following day, if at all. It was never a popular arrangement. By the 1860s, time guns could be fired electrically from an observatory and became precision signals.

2.5 Continuation After 1836

A letter published in The Nautical Magazine confirmed that the time ball service continued after closure of the Ladder Hill Observatory (Liddell, 1837). The Observatory time ball had been removed and the original repeater ball was not mentioned, but a repeater time ball on Ladder Hill was included in later Admiralty notices. The principal time ball was now located in Jamestown itself. There had also been changes to unhelpful regulations concerning ship movements.

> Perhaps it is not generally known to your maritime readers, interested in our eastern trade, that since the island of St. Helena was transferred from the Company's to the King's government, the port regulation, obliging ships before anchoring to send their boats to Bank's Battery, and others equally absurd, have been rescinded.
>
> On our homeward voyage last year, I wrote to the governor, General Middlemore, from Ascension, stating how unnecessary and annoying it was to the shipping; and I was glad to find, on our visit to the island in March last that this regulation, with others, preventing ships entering before sunrise, or departing after sunset, had been abolished many months before.
>
> It gave me also great pleasure to observe, that, although General Middlemore had brought out instructions to shut up the beautiful little observatory, and send home the valuable instruments so munificently provided by the East India Company, the admirable system of dropping a ball for regulating chronometers was still kept up as efficiently as ever. The ball has been removed from the hill to the town, and is now in charge of the master-attendant, Mr. Gulliver, late master of H.M.S. Thalia, whose attention to it is unremitting. It is dropped at two stated times every day, and oftener, if requested by any commander.

Liddell also encouraged the British government to provide time balls at Liverpool, Deal, Plymouth and Falmouth in England. It was many years before they were actually erected.

The photograph in Fig. 27.6 shows the lower level time ball on a tall mast. It was taken on Queen Victoria's birthday in 1866. By that time, the Ladder Hill Observatory had been long closed and this was the primary signal.

Fig. 27.6 The St Helena time ball in 1866. (Photo by John C. Lilley, St Helena Museum)

2.6 Admiralty List Entries for St Helena

The repeater time ball at the top of Ladder Hill was released from a yard arm on a flagstaff and had a much larger drop height than the primary ball. This may have been the original repeater time ball arrangement in 1834. The Time Office ball and the repeater time ball had been discontinued by 1908.

The Admiralty lists note that "Ladder Hill Observatory" time was used in St Helena. The observatory had ceased operation in 1836, but its latitude and longitude are likely to have been measured by Johnson when extensive instruments were available. Astronomical clocks had been retained after observatory closure and were probably regulated using a sextant and a nautical almanac. It has been shown recently that the longitude in use up to 1990 was in error by 732 m. The GPS latitude and longitude of the Observatory site are approximately 15° 55′ 30″ S, 5° 43′ 07″ W.

There was a fee of 5 shillings for provision of the time ball service in April 1859 (Russell, 1858, 343). This fee was noted by Findlay in 1867 and 1883 and was still extant in 1888 (notice in St. Helena Museum). It appears to have been accepted without complaint.

Fig. 27.7 Part of an 1840 drawing by Durand Brager, showing a flagstaff on Ladder Hill (St Helena Museum)

There are no known photographs of the repeater time ball on Ladder Hill, but a drawing of the rear of the observatory was made by Durand Brager, the artist who was present for the exhumation of Napoleon in 1840. Figure 27.7 shows the central part of the drawing, including a flagstaff which may have been used for dropping the repeater ball listed in Table 27.1.

3 Ascension

3.1 A Brief History of Ascension

Like St Helena, Ascension Island was discovered as an uninhabited island by the Portuguese in about 1502. There was then almost no vegetation on an extinct volcano that had risen from the Atlantic, about 1300 km from St Helena, 1600 km from Africa and 2200 km from South America. Its area is only about 90 km^2, with EW and NS dimensions of about 14 and 11 km. The extraordinary history and administration of Ascension are described in detail by Hart-Davis (2016).

Ascension became important as a victualling centre for ships on long voyages. Its flora and fauna were developed by importing species from many different countries. It was first colonised by the British in 1815, in order to discourage any attempt to rescue Napoleon from St Helena. After the death of Napoleon in 1821, it became an important centre for interception of slave ships sailing from the west coast of

Africa to South America. From 1815 to 1922, Ascension was run as a naval shore establishment with a detachment of marines. Its Commanding Officers were in either the Royal Navy or the Royal Marines (Hart-Davis, 2016, 233–234). Ascension became a dependency of the Crown Colony of St Helena in 1922.

Ascension was never strongly defended and its garrison was only a few hundred strong. The island was close to being abandoned in the 1860s, when the Admiralty found it hard to justify continued expenditure on the garrison and its facilities. It again became important in the twentieth century for communications, first as a telegraph station in 1901, then for radio communications during WWI and later for tracking of vehicles in Space. Its airfield was developed by the United States during WWII and further extended later. It continues to be an important military facility.

The islands of Tristan da Cunha and Gough Island were administered as dependencies of St Helena. No evidence has been found of visual time signals for mariners on those islands.

3.2 *The Ascension Time Ball*

Hart-Davis did not mention the existence of a time ball or associated observatory facilities on Ascension. He noted that David Gill (later Director of the Cape Observatory) and his wife had stayed there for several months in 1877 to observe the Opposition of Mars, but they brought their own equipment and set up a temporary observatory. Gill's observations were published in *Monthly Notices of the Royal Astronomical Society* (Gill, 1879).

Facilities for time determination in the 1860s were probably limited to sextants, chronometers and almanacs carried by naval vessels stationed there. Despite those limitations, there was a time ball service.

A notice concerning a time ball at Ascension was published in *The Nautical Magazine* (Time Signal Ball, 1860) and is transcribed below.

> *Time Signal Ball at Ascension.*—In order that vessels calling at Ascension may readily find the errors and rates of their chronometers, a Time Ball is dropped daily (Sundays excepted) from a flagstaff at the Master's cottage, precisely at one o'clock of Greenwich mean time. The Master's cottage is to the southward of Hayes Hill, and is the only one near it bearing a flagstaff. The Ball when hoisted ready to be dropped is at a height of ninety feet above the level of the sea, and may be readily seen from the anchorage. Considering the longitude of the flagstaff to be 14° 26′ 30″ W. or, in time, 0h. 57m. 42s; the ball falls at 2m. 18s. after noon of Ascension mean time, the first instant of falling being that which is to be noted by the chronometer. Vessels not intending to remain for the usual signal hour may have the ball dropped at any convenient period of the day for Greenwich time by applying for it to be done at the cottage.

Another, much shorter, notice was published in 1865, but the drop times had changed (The Time Ball, 1865).

> The Time Ball is now dropped from a staff on Hayes Hill at 8h. a.m. and 1h. p.m. mean time at Ascension. The corresponding mean times at Greenwich will be 8h. 57m. 42s. and 1h.

57m. 42s. respectively. Should the ball fall at the wrong time, the negative pendant will be hoisted, and the ball will be again dropped at ten minutes after the times stated.

In 1860, the time ball had been dropped at 1 pm Greenwich mean time. By 1865, the ball was dropped twice daily, first at 8 am and then at 1 pm Ascension mean time. Both notices pointed out its proximity to Hayes Hill. Other records of the time ball have not been found. It probably had a short life, given continuing budgetary pressures (Hart-Davis, 2016).

Figure 27.8 shows detail from a map of Ascension published in 1876. Hayes Hill is very close to George Town on the western side of the island. The GPS latitude and longitude of Hayes Hill are 7° 55′ 57″ S; 14° 24′ 52″ W. The longitude stated in 1860 and 1865 was therefore 1′ 38″ too far west. This corresponds to an error of about 2.5 km.

There were no entries for Ascension in Admiralty lists from 1880, as indicated in Table 27.1.

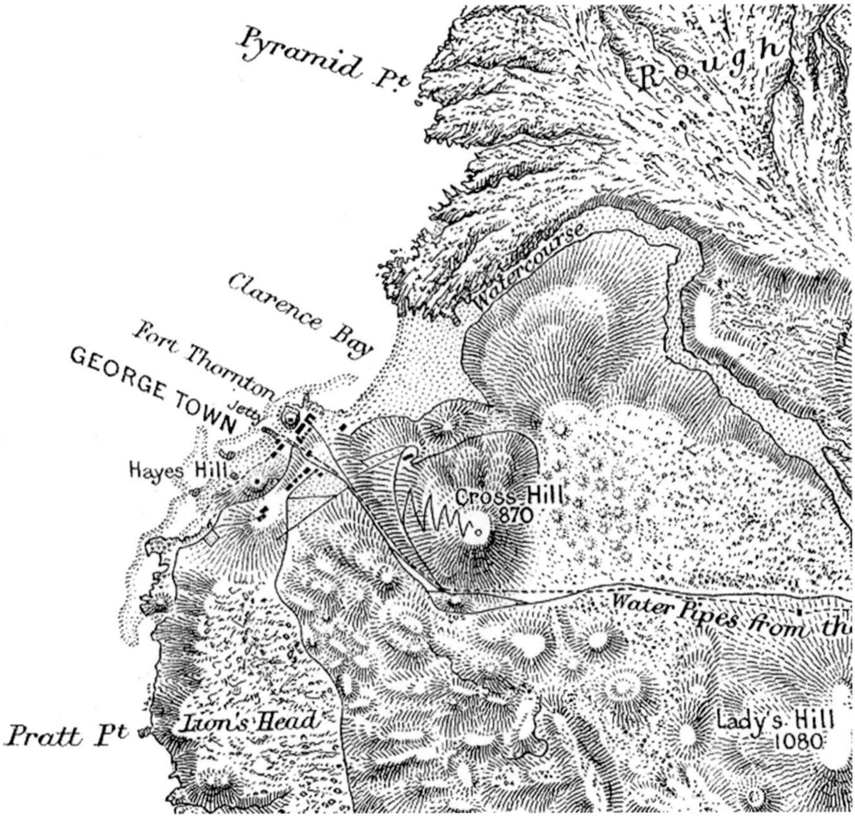

Fig. 27.8 Ascension NW coast. (Copyright, reproduced with permission from Science History Images ID: 2BE1090)

4 Cape Verde Islands

A telegraph signal at the Telegraph Office at Port Grande on São Vicente (St. Vincent) Island was first listed by the Admiralty in 1911. Visual signals were first listed in 1922. A telegraph signal from Greenwich was still noted, together with a blue flag at the "Captain of the port's office" and a time ball at the "fort on Monte Videa". New signals were introduced at Porto Grande in February 1922, after the Admiralty list had been published (Notices, 1922, 510). These were a time ball, a repeater signal with a diamond shape and a gun. All were at 10 am local time, corresponding to noon at Greenwich. The gun was fired on 3 days only. The complete notice is transcribed below.

> CAPE VERDE ISLANDS—São Vicente Island—Porto Grande Visual time signal—Information.—Since February 9, 1922, the time signal at Porto Grande has been made in the following manner from the signal mast of the meteorological station located in the building of the captain of the port:
>
> The ball (sphere) is hoisted at half mast at 9h 55m; mastheaded at 9h 58m, and dropped at 10h a. m., the time corresponding to that of the 30° meridian of longitude west of Greenwich.
>
> This signal is repeated from the semaphore mast at the fort with a black diamond-shape. The dropping of the diamond-shape is accompanied by the firing of a gun at the saluting battery on Mondays, Wednesdays, and Fridays. Approximate position of Porto Grande: latitude 16° 53′ N., longitude 24° 59′ W.

Subsequent Admiralty lists included the comment that the sphere was hauled down, rather than dropped. The sphere is shown as a time ball in Table 27.1.

The time ball and flagstaff signals were still in use in 1942, but the time gun had been discontinued (Sailing Directions, 1942, 273). No signal was listed in 1947.

5 Canary Islands and Madeira

There were no entries in Admiralty lists for either Madeira or the Canary Islands before 1908. Telegraph signals had become available to mariners at Madeira and Tenerife by 1911, but these could only be accessed by going ashore and are not shown in Table 27.1.

6 Azores

Clocks were available for chronometer calibration in the Azores at Punta Delgada by 1904 and at Horta by 1908, regulated from Lisbon and Hamburg respectively. These signals could not be seen from ships in harbour so are not included in Table 27.1.

Table 27.3 Time lights in the Azores

Location	Latitude Longitude	Signal	Additional details
Horta (1915): Tower of Meteorological observatory	38° 31′ 35″ N 28° 37′ 45″ W	Electric lights	The lights are switched on at 20 hours 57 minutes 00 second and extinguished at 20 hours 58 minutes 00 second standard time. After an interval of 1 minute the signal is repeated, the lights being extinguished at 21 hours 00 minutes 00 second standard mean time. Signals made daily when electric current is available.
Ponta Delgada (1914): Meteorological observatory	37° 44′ 20″ N 25° 40′ 30″ W	Electric lights	The lights are switched on at 14 hours 57 minutes 00 second and extinguished at 14 hours 58 minutes 00 second standard time. After an interval of 1 minute the signal is repeated, the lights being extinguished at 15 hours 00 minute 00 second standard time.
Terceira (1923): Meteorological observatory, Angra do Heroismo	38° 39′ 06″ N 27° 13′ 47″ W	Electric light	The light is switched on at 20 hours 57 minutes 00 second and extinguished at 20 hours 58 minutes 00 second standard time. After an interval of 1 minute the signal is repeated, the lights being extinguished at 21 hours 00 minute 00 second standard time.

From Admiralty list (1927)

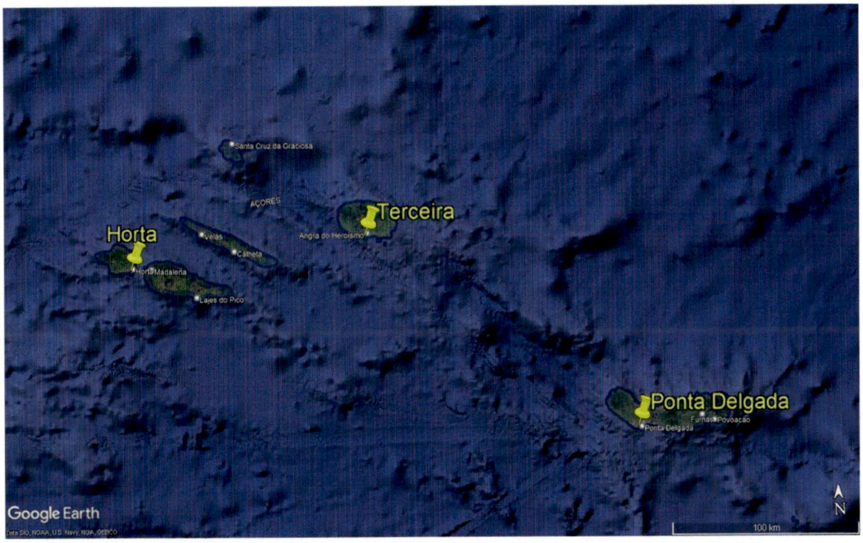

Fig. 27.9 Time lights in the Azores (Google Maps)

6.1 Azores Time Lights

Time lights were introduced at three locations in the Azores. The first was at Ponta Delgada in 1914, followed by another at Horta in 1915 and a third at Terceira in 1923. All were still listed in 1939, but not in 1947. Their descriptions in the 1927 Admiralty list are shown in Table 27.3. The Ponta Delgada signal was made 6 hours before the other two. Multiple lights were used at Horta and Ponta Delgada, but only a single light at Terceira. Figure 27.9 shows their locations.

Acknowledgements The research described in this paper makes extensive use of the excellent website *Saint Helena Island Info*, which includes a wealth of information about St Helena and its history from first discovery to the present day. Several illustrations have been taken from that website. I am grateful to John Turner, the website manager and to Nick Thorpe at the St Helena Museum in Jamestown, for additional images.

References

Editorial. (1835). The time-ball of St. Helena. *The Nautical Magazine, 4*, 658–660.

Findlay, A. G. (1867). *Sailing guide for the Ethiopic or South Atlantic Ocean* (5th ed.). Holmes Laurie.

Findlay, A. G. (1883). *Sailing guide for the Ethiopic or South Atlantic Ocean* (9th ed.). Holmes Laurie.

Gill, D. (1879). On the value of the solar parallax derived from observations of Mars made at Ascension Island during the opposition of 1877. *Monthly Notices of the Royal Astronomical Society, 39*(8), 434–437.

Hart-Davis, D. (2016). *Ascension, the story of a South Atlantic island* (2nd ed.). Merlin Unwin Books Ltd.

Johnson, M. J. (1835). *A catalogue of 606 principal fixed stars in the Southern Hemisphere*. Printed by the Royal Astronomical Society at the Expense of the Honourable East India Company.

Liddell, J. (1837). St. Helena time ball. Letter to the Editor dated 1 May 1837. *The Nautical Magazine, 1837*, 354–355.

Madras. (1841). Entry in *The Nautical Magazine and Naval Chronicle for 1841*, 363.

Notices to Mariners of 1922. (1922). Nos. 1 to 26. Washington Government Printing Office.

Russell, D. S. (Master of *Doctor Franklin*). (1858). Log of the whaling barque *Doctor Franklin*, 8 Nov 1856–1 Aug 1859. New Bedford Whaling Museum Catalogue Number: KWM 1033.

Sailing Directions for the West Coasts of Spain, Portugal and North-West Africa and off-lying islands (5th ed.). (1942). H. O. No. 134, United States Government Printing Office.

Saint Helena Island Info. (2023). http://sainthelenaisland.info/

St Helena. (1900, March 16). Article published in the *Inverness Courier*, 6.

The Time Ball. (1865). Nautical Notices in *The Nautical Magazine*, 161.

Time Signal Ball at Ascension. (1860). Nautical Notices in *The Nautical Magazine*, 499.

Warner, B. (1981). Manuel Johnson and the St. Helena observatory. *Vistas in Astronomy, 25*, 383–409.

Wauchope, R. (1836). Letter to the Editor of *The Nautical Magazine*, 2 April 1836, with supporting testimonials concerning time ball invention. *Nautical Magazine, 5*, 460–464.

Chapter 28
Concluding Remarks

Roger Kinns

Abstract Shore-based visual signals were the means used to check the accuracy of the time indicated by marine chronometers. The dropping time ball was the most popular visual signal for over a century, but there is still uncertainty about its original selection and key documents concerning its development have yet to be found. Admiralty lists of signals and postcard images show the recognised locations of signals for mariners, but there are errors of description and uncertainties about dates of introduction and withdrawal that need to be clarified. There also uncertainties about signal types, some arising from the use of "time-ball" as a generic term for time signals, while the specifications and suppliers of apparatus have yet to be identified in many cases.

Archives in different countries may reward close investigation and there is much scope for international historical research. This book provides an initial framework, indicating the number and variety of visual time signals that had existed in the nineteenth and twentieth centuries.

The marine chronometer was developed to allow accurate estimation of longitude at sea. Shore-based visual signals were the means used to check the accuracy of the indicated time. Two principal aims of this book have been to develop an understanding of how visual time signals for mariners came to exist and to provide a fresh perspective concerning their evolution up to the Second World War and their replacement by wireless signals. Another aim was to identify uncertainties that might be resolved by further research, noted in this final chapter.

The dropping time ball was the most popular visual signal for over a century after its first formal trial in 1829, but its history was not a simple progression. The 1833 installation at Greenwich showed how accuracy requirements could be met

R. Kinns (✉)
The Maudslay Society, Pembroke College, Cambridge, England

R. Kinns (ed.), *The World of Visual Time Signals for Mariners*, Historical & Cultural Astronomy, https://doi.org/10.1007/978-3-031-57334-7_28

using precise astronomical observations in combination with a reliable mechanical apparatus that gave a clear signal. Invention of the electric telegraph reinforced and extended application of visual time signals. The combination of astronomy, horology and telegraphy with mechanical and electrical engineering was an early precursor to what we now know as mechatronics.

1 Visual Time Signal Invention

It has been stated often that Robert Wauchope was the inventor of the time ball. He was undoubtedly the first to propose land-based signals under control by astronomical observations for rating of marine chronometers. He argued that it was essential for chronometers to be rated in situ, without the need to remove them to a shore location. As a Royal Navy Captain serving at the Cape of Good Hope he suggested a possible arrangement using a shutter signal in 1818 in his "remark book". Frustrated by inaction, he wrote again to the British Admiralty in 1824 suggesting that signals should be implemented at important ports. Newly discovered evidence shows that he was then proposing the hauling down of a flag at mean time noon as the time signal and persisted with that suggestion until at least 1827. A trial of his plan was initiated at Portsmouth in 1829, but it had been decided to exchange the flag at noon for a ball dropped at 1 pm. Careful examination of the wording in Wauchope's correspondence and particularly in a paper published in Edinburgh in 1830 indicates that the time ball idea is likely to have been introduced by the Admiralty. Wauchope never claimed the time ball as his invention, saying in his 1830 paper that "The accompanying signal by means of balls, is that which has been adopted at Portsmouth, and which appears to have answered perfectly". It was an improved signal which promised higher accuracy and better visibility than a flag and it was clearly based on Wauchope's plan for a controlled signal. Wauchope then adopted the time ball as his preferred signal, added his own design suggestions, and promoted it vigorously to the Admiralty and to authorities overseas. His persistence and continued belief in the importance of land-based signals played a vital role in provision of signals worldwide.

Wauchope published his design ideas in 1830, which has often been taken as proof that he was inventor of the time ball. There are, however, obvious differences between the notice of the trial published by the Admiralty and Wauchope's suggested design in 1830. The location of the time ball differs from the location which has been inferred previously from the 1830 paper. He proposed that the ball should be dropped slightly before the designated time and that the time when the ball had dropped by one diameter should be the signal. It was never implemented at Greenwich or elsewhere.

The 1833 design for Greenwich by Maudslay, Sons & Field for the Astronomer Royal, has always been regarded as a superior arrangement. Details of the Greenwich arrangement were published in 1835, establishing its reputation as a well-designed system that remained in service, largely unaltered, until 1915. Its subsequent development has been described. The service was discontinued during two world wars but the time ball is again in daily use.

Although Wauchope's correspondence included endorsements by senior naval officers that he indeed suggested land-based signals in 1818, key documents have never been found. The order from the Admiralty to the Superintendent at Portsmouth Dockyard to develop the trial apparatus is missing, as is a report on the trial which is known to have been prepared for the Lords of the Admiralty. Evidence of how long the first time ball remained in operation after the initial trial is also missing.

Perhaps we will never know who actually suggested the time ball idea. Wauchope undoubtedly deserves the credit for invention of the land-based signal and the ideas that underpinned its application.

2 Development of Other Visual Signals

Time ball familiarity is likely to have inhibited development of other devices that might have offered improvements in efficiency and reduced costs, although collapsible objects, rotating discs and semaphore signals were sometimes used. The hauling down of a flag was used as the time signal at several locations outside Europe, following Wauchope's original 1824 suggestion to the Admiralty that led eventually to the time ball.

Time discs, which rotated about a horizontal axis to simulate a disappearing ball, had been invented by Professor Frederik Kaiser, Director of Leiden Observatory. They were first used in the Netherlands in 1854 and were the preferred signal there and in Indonesia until they were rendered obsolete by wireless signals. The signals were not widely recognised outside the Netherlands and international notices about their existence, often as replacements for unsatisfactory time balls, were sometimes delayed for many years.

Time guns had long been used as approximate time signals but were only suitable for direct rating of chronometers if they were fired under control by an astronomical observatory. It was emphasised that the flash of the gun was the precise signal and warnings about the slow and variable speed of sound propagation were promulgated. Some guns were fired manually in response to the drop of a time ball. Other techniques included the firing of rockets and guns, with times of firing noted by an observatory for subsequent access by mariners. Much higher precision could be achieved after development of the electric telegraph.

A pioneering time light under observatory control had been introduced at the Cape of Good Hope in the 1820s by the Reverend Fearon Fallows, using an oil lamp behind a shutter that was closed as the signal. Electric time lights were introduced in the twentieth century, almost 90 years after the first lamp signal, when reliable electric power systems had been developed. Extinction of lights then replaced mechanical signals at many locations worldwide. They have been almost entirely ignored, but they played an important role in the history of visual time signals.

Electric time lights were first noted as supplementary signals for Hamburg in 1908 and 1911 in Admiralty lists of visual time signals, before being listed as principal signals there in 1915. They were favoured by Portugal and their earliest known introduction as principal signals was in 1909 at Lorenzo Marques in Mozambique. They were used widely in northern Europe, China and Japan. A unique arrangement of coloured time lights was introduced in New Zealand in 1912. Some replaced time balls while others offered a complementary or alternative service. Electric time lights were never listed for the British Isles or North America.

3 Early Time Balls

The locations of many early time balls before 1845 have been identified, but their detailed design and their accuracy and longevity have been difficult to establish, apart from the well-publicised arrangement at Greenwich. Many early records appear to have been lost and some anecdotes of doubtful veracity have gained currency, including a crude time ball arrangement for Table Bay in South Africa.

The first trial arrangement at Portsmouth had a limited early life that has still to be identified. It was not mentioned in considerations of the first Liverpool time ball that was erected in 1845 or in considerations of a time ball for Portsmouth after 1855.

The first time signal in Mauritius preceded the Greenwich time ball by 6 months, having been announced in April 1833 by John Augustus Lloyd, who had been appointed as Surveyor General and Civil Engineer in 1831. It was an extraordinary arrangement, using a stationary black ball or disc behind a shutter in a white-painted building. Its design was clarified in a paper prepared by Sir John Herschel when he was at the Cape of Good Hope, using information sent to him by Lloyd. It was presented on his behalf to the *Royal Astronomical Society* in 1836.

The first St Helena time ball, established in January 1834 at the Ladder Hill observatory, appears to have used Wauchope's ideas for a trigger arrangement, but the description of the apparatus published in 1835 is difficult to understand. Nevertheless, it appears to have operated satisfactorily in its short working life. This time ball was replaced by another nearer sea level using a different apparatus when the observatory was closed in 1836. The 1836 time ball at the Cape of Good Hope also used Wauchope's ideas for releasing the ball.

Evidence has been found in records of the Great Trigonometrical Survey of India, led by George Everest, of a time ball and repeater arrangement at Calcutta (now Kolkata) in 1835. There are mentions of an early time ball at Madras (now Chennai), but direct evidence has been limited to a time gun at Madras in 1841 and to a later time ball in the 1850s that was a private enterprise initiative. It appears at the moment that plans for an earlier time ball service may not have been implemented because of other priorities in Madras. There was, however, an 1840 time ball in Bombay (now Mumbai).

A time ball at Batavia (now Jakarta) was erected by the Dutch, and another was erected at the Cape Coast Castle (Ghana) in 1839. These all preceded the 1845

Liverpool time ball, which was the first to be erected in the British Isles after Greenwich. More were erected at other locations in the 1840s, some as private initiatives, including Valparaiso in 1843 and the US Virgin Islands. The first time balls in mainland Europe were erected on guardships in the Netherlands in 1846.

Although time balls were proposed initially for calibration and particularly rating of marine chronometers, they were also established at inland locations for domestic use. The first time ball in the United States, introduced at the Naval Observatory in Washington, DC in 1845, was a public time signal rather than a device for rating chronometers. At that time, The United States Navy purchased only chronometers of the highest quality and rated them carefully before issue to the fleet.

Time balls were always the dominant signal type after their first introduction as official signals. They took many different forms. Early time balls often used a thin skin over an internal wooden frame. Later time balls used spun metal construction. Others used an open frame. Their weight and diameter varied considerably. They were usually dropped at noon or 1 pm, but France preferred 10 am with a repeat drop 2 minutes later. Many other variations have been identified.

There was slow progress in the British Isles after introduction of the Liverpool time ball, but introduction of the electric telegraph in the 1850s allowed time signals to be operated using electromagnets and also the transmission of time signals to devices that no longer had to be visible from the observatory that originated the signal. A new type of apparatus, using a rack and pinion mechanism to raise the ball, was introduced by Maudslay, Sons & Field for Edinburgh and Deal. It retained the air spring arrangement that had proved so successful at Greenwich in 1833. The Deal time ball was dropped using the railway telegraph system over a distance of more than 100 km from its first operation at the beginning of 1855. Many different manufacturers provided systems of this type worldwide for several decades. Subsequently, improved manufacturing methods and materials allowed simplification of reliable apparatus, which was still being erected in the 1920s.

4 Admiralty Lists of Visual Time Signals

Admiralty lists contain details of latitude and longitude that have been used to create maps of signal locations. Tabulation of signals has shown clearly how certain types of signals were favoured by different countries.

Time guns were usually regarded by the British Admiralty as secondary signals, with the repeated qualification that it was the flash of the gun that should be used as the accurate time signal. Automatic firing of time guns by telegraph was introduced in the 1860s. The use of maps to show sound propagation times in average environmental conditions was promoted by time gun enthusiasts. Time guns for mariners were never listed for the United States but were favoured by Canada.

Several little-known or forgotten visual time signals have been found in the Admiralty lists, many of them short-lived. Examples are a time ball in the British

Isles at Immingham Dock on the River Humber and two in Australia at Port Pirie in South Australia and Devonport in Tasmania. Others were at Massawa in Eritrea, Makassar in Indonesia and Guaymas in Mexico. There was a time gun in Djibouti and another was introduced at Durazzo in Albania as late as 1939. The signals in Asia and South America were often innovative and merit deeper investigation, as do those in many European countries. The present work has identified where signals were located but collaborative work with authors in different countries is needed to establish how they were designed and operated.

Mistakes have occurred in published notices and these have been resolved as far as possible, but there may be others that have yet to be identified. Dates of signal withdrawal are often more difficult to establish than dates on introduction. There is some evidence that Admiralty lists, relying on receipt of accurate information from port authorities, included signals after they had ceased to be in commission. A few time discs were still listed by the Admiralty and other authorities after they had been dismantled.

There will be differences of opinion about signals that were not included in Admiralty lists but were noted in pilot guides and almanacs issued by different authorities. Their value depended on accuracy and reliability as well as location. Further work is needed to establish their design criteria. Inclusion in Admiralty lists usually required notice of the procedure to be followed in case of error in the time signal.

There are sometimes unexplained inconsistencies between descriptions in Admiralty lists and what were believed to be photographs of time signals. Two examples are the time balls at Gibraltar and Rangoon. There are also cases where dimensions such as time ball elevations and drop heights have been given incorrectly in published lists, possibly because changes to a signal arrangement had not been notified.

5 Finale

The provision of visual time signals, accurate to better than 1 second, was first achieved almost two centuries ago. This book has been a celebration of the achievements of visionaries, astronomers and engineers who developed apparatus of remarkable accuracy. Mechanical chronometers can be replaced now by much cheaper electronic watches. Introduction of the Geographical Positioning System has changed the world of navigation and achievable accuracy to an almost unrecognisable extent. The earlier techniques allow accurate navigation when electronic systems are unavailable. They deserve to be remembered.

Index

R. Kinns (ed.), *The World of Visual Time Signals for Mariners*, Historical & Cultural Astronomy, https://doi.org/10.1007/978-3-031-57334-7

T

U

V

W

Y

GPSR Compliance

The European Union's (EU) General Product Safety Regulation (GPSR) is a set of rules that requires consumer products to be safe and our obligations to ensure this.

If you have any concerns about our products, you can contact us on ProductSafety@springernature.com

In case Publisher is established outside the EU, the EU authorized representative is:

Springer Nature Customer Service Center GmbH
Europaplatz 3
69115 Heidelberg, Germany

Batch number: 10756373

Printed by Printforce, the Netherlands